Uncertain Judgements

Statistics in Practice

Statistics in Practice is an important international series of texts, which provide detailed coverage of statistical concepts, methods and worked case studies in specific fields of investigation and study.

With sound motivation and many worked practical examples, the books show in down-to-earth terms how to select and use an appropriate range of statistical techniques in a particular practical field within each title's special topic area.

The books provide statistical support for professionals and research workers across a range of employment fields and research environments. Subject areas covered include medicine and pharmaceutics; industry, finance and commerce; public services; the earth and environmental sciences, and so on.

The books also provide support to students studying statistical courses applied to the above areas. The demand for graduates to be equipped for the work environment has led to such courses becoming increasingly prevalent at universities and colleges.

It is our aim to present judiciously chosen and well-written workbooks to meet everyday practical needs. The feedback of views from readers will be most valuable to monitor the success of this aim.

A complete list of titles in this series appears at the end of the volume.

Uncertain Judgements: Eliciting Experts' Probabilities

Anthony O'Hagan*, Caitlin E. Buck*, Alireza Daneshkhah*, J. Richard Eiser*, Paul H. Garthwaite†, David J. Jenkinson†, Jeremy E. Oakley* and Tim Rakow‡
**University of Sheffield, UK* *†The Open University, UK*
‡University of Essex, UK

John Wiley & Sons, Ltd

Telephone (+44) 1243 779777

Email (for orders and customer service enquiries): cs-books@wiley.co.uk
Visit our Home Page on www.wiley.com

Other Wiley Editorial Offices

John Wiley & Sons Inc., 111 River Street, Hoboken, NJ 07030, USA

Jossey-Bass, 989 Market Street, San Francisco, CA 94103-1741, USA

Wiley-VCH Verlag GmbH, Boschstr. 12, D-69469 Weinheim, Germany

John Wiley & Sons Australia Ltd, 42 McDougall Street, Milton, Queensland 4064, Australia

John Wiley & Sons (Asia) Pte Ltd, 2 Clementi Loop #02-01, Jin Xing Distripark, Singapore 129809

John Wiley & Sons Canada Ltd, 6045 Freemont Blvd, Mississauga, ONT, L5R 4J3, Canada

Wiley also publishes its books in a variety of electronic formats. Some content that appears in print may not be available in electronic books.

British Library Cataloguing in Publication Data

A catalogue record for this book is available from the British Library

ISBN-13: 978-0-470-02999-2 (HB)
ISBN-10: 0-470-02999-4 (HB)

Typeset in 10/12 Times by Laserwords Private Limited, Chennai, India
Printed and bound in Great Britain by TJ International, Padstow, Cornwall
This book is printed on acid-free paper responsibly manufactured from sustainable forestry in which at least two trees are planted for each one used for paper production.

Contents

Preface

This book arises from a multi-disciplinary research project commissioned by the R&D Research Methodology Programme of the National Health Service in the United Kingdom. The BEEP (Bayesian Elicitation of Experts' Probabilities) project brings together expertise in statistics, psychology and health. Although based in the Department of Probability and Statistics at the University of Sheffield, BEEP includes researchers from two other departments at Sheffield and from the Open University and the universities of Essex and Leeds. The first task within the BEEP research programme was to produce an authoritative review of the diverse literature on elicitation. As our perspective over this broad field grew, it became clear that what we were compiling was more than a review. The reader will find herein an extensive coverage of the field of elicitation, with very many references to the literature, but will also find commentary and synthesis. The result is a book that sets out our view of what elicitation is, what constitutes the best current practice in elicitation and the outstanding areas where research is needed, and it also provides detailed citations of the existing literature.

It has been written by the following BEEP team members:

- University of Sheffield, Department of Probability and Statistics: Professor Tony O'Hagan, Drs Caitlin Buck, Alireza Daneshkhah and Jeremy Oakley.
- University of Sheffield, Department of Psychology: Professor Dick Eiser.
- Open University: Professor Paul Garthwaite and Mr David Jenkinson.
- University of Essex: Dr Tim Rakow.

The authors have also benefited from valuable comments and contributions from other team members and the BEEP international advisory group. We are particularly grateful for contributions from Dr Gaëlle Villejoubert (University of Leeds) and Professors Roger Cooke (Delft University of Technology), Baruch Fischhoff (Carnegie Mellon University), Nigel Harvey (University College London), Colin Howson (London School of Economics), Jay Kadane (Carnegie Mellon University), Larry Phillips (London School of Economics) and Bob Winkler (Duke University).

Further information about BEEP can be obtained from the project website <www.shef.ac.uk/beep/>

Scope of the book

A number of reviews of elicitation have been published, including Chaloner (1996), Cooke (1991), Garthwaite et al. (2005), Hogarth (1987), Kadane and Wolfson (1998) and Wallsten and Budescu (1983). With the exception of Garthwaite et al. (2005), where the emphasis is on the statistics literature, these are all now out of date. We have built on these works and made use of our own knowledge of the literature, but have also carried out extensive new searches.

- We searched the ISI Science, Social Science and Arts and Humanities Citation Indices under the terms 'probability assessment', 'subjective probability', 'elicitation' and 'expert opinion'. More than 5000 references were found and abstracts were checked where available. More than 1000 articles were identified as relevant and investigated further.

- In keeping with the BEEP project's funding source, we made particular efforts to identify articles in the medical literature, searching the MEDLINE database under the same four terms. More than 8000 references were found and some 900 were identified for further investigation.

- A hand search of *Organizational Behavior and Human Decision Processes* between January 1990 and January 2004 was also conducted. There is a considerable amount of material in this journal that relates to the psychological issues around human judgement. Approximately 100 further papers were identified in this way as being relevant to the book.

- Of the 2000 or so references identified in these ways for further investigation, careful reading of abstracts led to the selection of more than 400, whose full text was retrieved and read.

- We asked each of our international advisory panel members to give us a short list of seminal papers in their own area of expertise; they also checked the final content of the book.

In this way, we hope that this book will be comprehensive in its scope, authoritative and up to date. Inevitably, there will be omissions and the discussion and emphasis will reflect the judgements of the authors about which ideas are the most useful and important.

Target audience

We hope that the book will be found useful by a wide range of people. Those who need to take decisions in the context of substantial uncertainty and risk know that they must often rely on expert judgement. Decision makers should find in this book a valuable introduction to how the elicitation of those judgements in probabilistic form can be achieved, and an overview of the current state of the art. Researchers in psychology (particularly the psychology of judgement and decision-making), decision theory, risk assessment and statistics (particularly Bayesian statistics) should find valuable synthesis of parts of their own research areas, with indications of the gaps that we feel need to be filled by new research. More importantly, they will see how the work in their own area relates to research in other disciplines and will gain a more complete appreciation of the field of elicitation.

With such a diverse readership in mind, there is a risk that the technical jargon of one discipline will be impenetrable to those whose background is in another. We have tried to introduce ideas in a simple manner wherever possible. We have also provided an extensive Glossary, covering the major concepts in both statistics and behavioural psychology. Even the reader who skips over the most technical parts of unfamiliar subjects should find enough less technical discussion to obtain a general understanding.

Outline of the book

The book is structured as follows:

- Chapter 1 sets out some fundamental concepts regarding elicitation, probability and expert judgement.
- Chapter 2 outlines the elicitation process in general terms, highlighting the differing perspectives of statistical and psychological research in elicitation.
- Chapters 3 and 4 are concerned with psychological theories and experimental evidence about expert judgement of uncertainty and with how these findings relate to the practical elicitation of probabilities.
- Chapters 5 to 7 deal with eliciting probability distributions, primarily from a statistical perspective.
- Chapter 8 discusses ways to evaluate the accuracy of elicited probabilities and distributions.
- Chapter 9 deals with issues when eliciting from multiple experts.
- Chapter 10 reviews a selection of applications, to give a flavour of the variety of methods that are actually in use.
- Chapters 11 and 12 collect together the main findings in Chapters 2 to 9, with regard to best elicitation practice and areas where further research is needed.
- The book ends with an extensive Bibliography of cited references, and a Glossary that explains some common terms in psychology and statistics.

Chapter 1

Fundamentals of Probability and Judgement

1.1 Introduction

This book concerns the elicitation of expert knowledge in probabilistic form. Before we can discuss what this means and the techniques for doing it, we need to explore some fundamental facts about probability and the way in which people formulate judgements of probability. This chapter begins with an introduction to probability and elicitation. It continues with a discussion of the nature of probability, arguing that, for the kinds of uncertain quantities for which expert opinion is typically sought, the usual understanding of probability in terms of long-run repetition of events is inadequate. We then consider how experts construct probability judgements, and find that probabilities are not pre-formed numbers just waiting to be expressed. On the contrary, psychological research tells us that judgements are formed 'on the fly' in response to questioning about uncertain quantities and are likely to be highly context dependent. Finally, we ask how such probability judgements might relate to the normative theories that underpin the interpretation and use of probabilities in statistics, decision theory and risk analysis.

1.2 Probability and elicitation

1.2.1 Probability

The probability of an event is a measure of how likely it is to occur. Probability 0 means that the event is certain not to occur, whereas probability 1 means that it is

Uncertain Judgements – Eliciting Experts' Probabilities A. O'Hagan, C. E. Buck, A. Daneshkhah, J. R. Eiser, P. H. Garthwaite, D. J. Jenkinson, J. E. Oakley and T. Rakow

certain to occur. Values from 0 to 1 describe increasing chances that the event will occur. The central value, 0.5, represents an event that is as likely to occur as it is not to occur. Events with probabilities above 0.5 are more likely to occur than not to occur, and conversely events with probabilities below 0.5 are more likely not to occur than to occur.

The symbol that is almost universally adopted to denote probability is P. Thus, if E is an event, then $P(E)$ denotes the probability of that event. For example, if E is the event of getting the result 'Heads' in a toss of an ordinary coin, then we can say $P(E) = 0.5$, because it is generally agreed in this situation that 'Heads' and 'Tails' are equally likely to occur. Similarly in the roll of a die ('die' here is the singular of 'dice'), there are 6 equally likely results and if S is the event of getting a 6 then $P(S) = \frac{1}{6}$.

The theoretical study of probability is a branch of mathematics that deals with laws and theorems about how probabilities behave and combine. For instance, suppose that events E and F are mutually exclusive. The term 'mutually exclusive' is defined in the Glossary; it simply means that E and F both cannot occur. If one occurs, then the other cannot. Let $EorF$ be the event that either E or F occurs. Then one of the fundamental laws of probability theory (the Addition Law) is that $P(EorF) = P(E) + P(F)$.

The statement that E and F are mutually exclusive implies that if I know that E has occurred, then the probability that F occurs must be zero. The occurrence of E changes the probability of F. For the inexperienced observer, one of the most difficult aspects of probability (and the source of some perplexing paradoxes) is the manner in which the probability of an event is affected by other events or other information that we might have. In this case, we need to distinguish between the probability of F when we do not know whether E has occurred and its probability when we do have that information. The first is just $P(F)$, and is called the *unconditional* probability of F. But if we know that E has occurred we have $P(F \mid E) = 0$, and this is the *conditional* probability of F *given* E. Another example of conditional probability can be found in the toss of the die. If E denotes the event that the result is an even number, then $P(S \mid E) = \frac{1}{3}$; that is, given that the result is an even number (2, 4 or 6) the probability of getting a 6 is one-third.

A more complex example is the probability that a specified person is killed in a road accident in the next 12 months. If we know nothing about the person except that he/she lives in England, then we could assess that probability as about one in 20,000 (because, although figures are not readily available for England alone, about 3000 people are killed on British roads each year). If, however, we know that the person is aged between 17 and 21, then the probability is larger, because this age group has more accidents. If we also know that the person is male, the probability increases again. A person's chance of being killed on the road varies with their age and gender, where they live in England, their occupation, whether they are married, and so on.

Pursuing this example further, what is the probability that I will be killed in a road accident in the next 12 months? If we consider all the relevant conditioning factors – my age, gender, location, marital status, the model of car that I drive, the number of miles that I drive each year, and so on – then there is nobody else in England (and never has been) with exactly the same characteristics. There will therefore be no data on which to assess that probability, and it is even questionable how to define it. We will explore these issues more thoroughly in Sections 1.3.2 and 1.3.3, but it is already clear why probabilities can be confusing for ordinary people. One reason why road safety advice (such as to wear seat belts, not to use mobile phones while driving or to drive more slowly in conditions of poor visibility) often has limited effect is because people do not see it as necessarily applying to them personally. They can believe that using mobile phones is dangerous in general, but that they personally can do so safely. It may not be rational to believe that they are special in this way, but it is certainly true that each person's individual characteristics will condition the probability and make them more or less at risk than the average person.

1.2.2 Random variables and probability distributions

Uncertainty about a single event is quantified by its probability. We are often interested, however, not so much in an uncertain event as in an uncertain quantity. An uncertain quantity is usually called a *random variable*. An event either occurs or does not occur, whereas a random variable may take any value of some collection of possible values. If the variable may take any value within some range, then we call it a *continuous* random variable. An example is the weight of the Great Pyramid at Giza, Egypt, which could, in principle, be any positive value (although clearly it would be very many tons). In contrast, a *discrete* random variable can only take certain distinct values, and cannot have values between these. An example is the number of stones in the Great Pyramid, which, in principle, could be 0, 1, 2 or any other positive integer value (although again it is clearly a large number), but not, for instance, any value between 0 and 1 or between 2,000,000 and 2,000,001.

Uncertainty about a random variable X is described by specifying the probability $P(X \leq x)$ for any x. So, although we can no longer characterise uncertainty about a random variable with a single probability, the description is still in terms of probabilities. (Note that $X \leq x$ is an event, the event that the true value of X is less than or equal to x.) If we think of $P(X \leq x)$ as a function of x, then it is called the (cumulative) *distribution function* of X. Examples of these functions for both discrete and continuous random variables are shown in the Glossary. Note that we can also have conditional distributions. For instance, the conditional distribution of X given some event E is specified by the probabilities $P(X \leq x \mid E)$ for any x.

While the distribution function is formally the way to define the probability distribution of a random variable, there are alternative formulations that are more intuitive, but which differ for continuous and discrete random variables. For a discrete random variable, it is more natural to use the set of probabilities $P(X = x)$ that give the probability that X will take each of its possible values. This is called the *probability mass function* of X. Figure 1.1 shows the probability mass functions

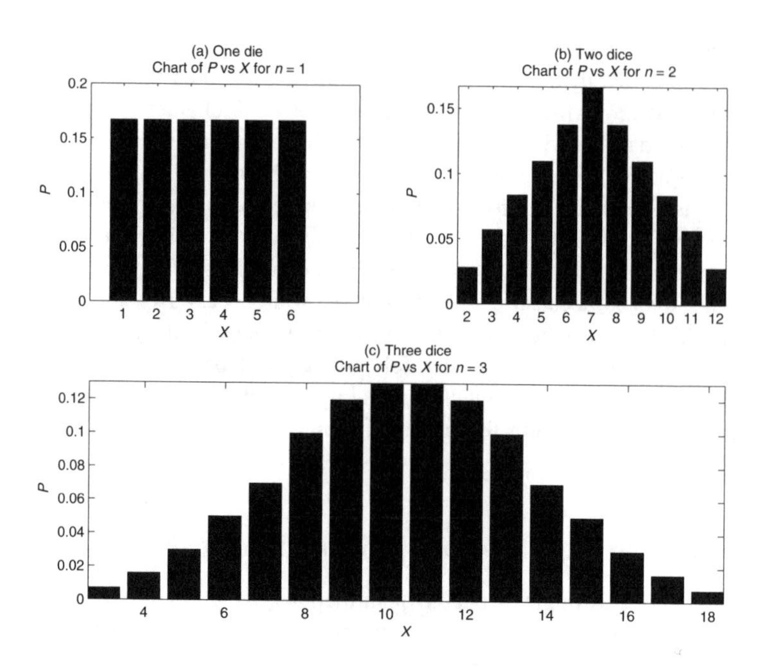

Figure 1.1: Probability mass functions for total score on n dice.

for three random variables. In Figure 1.1(a), we see that of the score on the toss of a single die. Figure 1.1(b) shows the probability mass function for the total score on tossing two dice and Figure 1.1(c), the same for the total of three dice.

The score on one die is equally likely to be 1, 2, 3, 4, 5 or 6, and this appears in Figure 1.1(a) as a distribution of uniform height. The same for two dice in Figure 1.1(b) has a triangular shape, while that of three dice in Figure 1.1(c) climbs, flattens out and then falls smoothly.

For continuous random variables, a similar picture is shown by the *probability density function* (pdf). Figure 1.2 shows some typical pdfs.

Both density functions are *unimodal*, meaning that they rise to a single peak (mode) before falling again. The density in Figure 1.2(a) is *symmetric* (like those in Figure 1.1), while that in Figure 1.2(b) is *skewed*. It should be noted that, whereas the heights of the bars in the probability mass function plots in Figure 1.1 are actual probabilities, the heights of the pdf curves in Figure 1.2 are not probabilities. Instead, it is the area under the curve between any two points, say x_1 and x_2, that is a probability; specifically, this area is $P(x_1 \leq X \leq x_2)$, the probability that X lies between x_1 and x_2.

The distributions in Figure 1.2 are examples of the many families of distributions that are used in statistics. Figure 1.2(b) is an example of a *beta* distribution; specifically it is the beta distribution with parameters 2 and 4. Figure 1.2(a) is a *normal* distribution; specifically it is the normal distribution with parameters 0

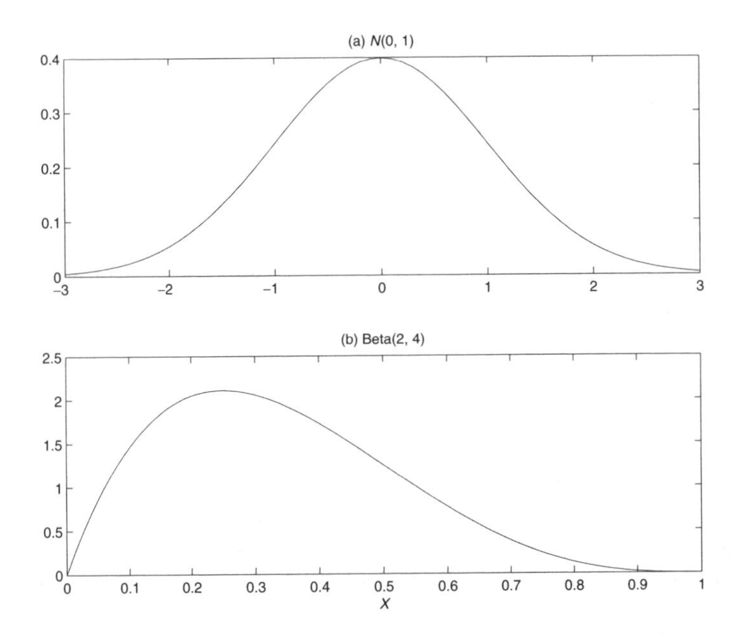

Figure 1.2: Two probability density functions.

and 1, also known as the *standard normal distribution.* Several other widely used distribution families are described in the Glossary.

1.2.3 Summaries of distributions

Although a probability distribution is defined by its distribution function, or equivalently by its probability mass function or probability density function, it is useful to have other kinds of descriptions that capture particular features of a distribution. So there are many different *summaries* that are used in statistics to present information about a distribution. Since these are also widely used in elicitation, the most important ones are listed here.

- *Probabilities.* Individual probabilities, such as $P(X = 2)$, $P(1 \leq X \leq 2)$ or $P(X \leq 10)$, are often used as summaries in their own right.

- *Quantiles.* The qth *quantile* of the distribution of X is the value x_q such that $P(X \leq x_q) = q$. The most widely used quantiles are the percentiles, median and quartiles. The nth percentile is $x_{0.01n}$. The 50th percentile, $x_{0.5}$, is known as the *median*, and it divides the range of X into two equally probable ranges (with probabilities 0.5); X is equally likely to lie above $x_{0.5}$ or below $x_{0.5}$. The lower *quartile* is the 25th percentile and the upper quartile is the 75th percentile. Together, the quartiles and the median divide the range of X

into four equiprobable regions (with probabilities 0.25). Of course, all the percentiles together divide the range into 100 equiprobable regions (all with probability 0.01). We sometimes refer to the *tertiles*, which divide the range into three equiprobable regions.

- *Intervals*. For any $s > t$, there is a probability $s - t$ that X lies in the interval (i.e., range of values) from x_t to x_s, written as $[x_t, x_s]$. It is often referred to as a $100(s - t)\%$ probability interval (or *credible interval*) for X. An interval like this with a suitably high probability, such as 90 or 95%, provides a range of values in which the true value of X will 'probably' lie.

- *Location measures*. These measures try to represent in some sense a typical, or representative, value of X. The median is a location measure, being the central value (such that X is equally likely to be higher or lower). Another measure is the *mode*, defined to be the value of X at which the probability mass function or pdf reaches its maximum. In the case of a discrete random variable, the mode is the *most probable* value for X. For a continuous random variable, this interpretation suffers from some technical ambiguities, but is still the usual way to explain the mode. The location measure that is most often used in statistical analysis is the *mean*. The mean, or expected value, of X is interpreted as the average value, or more formally, if we were able to observe the values of many random variables all with the same distribution as X, then the average of these values would be the mean. The mean has its own notation, $E(X)$ (standing for the *expectation* of X).

- *Measures of scale or dispersion*. These measures represent in different ways how far from its mean (or some other location measure) the random variable X might be. They can be seen as descriptions of how much uncertainty there is concerning X, since a large value of any of these measures implies that X may be far from any typical or representative value. The simplest is the inter-quartile range, which is the difference between the upper and lower quartiles. The most widely used measure in statistical analysis is the *variance*, which is the expected squared distance of X from its mean. Formally, we write this as $E\{(X - E(X))^2\}$. The square root of the variance is known as the *standard deviation* and is often more useful as a measure of dispersion because it is on the same scale as X.

- *Measures of shape*. Qualitative measures of shape include describing the density as unimodal, bimodal (rising to two distinct maxima, with a dip between) or multi-modal (having three or more maxima). We can also say that it is symmetric (as in Figure 1.2(a)), skewed to the right (as in Figure 1.2(b)) or skewed to the left (as in the mirror image of Figure 1.2(b)). However, there are also quantitative measures of skewness (where a symmetric distribution has value 0, a distribution skewed to the right has a positive value and a distribution skewed to the left has a negative value), and kurtosis (the tendency for the mode to be more or less sharply curved).

In order to describe a distribution effectively, a statistician will often use several summaries.

The reader may have encountered many of these summaries in a different guise, as summaries of a set of data. For instance, the mean of a sample is the average of all the values. There is a natural correspondence between the summaries of samples and the summaries of distributions, and a sample can often usefully be thought of as an *empirical* distribution.

1.2.4 Joint distributions

If we have two random variables, say X and Y, the uncertainty about them is not completely described by giving their separate distributions. The distribution of X gives us the value of $P(X \leq x)$ and that of Y gives us the value of $P(Y \leq y)$, but this is not enough to determine the *joint* probability $P(X \leq x, Y \leq y)$, which is the probability that both events, $X \leq x$ and $Y \leq y$, occur. The reason is that the occurrence of one event may change the probability of the other, in the way considered in Section 1.2.1.

Another of the laws of probability theory (the Multiplication Law) is that for two events E and F the joint probability $P(E, F)$ equals the product of the unconditional probability $P(E)$ and the conditional probability $P(F \mid E)$, that is, $P(E, F) = P(E)P(F \mid E)$. Equivalently, $P(E, F) = P(F)P(E \mid F)$. If the occurrence or non-occurrence of F does not change the probability of E then $P(E \mid F) = P(E)$ and we have $P(E, F) = P(E)P(F)$. In this situation, we say that E and F are *independent*. Notice now that this also implies that $P(F \mid E) = P(F)$: independence is a symmetric relationship, and if the occurrence or non-occurrence of F does not change the probability of E then the occurrence or non-occurrence of E does not change the probability of F.

The same ideas apply to probability distributions. If X and Y are two discrete random variables, then their joint probability mass function comprises the probabilities $P(X = x, Y = y)$ for all possible values x of X and y of Y. They are said to be independent if $P(X = x, Y = y) = P(X = x)P(Y = y)$ for all x and y. Two continuous random variables are said to be independent if their joint pdf is the product of their separate pdfs. If random variables are independent then knowing their separate probability distributions *is* enough to know all about their joint uncertainty. But otherwise we need to consider that the occurrence of some particular value of X may influence the distribution of Y or, conversely, that the occurrence of any particular value of Y will influence the distribution of X.

If X and Y are not independent, we will need to consider the *conditional* distributions of X given $Y = y$ (for all possible values y) and/or the conditional distributions of Y given $X = x$ (for all possible x). Therefore, the joint uncertainty of two (or more) random variables is potentially a complex thing, and may require new kinds of summaries to describe it.

- *Measures of correlation.* These measures describe the degree to which the value of one variable influences the value of another. They take the value 0 when random variables are independent and ± 1 when they are totally dependent, meaning that as soon as we know the value of one variable there will be no uncertainty about the value of the other. In the case of total dependence, the sign of the correlation coefficient indicates which of two forms of total dependence applies. Correlation of $+1$ means that as X increases so does Y, whereas if the correlation is -1 then as X increases, Y decreases. Values between these extremes indicate greater or lesser degrees of dependence, with a positive sign indicating that higher values of one tend to be associated with higher values of the other (and negative sign meaning that higher values of one tend to be associated with lower values of the other). The usual correlation coefficient is formally known as the *Pearson correlation coefficient*. It only takes the value ± 1 if the variables are totally dependent in a linear relation (increasing X by one unit always causes Y to increase by the same amount, regardless of the original value of X). Other correlation coefficients exist that measure *rank* correlation, and give values ± 1 whenever each variable is totally dependent on the other.

Also, just as individual probabilities are used as summaries for a single variable, we may use joint or conditional probabilities to summarise the features of a joint distribution.

1.2.5 Bayes' Theorem

An important consequence of the asymmetry in the Multiplication Law of probabilities is Bayes' Theorem (named after an eighteenth-century mathematician and clergyman called Thomas Bayes). In its simplest form it states that

$$P(E \mid F) = \frac{P(E)P(F \mid E)}{P(F)}.$$

The reason this is an important result is that it provides a recipe for learning from experience. In this context, we interpret E as an uncertain event of interest and F as a piece of new information that we obtain (we learn that the event F occurs). Then Bayes' Theorem explains how to convert from the *prior probability* of E, which is $P(E)$, to the *posterior probability* $P(E \mid F)$. The words 'prior' and 'posterior' here refer to the state of knowledge before and after learning that F occurs. The conversion consists of multiplying by $P(F \mid E)/P(F)$.

What is not apparent from this simple description, but would take too much space here to explain more fully, is how this 'recipe for learning' can really be applied in practice. However, this simple result underpins a philosophy of statistical inference known as the Bayesian approach, which is characterised by using the data and a form of Bayes' Theorem to update an initial state of knowledge (the prior distribution) to a new state of knowledge (the posterior distribution).

1.2.6 Elicitation

This book concerns the elicitation of experts' knowledge about one or more uncertain quantities in probabilistic form, and we are now in a position to appreciate what this 'probabilistic form' is. It is a (joint) probability distribution for the random variable(s) in question. The purpose of such elicitation is to construct a probability distribution that properly represents the expert's knowledge/uncertainty. The person whose knowledge is to be elicited is usually referred to as an 'expert', and while in principle there is no particular reason for them to have special knowledge or expertise, the fact that someone deems it worthwhile to carry out the elicitation implies that the expert's knowledge and judgements are worth having.

Elicitation is an important activity in a variety of fields. It has been widely practised in the design and management of large, complex engineering projects. Such projects are often essentially unique, so that there is very limited experience about the performance of components individually and in combinations. It is natural then to draw on expert judgements. In particular, there has been extensive use of elicitation in connection with nuclear installations.

Similarly, elicitation has played an important role in complex decision-making. The most difficult decisions are those where the consequences are subject to substantial uncertainty, and where those uncertainties are themselves not easy to judge. The use of expert elicitation to quantify the uncertainty in key variables then feeds directly into the decision itself.

Two statistical contexts also call for elicitation. One is the design of experiments. The purpose of experiments is to gain information regarding variables about which there is substantial uncertainty. Paradoxically, however, it is important to be able to use what knowledge one has about those variables in order to plan efficient experiments.

The other statistical context is in the Bayesian approach to statistics, a vital component of which (as is suggested in Section 1.2.5) is the use of prior information to augment the information from the statistical data. See, for example, O'Hagan and Forster (2004, Chapter 6). Elicitation of prior information is accepted as having a fundamental role in Bayesian statistics. In other areas in which elicitation is practised, the expert's knowledge feeds directly into the analysis of the underlying problem and will typically influence the outcome of that analysis strongly. In Bayesian statistics, however, it will often be the case that the statistical data will contain far more information than the prior knowledge, so the prior information may not be influential. Formal elicitation of prior distributions in Bayesian statistics has been used only in situations where prior information is appreciable and the data limited.

Numerous examples of all these contexts for elicitation will be found in Chapter 10.

1.3 Uncertainty and the interpretation of probability

1.3.1 Aleatory and epistemic uncertainty

The essence of elicitation is to capture an expert's knowledge about some uncertain quantity in a probability distribution that appropriately recognises the degree of uncertainty. It is useful to identify two different kinds of uncertainty that are sometimes known by the terms *aleatory* and *epistemic* uncertainty.

Aleatory uncertainty is induced by randomness. The word 'aleatory' derives from the Latin *alea*, meaning a die (singular of 'dice', readers may know the Latin quotation *alea jacta est* – the die is cast – attributed to Julius Caesar on crossing the Rubicon). Wherever we are interested in characterising uncertainty in one or more instances of a random process, then aleatory uncertainty is present. Epistemic uncertainty is due to imperfect knowledge about something that is not in itself random and is, in principle, knowable. The word 'epistemic' is Greek and means 'pertaining to knowledge'.

Consider, for example, the improvement in lung function that might be produced by a drug for asthma sufferers. The most widely used measure of lung function is FEV_1, which is the amount of air that the patient can expel in one second with maximum effort. If we ask an expert to assess the FEV_1 value that an individual patient will achieve using the drug, then she will have uncertainty about this value for a variety of reasons. (Note that we are adopting a convention here, which is explained in Section 2.3, that the expert is female.) First, there is aleatory uncertainty due to the fact that an individual patient will produce different FEV_1 readings when given repeat lung function tests. This is unavoidable random variation. Second, if we suppose that the expert is being asked about an unspecified, randomly chosen individual, then there is also aleatory uncertainty due to variability between patients. In addition, there is epistemic uncertainty because of various things that the expert has imperfect knowledge of. These may include uncertainty about how much within-patient variability there is in repeated FEV_1 measurements or uncertainty about how FEV_1 varies between patients. Even if the expert has enormous experience of both between- and within-patient variability of FEV_1 readings, she is likely to be uncertain about the extent of improvement that is achieved by the drug.

Statisticians usually separate the two kinds of uncertainty in the statistical models that they build. For the above example, we could characterise the single FEV_1 reading y as

$$y = \mu + \alpha + \tau p + \sigma e,$$

where μ is the mean level of FEV_1 for untreated patients, α is the effect of the drug in terms of the mean increase in FEV_1 that it produces, τ is the standard deviation of between-patient variability, σ is the standard deviation of within-patient (i.e., between measurements) variability, and p and e are zero-mean, unit-variance random variables that we might assume to be normally distributed. In this expression, it is p and e that represent the aleatory uncertainties. These give y a random addition (positive or negative) for the individual patient and the individual measurement. The other symbols, μ, α, τ and σ, represent epistemic uncertainties. Unless the expert has considerable practical experience, they will all be uncertain, but, in principle (given enough data), they are knowable. Statisticians refer to μ, α, τ and σ as *parameters*. A statistical model can be viewed as a representation of data in terms of (aleatory) probability distributions and (epistemic) parameters.

1.3.2 Frequency and personal probabilities

The distinction between aleatory and epistemic uncertainties is paralleled by the distinction between two different definitions of probability. *Frequency* probability is the definition that almost all people learn when they first encounter theories of probability and statistics. According to the frequency definition, the probability of an event is the proportion of times that it occurs if we conduct a long sequence of repetitions. Thus, the probability of obtaining 6 on a single toss of a die is defined to be the proportion of times that 6 would occur if we tossed it an infinite number of times. This definition is essentially only applicable to aleatory uncertainties, because it requires events to be repeatable in a process having intrinsic randomness. This is obviously true of tossing a die and is also true of making repeated measurements of FEV_1 on an individual patient or FEV_1 measurements on a series of randomly chosen patients. The definition cannot, however, apply to the effect of the drug. We cannot imagine this to be repeatable, since this is a specific drug and would not be completely equivalent to any other.

Epistemic uncertainties are typically associated with one-off, unrepeatable things. The same is almost always true of parameters in statistical models. If we wish to express epistemic uncertainty through probabilities, we must find another definition.

The answer is to use *personal* probability, also sometimes called subjective probability. According to this definition, probability represents someone's *degree of belief* in an uncertain proposition. This applies to both aleatory and epistemic uncertainties. I have, for instance, a degree of belief in whether a toss of a die will yield a 6, and I can have a degree of belief in the proposition that a particular drug will increase FEV_1 for asthma patients on average by 100 ml or more.

It is clear that the terms 'personal' or 'subjective' are appropriate because my degree of belief in one of these propositions may be different from yours. This may not be true for something as simple as a toss of a die but for epistemic uncertainties (which are associated with imperfect knowledge) probabilities will always depend on what knowledge a person has. In everyday usage, the word 'subjective' has unfortunate connotations of opinions contaminated by personal bias, prejudice and even irrationality or superstition. It is important to recognise that the objective of good elicitation is to eradicate such elements and to structure the process of elicitation in such a way as to assist the expert in rational and thoughtful evaluation of her knowledge and experience. The expert inevitably has different knowledge from others, so her probabilities are personal, but they should not be 'subjective' in any of those pejorative senses.

1.3.3 An extended example

To help clarify the distinctions and ideas in Sections 1.3.1 and 1.3.2, it is useful to consider another example in some detail.

Suppose that a timber company is considering planting a species of tree that it has not previously used. It asks an expert for her judgement of what yield it will get if it plants this species. (For the purposes of this example, we will define the yield to be the volume of usable timber per tree, although in forestry the more usual definition is volume per hectare per year.) An important first distinction is between the yield of a single tree and the average over all trees that the company might plant. This is known in statistics as the distinction between an *individual* sampled observation and the underlying *population* mean. In this case, the population is the collection of all the trees that the company will grow if it decides to use this species, and an individual tree will usually be regarded as being randomly drawn from this population. The yield of an individual tree therefore has aleatory uncertainty that is described by the *distribution* of yields in the population. For instance, if 30% of trees in the population yield more than $50\,m^3$ of timber, then there is a probability of 0.3 that an individual tree will yield more than $50\,m^3$. The aleatory uncertainty is completely described if we know this distribution of yields in the population.

However, there is another source of uncertainty – that this distribution is not known. The yields of trees of this species will have been observed in other places where it has been grown, and this is likely to form a part of the expert's knowledge, but there is uncertainty about how well the species will grow on this company's land. Furthermore, it is this distribution, and particularly the mean of the distribution, that is of interest to the company. It is the mean yield, the average over all the trees, that relates directly to the profitability of this species, and to the decision whether to plant it. It is this mean yield that the expert is asked to assess, not yields of individual trees.

Is the expert's uncertainty about mean yield aleatory or epistemic? We could think of the company's land as just one of the many sites where this species has been and might be grown. There is then another level at which we can conceive a population, the population of sites, and a distribution of mean yields over these sites. So if 25% of sites have mean yields of more than $45\,m^3$ per tree, then we might suggest that the probability of the company's site producing a mean yield over $45\,m^3$ per tree is 0.25. This presents the uncertainty about mean yield as aleatory, but such an interpretation is *not* appropriate. The company's site cannot be regarded as randomly drawn from the population of sites. We know its latitude, its altitude, the nature of the geology and topography, all of which make this site different from others. It is factors such as these that the expert will be expected to take into account, in addition to any knowledge about the yield of this species at other sites.

The uncertainty about mean yield in this site is predominantly epistemic because it is *not* a randomly chosen site. The uncertainty derives from lack of knowledge about how the specific features of this site will affect the mean yield. Whereas the frequency interpretation of probability is adequate to describe the distribution of yields in a population of trees, it cannot apply to the mean yield of a specific site, because that site is a 'one-off'. There is no other site that is exactly like it, and when we use probability to describe the expert's uncertainty about the mean yield, the only meaningful interpretation for those probabilities is the personal or subjective interpretation.

Note that if the expert were to be asked about the yield of an individual tree on this site, her uncertainty would be a compound of the aleatory and epistemic uncertainties. It would be purely aleatory *if* she knew the distribution of yields in the population of trees that might be grown on that site, but this distribution is *not* known. In particular, its mean is unknown. Uncertainty about features of the population is epistemic and will also contribute to the uncertainty about an individual tree. In statistics, features of populations, such as means and variances, are generally referred to as *parameters* (like the parameters μ, α, τ and σ in the medical example of Section 1.3.1). The theory of statistics is concerned with ways to make inferences about the unknown parameters, using the available data. The things that we wish to ask experts about are very often what statisticians would call parameters. Uncertainty about them is always epistemic because the population is unique, and elicitation is always concerned with the expert's personal probabilities.

There is controversy in the world of statistics about the use of personal probability. The most widely taught theory of statistical inference is the frequentist theory, in which parameters are regarded as unknown but fixed. In frequentist statistics, it is not legitimate to express probabilities about parameters, because only the frequency interpretation of probability is admitted. The rejection of personal probability as a basis for scientific reasoning is one of the differences that distinguishes most followers of frequentist statistics from most advocates of Bayesian statistics, the latter generally embracing personal probability in their methods. However, in the practical elicitation of expert knowledge, this controversy does not arise. The focus of attention in practice is *always* on variables for which there is at least a component of epistemic uncertainty, and expert judgements are therefore always personal probabilities.

In the light of the timber yield example, we can now refer back to the problem in Section 1.2.1 of assessing the probability that I will be killed in a road accident in the next 12 months. It was noted there that the combination of relevant factors – my age and gender, where I live, the kind of car I drive, and so on – make me unique, so that it is no longer possible to ascribe a probability by referring to road accident data. This is clearly analogous to the uniqueness of the timber company's site. It may be entirely natural to ask about the probability that I will be killed on the road in the next 12 months, but it is not possible to give a frequency interpretation to such a probability. The only sense in which we can discuss it meaningfully is within the personal probability framework. The fact that people, in general, are willing to talk about unique events as having probabilities emphasises the importance of personal probability.

1.3.4 Implications for elicitation

Most people are familiar with probability only in terms of repeatable, random events, and this has important implications for the process of elicitation. If an expert is asked to express her probability for the proposition that the asthma drug will increase FEV_1 by an average of 100 ml or more, or that the mean yield will exceed 45 m^3 per tree, we are asking for a personal probability. In trying to answer the question, she cannot appeal to any experience of repetitions since the events she is being asked about are unique and repetition is impossible. Nevertheless, the familiar ideas of frequency probability are a valuable guide.

First, when explaining to the expert what is needed, it is usual to draw analogies between personal probabilities and frequencies. The expert will be advised that she should give a probability of one-sixth if she has the same strength of belief in the proposition as in throwing a 6 with a die. Well-known frequency probabilities associated with familiar gambling devices such as dice, coins, roulette wheels and cards help the expert assign personal probabilities to one-off propositions.

Second, experience with frequencies of related things may suggest a probability. For instance, the medical expert may know that six out of seven asthma drugs claim to increase FEV_1 by at least 100 ml. This is not really repetition because the drugs are all unique, but it gives the expert a sense of how realistic it is for a new drug to reach that level of effect. In the same way, the forestry expert will use the yields of the tree species in other places to indicate a probability, but must also account for the unique features of the specific site. The knowledge that an expert draws on is often a kind of quasi-repetition, moderated by a judgement of how much the proposition in question is representative of those quasi-repetitions.

1.4 Elicitation and the psychology of judgement

When we talk of 'elicitation', we imply that our respondents have some kind of knowledge or beliefs 'in their heads' and it is our task to devise the right kind of questions to 'extract' this information from them. But is this picture correct? Do people have ready-formed beliefs waiting to be extracted in this way? And even if they do, if such beliefs concern uncertain events or prospects, are they represented subjectively in terms of numerical probabilities? To start answering such questions, we need to go back a bit into history to remind ourselves of the concerns that have guided the development of theory and method in psychology.

1.4.1 Judgement – absolute or relative?

Psychology is, very largely, the scientific study of how human beings think and feel and act on the basis of their thoughts and feelings. So one of the first questions is how we can tell what someone is thinking or feeling. This, of course, was, and still is, a fundamental question of philosophy, but what distinguished the aspirations of early experimental psychologists was a conviction that thoughts and feelings were, in principle, *measurable*. Those who initiated this programme of work in the mid-nineteenth century referred to their field as *psychophysics*. Partly, this signified an intention to bring the methods of the physical sciences to bear on the subject matter of psychology, but it had a more precise sense too. The search for the 'psychophysical law' involved attempting to specify, in exact mathematical terms, a function that described the relationship between 'psychological magnitudes' on the one hand and 'physical magnitudes' on the other.

The context in which this enterprise was undertaken was that of sensory perception. Hence, the 'psychological magnitudes' studied were sensations – of the loudness of tones, the brightness of lights, the length of lines and (particularly in the early days, since such stimuli could be manufactured easily) the perceived heaviness of different weights. Consider the last of these. Participants are presented with a series of small brass cylinders, made to be identical visually but of different actual weights. The 'physical magnitude' of these stimuli is simply their weight in grams and their 'psychological magnitude', is, how heavy they feel. Obviously, these two sets of magnitudes will be related to each other. A weight of 200 g will feel heavier than one of 100 g, but will it feel exactly twice as heavy? And will the difference between these two weights feel the same as that between two weights of 200 and 300 g? The broad answer is no. For many sensory continua – at least those that involve changes in perceived intensity rather than in quality (colour is an example of the latter) – sensitivity to differences is reduced as the stimulus intensity increases.

The so-called Weber–Fechner law (Fechner, 1860) proposed that the 'difference threshold' or 'just noticeable difference' (JND) for any stimulus – the amount by which the magnitude of the stimulus would need to be increased for the difference to be just detectable – is a logarithmic function of the distance of that stimulus above 'absolute threshold' (i.e., the smallest detectable stimulus intensity). It was also assumed that the perceived difference between any pair of stimuli is a direct function of the number of JNDs by which they were separated. This law remained intact for a century until Stevens (1957) proposed its replacement by a power function. Even then, the differences between the predictions of these two versions of the 'psychophysical law' are quite subtle over many ranges of stimulus intensity.

In the course of their search for this precise mathematical function, however, psychophysicists were quickly confronted by another phenomenon. Judgements are highly dependent on context. Perhaps the most common context phenomenon is termed the *contrast effect*. Here is a typical experiment. In a control condition, participants have to lift a series of weights ranging between, say, 200 and 400 g and rate how 'light' or 'heavy' they feel (e.g., on a scale with, say, 11 response categories labelled 'extremely light' at one end and 'extremely heavy' at the other). In other conditions, an 'anchor' or standard weight (that need not be judged) is added on alternate trials between each of the original variable weights. If this anchor is much heavier than the original series (say 900 g), the remaining weights will be judged lighter; if the anchor is lighter than the original series (say 50 g), they will be judged heavier.

Interestingly, the original reason for introducing such 'anchors' was, as the term suggests, to stabilise participants' use of the rating scale so that the relationship between the physical magnitudes of the stimuli and how they were rated remained relatively unchanged across the course of the experiment. Sometimes these anchors could fall within the range of the stimuli to be judged and sometimes at or beyond the extremes of the stimulus distribution. Sometimes participants were instructed that the anchor or anchors corresponded to a specific category or score on the response scale; sometimes this was left unstated or implicit. Either way, this procedure did generally achieve the desired effect. In other words, the resulting judgements tended to be more consistent in terms of their rank ordering and defensibly interpretable as constituting an interval scale (as required for any test or application of the 'psychophysical law'). However, participants' judgements, while internally consistent, were entirely relative to the context in which they were presented. Although a 350-g weight would always receive a 'heavier' rating than one of 250 g *in the same experimental condition*, there is no guarantee at all that a 350-g weight in one condition would be rated as 'heavier' than a 250-g weight *in a different condition* (as for instance, if the stimuli in the first condition ranged from 200 to 800 g and those in the second condition, from 50 to 400 g). In short, such judgements are always relative, not absolute.

Contrast effects are easy to demonstrate, but less easy to explain definitively. A major ambiguity is whether such effects reflect changes in sensation (or 'psychological magnitude') resulting from perceptual adaptation (as when an indoor room initially seems dark after coming in from bright sunshine) or merely a 'semantic shift' in terms of participants defining descriptive terms such as 'light' and 'heavy' to match the range of stimuli actually presented. According to the latter interpretation, a weight of 400 g would not necessarily *feel* any lighter in the context of a 900 g anchor. It is just that the term 'extremely heavy' now has to allow for weights of at least 900 g, whereas in the control condition, it could be used for weights of just 400 g. Probably both perceptual adaptation and semantic shifts are present in this example, but separating out their effects is enormously difficult with methods such as those described. Of more general significance than debates over the role of perceptual adaptation, however, is the fact that relativity to context pervades all kinds of judgements, not just those involving sensory perception. Judgements of personality, of the seriousness of offences and of political and social attitudes all show similar effects (see, for example, Eiser, 1990; Eiser and Stroebe, 1972).

What are the implications of this piece of history for the elicitation of experts' probabilities? The main messages are, firstly, that subjective perceptions and sensations are, in principle, measurable – and with some precision – but such measurements can only be interpreted relatively and not absolutely. Secondly, in recognition of such relativity, the psychology of judgement needs to take on issues that extend beyond the original aim of the psychophysicists in relating psychological and physical continua to each other. Simply stated, there is a logical distinction between people's sensations, or subjective representations, and their *descriptions* of such sensations in terms of any response scale. In other words, whereas classical psychophysics was primarily concerned with the relationship between continua of psychological and physical magnitudes, more modern judgement research addresses the relationship between psychological and response continua.

Translating this into the question of eliciting probabilities, we therefore need to remember two basic distinctions. The first distinction, analogous to that between psychological and physical magnitudes, is that between people's subjective representations of how probable things are and the objective events or evidence that provide grounds for such representations. The second distinction, corresponding to that between psychological and response continua, is that between people's subjective representations of how probable things are and the manner in which they express such representations on any given response scale. For example, just because respondents rate something as '90% certain', we cannot assume that they really *mean* 90% rather than 95%, 85% or even 65% or that what they think of as 90% equates to what anyone else would mean by 90%. However, we *can* assume that they mean it is more probable than something else they have rated as '80% certain'. In other words, just because respondents may seem happy to use numerical probabilities to *express* an estimate, we cannot assume that they represent such estimates numerically to themselves (though they may) and, even less, that they have arrived at any given numerical estimate through a process of normatively correct statistical reasoning (though with training they sometimes may).

The argument, then, is that judgements of probability, however elicited, are just that – *judgements*. Knowing what we do about the sensitivity of all judgements to context effects, we should always be wary of interpreting them in absolute rather than relative terms. Nonetheless, judgements of probability at least *appear* more absolute than, say, judgements of heaviness or loudness. This is largely because the meaning of different probability values is well defined (i.e., absolute), is assumed to be well known and, importantly, the scale on which they fall is bounded at both extremes (0 and 1). In comparison, many scales of psychological measurement are not bounded at all, and some (e.g., perceptual judgements) only have a minimum or lower bound (i.e., absolute threshold) but no upper bound. In this sense, probability judgements on a scale of 0 to 1 are already 'anchored' at the two extremes. So does this remove the problem of relativity? Not quite.

The first reason is conceptual. People's perceptions of how probable things are may not correspond to any formal definition of probability. By analogy, we know that the loudness of sounds can be measured in decibels, but asking people (even expert sound engineers) to *estimate* the decibel value of a given sound in no way guarantees that their estimates will be accurate. Eliciting probabilities from experts is not a straightforward matter, and context effects may be expected to contribute to errors of judgement. The second reason is more empirical and relates to the actual distribution of probabilities which judges are asked to estimate when required to make a series of judgements, for instance, whether the distributions are positively or negatively skewed and how much of the range from 0 to 1 they cover. A well-established principle, related to the contrast effects previously described, was first identified by Parducci (1963) and is termed the *range-frequency compromise*. This involves two tendencies. The first is for judges to use the different regions (or categories) of a response scale to cover broadly equal intervals or fractions of the total range between the smallest and the largest stimulus presented. The second is for judges to use the different regions or categories with broadly equal frequencies. Of course a scale such as 'extremely light' to 'extremely heavy' can be assumed to be more vulnerable to such effects than a probability scale from 0 to 1. However, the influence of such a principle cannot be ruled out in situations where the actual probabilities being judged are heavily skewed. Even an association that is very strong in epidemiological terms (e.g., the conditional probability of contracting lung cancer if one is a smoker, about 0.1) can fall very close to the bottom end of the 0 to 1 scale. If judges are asked to make comparative estimates, on this scale, of various mathematically small probabilities, it could well be the case that they would achieve 'better' differentiation by overestimating some of the less improbable events. Another way in which the range-frequency principle might have an effect could be when judges are asked to provide estimates of distributions, rather than single probabilities. A speculation implied by research in other areas is that, depending on the elicitation procedure, judges may produce distributions that are less skewed, and possibly less peaked, than they should be.

1.4.2 Beyond perception

A possibly less helpful legacy of classical psychophysics has been the tendency to conceive judgement prototypically as a form of perception. It is as though participants are told, "Here's a stimulus, please look at it (or listen to it or touch it) and tell me what you see (or hear or feel)." Sometimes this prototype is quite appropriate, as when a stimulus is physically present and can be directly perceived. But not all judgements are like that. Frequently, we are asked to make judgements of concepts or hypothetical events and consequences. Examples of the latter include judgements of political preference and trust in politicians, elicitation of probabilistic estimates of the outcomes of medical treatment or of the susceptibility of members of a given population to a particular disease. Here, there is no specific thing that has a determinable 'physical magnitude'. Even where the question is one of aleatory uncertainty and there is a correct answer determinable in advance (e.g., the prevalence of disease D in population P), this fact is not known as such to the respondents (or there would be no point in asking for their judgements). So they are not 'perceiving' the prevalence and then describing their 'perception'. They are thinking about the problem, forming some notion of what the prevalence might be and then expressing this notion in terms of the particular response format presented to them by the researcher. Such 'thinking about' involves a much more deliberative and effortful process than that typically implied by more automatic perceptual responses. The challenge, then, is to determine what this 'thinking about' comprises.

The starting point for an analysis of this problem is to realise that this 'thinking about' is a process that is set in motion by the researcher's interrogation. This process leads to the respondent arriving at a notion of a possible answer and then translating this notion into the response language provided. Within this process, there is lots of room for conceptual slippage. Does the respondent understand the researcher's question in exactly the way the researcher intended? Does the respondent then continue to think only (and wholly) about the question asked or does he or she start thinking about other associations that are not strictly part of the problem itself, while selectively failing to properly attend to all the features of the problem that are relevant? (Note the discussion of cognitive heuristics in Chapter 3.) And once the respondent has arrived at a notion of how to answer the question, will the answer then offered be interpreted by the researcher exactly as intended?

It is now acknowledged that an extremely important influence on this process of 'thinking about' is memory. There are a number of reasons why memory is so important for judgement. One reason is that memory is selective. Suppose that our expert is asked to estimate the prevalence of condition or disease D in population P (e.g., the proportion of children under five in an East African state who are likely to die from malnutrition). Our respondent (who, let us assume, has never been to Africa) will try to remember relevant sources of evidence, from news reports, from articles in professional journals, from conversations with colleagues, and so on. Some of this information may be easier to recall because of the vividness of news reports. In other words, the information on which the respondent's judgement is based may be only a sub-sample, or selection, of what is potentially relevant and available. Judgements can also be influenced by memory associations, that is, by thoughts triggered by irrelevant (or non-diagnostic) features of the problem, or by over-generalisation from inferences based on the membership of a broader category.

To continue our example of malnutrition in Africa, let us suppose that, whereas many states in that part of the world have had years of poor harvests, the state in question has remained relatively prosperous. If so (and particularly if this has not been considered newsworthy by western media), our respondent may overestimate the mortality rate of the specific country as a consequence of a broadly accurate but undifferentiated association between Africa, poverty and famine. Thus, through the combined effect of selectivity and associative memory, our respondent may fail to recall some useful information, and let other, misleading or less relevant, information intrude. The point is that our respondent's prevalence estimate is not something ready formed and 'sitting there' in some memory store just waiting to be retrieved. It is something constructed from the ideas and associations that come to mind while the respondent thinks about how to answer the question.

In short, memory can involve the reconstruction of meaning and not simply the recall of facts or events. Related to this is the fact that we frequently have little control over what memory associations do come to mind and little insight into why we remember what we do or what has triggered particular thoughts. There is a considerable body of recent literature on how 'automatic' memory associations can influence attitudes, judgements and decisions, with many of these processes occurring below the level of conscious awareness. See, for example, Bargh et al. (1996), Bargh and Ferguson (2000) and Fazio (2001).

1.4.3 Implications for elicitation

We can draw from this generic psychological research some important conclusions about the process of eliciting experts' probabilities. First, an elicited probability is a judgement, and we should expect in principle that generic findings about judgements will apply. In particular, research has shown that judgements of stimuli such as weights are intrinsically relative. Even when anchors are provided, changing the anchors changes the context and is likely to influence the respondents' judgements. However, it has already been remarked that probability judgements are different with respect to having natural limits of 0 and 1. These act as absolute and unvarying anchors. Furthermore, it is usual in eliciting probabilities to give the expert additional absolute anchors in the form of reference events having specific probabilities. For instance, the probability of 0.5 is explained as corresponding to a proposition that is equally likely to be true or false, equivalent to the event of drawing a red ball from a bag containing one red and one white ball. These practical measures should mean that probability judgements suffer much less from relativity effects than those found in experiments using judgements of other stimuli.

We should, nevertheless, expect to find that the ways that people make such judgements lead to biases, such as through the range-frequency compromise. A number of such bias-inducing heuristics are discussed in Chapter 3.

It is also important to recognise that experts construct probability judgements in response to the stimulus of questioning; their probabilities are not pre-formed values simply waiting to be expressed. The role of memory in this process, and the effect that different forms of questions can have on which memories are accessed, also underlies some of the sources of bias in probability judgements that are discussed in Chapter 3.

1.5 Of what use are such judgements?

The purpose of elicitation is to obtain a formal expression of the expert's knowledge regarding an uncertain quantity (or quantities). Usually, the resulting probability distribution will be used as part of some analysis (for instance a risk analysis) or to aid in making a decision. It is therefore important to consider the extent to which probability judgements elicited from an expert have the status and interpretation that is expected of them in these applications, particularly bearing in mind the preceding discussion about how those judgements are constructed in practice.

1.5.1 Normative theories of probability

The probabilities that we wish to elicit (and which statisticians have often implicitly assumed are indeed elicited) are those that are implied by normative theories of decision-making under uncertainty. The relevant theory was first fully developed by Savage (1954) and further expounded in the classic textbook of DeGroot (1970). According to this theory, in order to make decisions which satisfy some natural axioms, persons must behave as if they (a) have a probability distribution for all relevant uncertain quantities, (b) have a utility function expressing the value of making any given decision conditional on the true values of those quantities, and (c) choose an optimal decision as the one that maximises expected utility. In principle, the expert's probability distribution can be revealed by offering enough different decision options and rewards and then observing what decisions she makes.

In another seminal work, de Finetti (1974) formally defined probability to be the decision made in response to a quadratic scoring rule (see Section 8.2). Specifically, suppose that the expert states her probability for an event E to be q. Then when it is determined whether E does occur, she receives a reward $1-(1-q)^2$ if E occurs and $1-q^2$ if it does not (in some appropriate monetary units). If the expert actually judges the probability for E to be p, then her expected reward in stating it to be q is

$$\begin{aligned} p\{1-(1-q)^2\}+(1-p)\{1-q^2\} &= 1-p(1-q)^2-(1-p)q^2 \\ &= 1-p(1-p)-(p-q)^2. \qquad (1.1) \end{aligned}$$

This expected reward is maximised by the expert stating a probability, q, that is equal to her actual assessment p. So de Finetti's scheme encourages the expert to assess her probability accurately. However, it assumes that she is able to balance probabilities and rewards appropriately. Under this assumption, it is proved that the expert's probabilities will behave according to the laws of probability theory and will be appropriate for use in subsequent analyses or decision-making.

These normative theories state that probabilities are the uniquely scientific way to represent uncertainty. Furthermore, probabilities defined according to such theories are what is needed for use in applications such as risk analysis or decision-making.

1.5.2 Coherence

Whatever interpretation we place on probability, frequentist or personal, it is agreed that probabilities should obey the laws and theorems of probability theory (such as the Addition Law of Section 1.2.1 and the Multiplication Law of Section 1.2.4). A set of probability judgements that follow all these laws and theorems are called *coherent* (see Section 8.3 for more discussion of coherence). In developing their theories of personal probability, Savage, DeGroot and de Finetti took care to show that the probabilities that would be obtained must be coherent. For example, de Finetti (1974) shows that a person who assigns a series of probabilities according to the reward scheme illustrated in (1.1) will necessarily expect to obtain a lower reward if her probabilities are not coherent. The assumption that she is able to combine accurately the probabilities and rewards (and that she wishes to maximise her expected reward) would imply that she would not make this mistake, and hence her probabilities must be coherent.

In practice, however, we know that people do assign probabilities non-coherently. They make errors of judgement that are assumed not to happen in the normative theories. This raises the fundamental question of the nature of elicited probability judgements and the extent to which they can be treated as having the interpretation that is required for practical risk assessment and decision-making.

1.5.3 Do elicited probabilities have the desired interpretation?

Both de Finetti and Savage considered the process of obtaining expert responses to choices with rewards such as (1.1) to be elicitation; see, for instance, Savage (1971). In general, though, psychologists would regard such choices as cognitively more complex than asking directly for an assessment of probability. It seems unlikely that experts would perform better in such tasks. If, in order to determine q, the expert follows the above reasoning and decides that her answer ought to be her probability p, then she still has to determine p with all the difficulties outlined above and in Chapters 3 and 4. If a less analytical approach is used, without explicit assessment of probability but with the expert making a more intuitive choice in the face of the reward scheme, then this seems likely to lead to a less accurate judgement through imperfect appreciation of the implications of the reward scheme.

Research and the practice of elicitation have since concentrated on the direct elicitation of probabilities. However, this compounds the question of what interpretation we can place on the elicited probabilities. If they have not been obtained according to their formal constructions in the theories of Savage, DeGroot or de Finetti, and if they may, in practice, be non-coherent, what status do they have?

Winkler (1967, p. 778) writes

> "The assessor has no built-in prior distribution that is there for the taking. That is, there is no 'true' prior distribution. Rather, the assessor has certain prior knowledge which is not easy to express quantitatively without careful thought. An elicitation technique used by the statistician does not elicit a 'true' prior distribution, but in a sense helps to draw

> out an assessment of a prior distribution from the prior knowledge. Different techniques may produce different distributions because the method of questioning may have some effect on the way the problem is viewed."

Winkler seems to regard the different distributions that result from different elicitation techniques as equally valid, but this would deny the considerable research in the psychology literature (see Chapter 4) that demonstrates how some forms of questioning lead the expert to view the problem inappropriately, in the sense that they do not utilise the available information fully and accurately.

On the other hand, O'Hagan (1988) explicitly defines 'true' probabilities as those that would result if the expert were capable of perfectly accurate assessments of her own beliefs. He shows that such 'true' probabilities will satisfy the coherence requirements of the normative theories. O'Hagan regards different 'stated' probabilities, that might result from different elicitation methods, as more or less inaccurate attempts to specify the expert's underlying 'true' probabilities.

It is clear that whether or not we believe that experts' knowledge is representable by unique, *true* probability distributions there are ways in which the expert might give poorly judged assessments. So not all elicitation techniques will lead to equally valid results. It is important that the expert should view the problem from as complete a perspective as possible, utilising all the relevant information in an unbiased way. If this were achievable, taking care, in particular, to avoid the sources of poor judgement and bias that have been identified by psychological research, then the elicited probabilities and distributions would be coherent. We could call such a set of probabilities or probability distributions *good*.

In practice, an elicited probability distribution can be seen as an approximation to such a 'good' distribution. O'Hagan (1988) and the earlier developers of personal probability implicitly assumed that there would be a unique 'good' distribution, which can then be called 'true', but this is an open question. Are people's probabilities different only because no two people have identical knowledge and experiences, or if we could ever find two people who are identical in this respect might they legitimately have different probabilities? In a similar vein, if two elicitation techniques were both so perfect that they yielded 'good' distributions, could they, nevertheless, produce different distributions? There are theories of probability that take an 'objective' view that there is a unique probability distribution associated with any state of knowledge. They are associated, in particular, with the debate about probabilistic representation of ignorance (Jeffreys, 1967; Kass and Wasserman, 1996).

We take the view that the purpose of elicitation is to represent an expert's knowledge and beliefs accurately in the form of a 'good' probability distribution. In later chapters, we may refer to this as a true distribution, but the reader should be aware that this does not necessarily imply that the true distribution is unique.

1.6 Conclusions

Each chapter of this book will conclude with a summary of its findings relating to good elicitation practice and areas of research need. These are collected in Chapters 11 and 12.

1.6.1 Elicitation practice

- The distinction between aleatory and epistemic uncertainty is important for elicitation practice. Elicitation usually focuses on uncertainties that are either purely epistemic or have an epistemic component. However, people are most familiar with the concepts of probability in the context of aleatory uncertainties.
- It is important to remember that elicited statements of probability are judgements made in response to the facilitator's questions, not pre-formed quantifications of pre-analysed beliefs. The psychophysics literature suggests that all such judgements are intrinsically relative.
- The range-frequency compromise suggests that in some situations experts will tend to distribute their elicited probabilities evenly over (the whole or part of) the probability scale.
- Elicited probabilities may suffer from biases and non-coherence in practice, but the goal of elicitation is to represent the expert's knowledge and beliefs as accurately as possible.

1.6.2 Research questions

- To what extent does the existence of an absolute scale (0 to 1) for probabilities, and the way that training usually gives the expert other anchors or landmarks on the scale, allow absolute (rather than relative) judgements?
- What are the implications of the range-frequency compromise in the context of probability elicitation?
- Does elicitation using proper scoring schemes (as propounded by Savage and others) lead to less accurate assessments than the direct elicitation of probabilities?

Chapter 2

The Elicitation Context

2.1 How and who?

2.1.1 Choice of format

This chapter concerns the whole process of setting up and carrying out an elicitation of the beliefs of one or more experts. Whereas later chapters deal almost exclusively with the details of what questions might be asked, and how the expert's replies might be interpreted, this chapter deals with the wider context in which that questioning takes place.

First we consider two issues that arise in planning an elicitation.

- *One or more experts*. When expert opinion concerning some uncertain parameter is sought, it is perhaps unusual for the expertise to reside in a single expert. Often, we wish to gather opinions from several experts. Often, we wish to gather opinions from several experts and to synthesise their knowledge and beliefs into a single probability distribution for the unknown quantities of interest. If we elicit the judgements of each expert separately, then it is usual to apply an algorithm to combine their separate distributions, a process known as *mathematical aggregation*. Alternatively, we might elicit their views as a group by bringing them together and treating the group as a single 'expert'. This is called *behavioural aggregation*.

- *Interview versus questionnaire*. Elicitation by a face-to-face interview between the facilitator and the expert involves substantial commitment of human resources. Particularly when elicitation is sought from many experts, a postal or Internet-based questionnaire is an appealing alternative on the grounds of cost. Using a questionnaire has the merit of ensuring that each

Uncertain Judgements – Eliciting Experts' Probabilities A. O'Hagan, C. E. Buck, A. Daneshkhah, J. R. Eiser, P. H. Garthwaite, D. J. Jenkinson, J. E. Oakley and T. Rakow

> expert receives exactly the same information and answers the same questions. On the other hand, it is extremely difficult to formulate a questionnaire in such a way that all experts will clearly understand the questions and the tasks that are required of them. Also, the choice of later questions may depend on the expert's answers to earlier questions. Software has been used in various ways to support elicitation and, at one extreme, may play the part of a questionnaire by enabling an expert's judgements to be elicited without the intervention of a facilitator (see, for instance, O'Hagan, 1998).

For a single expert, there is no doubt that a face-to-face interview is the best approach. It is also likely to be the most cost-effective, since the effort of constructing a questionnaire could rarely be justified for a single respondent. A telephone interview may also be effective, but can be hampered by the impossibility of certain kinds of interactions, such as drawing graphs.

When eliciting from many experts, the administrative ease and low cost of questionnaires is attractive, but the quality of responses is likely to be severely compromised. Experts may misunderstand the questions, and there is the perennial problem of low response rates. Interviews are to be strongly preferred unless a high level of commitment from the experts can be assured and the questions to be asked are relatively simple.

Given that interviews are preferred, there are again administrative and cost benefits from separate elicitations as opposed to interviewing the experts in a group. Genuine experts are busy people, and finding a time and location at which a group could assemble can be a very difficult task. Furthermore, group elicitations tend to take substantial amounts of time, because of the interaction and discussion that takes place between the experts in sharing their knowledge and agreeing on responses to questions. So the cost, in numbers of person-days of expert time to be paid for, can be substantially lower for separate elicitations. However, the group elicitation has the benefit of allowing experts to share their knowledge and to reach a consensus (if they are able and willing to do so), instead of requiring separately elicited beliefs to be combined afterwards.

Methods of eliciting and aggregating the beliefs of several experts are discussed in Chapter 9.

2.1.2 What is an expert?

Our use of the word 'expert' deserves some clarification. In general, elicitation can be used to quantify the knowledge and judgement of any person(s), whether they would commonly be regarded as expert in the relevant topic. Wherever we use the word, 'expert' may, in principle, just mean the person whose judgements are to be elicited, whatever their actual degree of expertise.

In practice, much of the literature is concerned with situations where major decisions are to be made in the presence of substantial uncertainty and where genuinely expert judgement is deemed essential to minimise and characterise the uncertainty. It is then common to elicit the judgements of several experts, and the choice of experts is one of the most important stages of the elicitation process. The success of such elicitation depends on the personality, experience and technical background of individual experts.

The question of what makes an expert is an interesting one. A simple conception is that an expert is someone who has great knowledge of the subject matter. However, expertise also involves how the person organises and uses that knowledge. For instance, Wood and Ford (1993) described four ways in which an expert's approach to problem solving differs from that of an amateur: expert knowledge is grounded in specific cases, experts represent problems in terms of formal principles, experts solve problems using known strategies, experts rely less on declarative knowledge (facts) and more on procedural knowledge (relationships). For good elicitation, we also wish that the expert will be able to assess and express her uncertainty accurately, and expertise in judgement and decision-making (normative expertise) under uncertainty is different from subject-matter expertise (substantive expertise). Some psychological theories of expertise are discussed in Section 3.5.

2.2 The elicitation process

The literature on the elicitation of probabilities suggests various elements as being critical to a good elicitation process. Clemen and Reilly (2001) suggest that any assessment protocol should contain the seven steps: background, identification and recruitment of experts, motivating experts, structuring and decomposition, probability and assessment training, probability elicitation and verification, aggregation of experts' probability distributions. Once the experts have been identified and recruited, Phillips (1999) suggests a four-stage process: introduction and training, motivation, conditioning, encoding. The final stage includes the elicitation of a consensus distribution from the experts. Note that both articles assume that elicitation is required from several experts, but whereas Clemen and Reilly's protocol involves separate elicitations, Phillips' uses group elicitation.

Walls and Quigley (2001) also give an elicitation process, with five main stages and a flow chart giving greater detail. The main components are similar to those described by Clemen and Reilly. Similarly, Shephard and Kirkwood (1994) present five stages via a transcript of an actual elicitation interview (see Section 10.2 for an extended account of this work).

Garthwaite et al. (2005) suggest a four-stage method for the actual elicitation stage: set up, elicit summaries of the expert's distributions, fit a (joint) probability distribution, assess the adequacy of the elicitation. If the elicitation is deemed to be inadequate, the assessor should return to the elicitation stage and repeat the process until adequacy is achieved. (Chapter 8 discusses how the adequacy of an elicitation can be assessed.)

There is broad consensus as to the components of the elicitation process but there are differences in detail, for instance regarding the order in which the components should be carried out. Section 2.2.2 suggests a model for the elicitation process, based on the schemes described above. But first it is helpful to clarify the different roles involved in the whole process.

2.2.1 Roles within the elicitation process

We can identify the following four distinct roles.

1. The decision maker or the person or group that needs to use the results of the elicitation – the client.
2. The substantive expert(s), having relevant knowledge about the uncertain quantities of interest.
3. The statistician, a normative expert, who can give probabilistic training, validate the results and provide feedback.
4. The facilitator, an expert in the process of elicitation, who will manage the dialogue with the expert.

In practice, one person may play different roles, and it is common for the facilitator and statistician to be combined. Similarly, the client may be a substantive expert.

The distinction between the expert(s) and the facilitator is important, although if the substantive expert is also knowledgeable about statistics, one might consider their roles being combined, too, into a single person who acts as the facilitator to elicit his own probabilities. Savage (1971) discusses the difficulties of self-elicitation, mainly vagueness, and proposes using proper scoring rules (see Section 8.2) and practising with others who are eliciting probabilities as ways of overcoming this problem.

2.2.2 A model for the elicitation process

1. Background and preparation

   Clemen and Reilly (2001) suggest, "This first step is to identify those variables for which expert assessment is needed". This is a simple statement, but the task is enormously important and can be difficult. If, through some misunderstanding, the elicited probability distributions do not meet the client's needs then the whole exercise is wasted. For instance, if expert knowledge is to be elicited as prior information for a subsequent Bayesian analysis with new data, then a clear idea of the structure of those data and the appropriate statistical model for them is essential before the parameter(s) whose prior distribution is to be elicited can be specified. Identifying the variables to be assessed would be done by the client, perhaps with involvement from the statistician. This is the stage at which the statistician/facilitator can gain enough substantive expertise to be able to converse with the expert(s) about their field to the required level of competency. This stage also includes planning the elicitation session and preparing any required documents (such as background information for the experts or a pro forma for recording a transcript of the elicitation).

2. Identify and recruit expert(s)

   There may be obvious candidates, or perhaps the client will want to look for experts with alternative points of view. It may be worth investigating the level of statistical understanding, particularly of probability and distributions, that each expert has at this stage, since it may be possible to provide some of the training (see stage 3, below) before the elicitation event. The selection of experts is covered by, amongst others, Hora and von Winterfeldt (1997) in their work in the field of nuclear waste. They recognise the need for the recruitment process to be open to scrutiny, especially when there is considerable public interest. For situations such as nuclear waste, which have high levels of controversy and wide-ranging views attached to them, they recommend the use of a nomination system that includes public interest groups as well as professional organisations in order to gain a balanced perspective. They list six criteria for the experts:

   (a) Tangible evidence of expertise
   (b) Reputation
   (c) Availability and willingness to participate
   (d) Understanding of the general problem area
   (e) Impartiality
   (f) Lack of an economic or personal stake in the potential findings.

   Where it is not practical to recruit experts who satisfy criteria (e) and (f), it is important to record any potential conflict of interest that an expert may have.

3. Motivating and training the expert(s)

   At the start of the elicitation session, it is advisable to explain to the experts why their judgements are required and how what they say can be used. It is also advisable to keep a record of the proceedings, and its purpose and use should be made clear to the expert(s). Clemen and Reilly (2001) note that when the experts are scientists, they may find it outside of their normal practice to make assessments of probability. That their opinions may not be 'correct', in the scientific sense, may cause them to be hesitant to express those opinions. The experts should be assured that uncertainty is natural and that the objective of the elicitation is to capture their knowledge in a form that expresses neither too much nor too little uncertainty. Clemen and Reilly (2001) also point out that "it is important to establish rapport with the experts and to engender their enthusiasm for the project". Fischhoff (2000) proposes obtaining informed consent from experts as an aid to ensuring quality of communication between the facilitator and the expert.

For effective elicitation, the expert should have both general numeracy and some understanding of probability. An expert who is a scientist, or a social scientist used to quantitative work, is highly likely to be numerate but may not have as good an understanding of probability as the statistician would like. In their discussion on eliciting assessments of benzene exposure, Walker et al. (2001) note that "for all of the experts, elicitation of their subjective judgements about variability and uncertainty in a probabilistic form was a new and, for most, a discomforting experience". The probability training of the experts should have three parts:

(a) Probability and probability distributions

(b) Information about the most common judgement heuristics and biases, including advice about how to overcome them

(c) Practice elicitations, particularly using examples where the true answer is known, but unlikely to be known by any of the experts. An example of such a question is the distance between two towns, as used by Garthwaite (1983, p. 167) and O'Hagan (1998).

4. Structuring and decomposition

In his response to three articles on the subject of probability elicitation, Smith (1998) states that in his experience "it is paramount to spend a significant proportion of my time eliciting structure: dependencies, functional relationships and the like". Depending on the field in which the elicitation takes place, some of this could be done by the client, perhaps with some help from the statistician during the preparation stage. But it seems only sensible that such relationships should be agreed to by the expert before they are used: the expert may wish to modify them and will, almost certainly, have new insights of her own. In the same comment, Smith also stresses the need for the result of any elicitation exercise to be owned by the expert(s), in the sense that they feel it adequately represents their understanding and beliefs and that they could give a convincing explanation to a colleague. One of the problems of conducting separate elicitations with a number of experts is that they may not all find the same structuring of the problem natural or helpful.

The quantities that are to be elicited must be defined precisely, including a specification of their units of measurement. It is important to review with the experts the available evidence upon which they will draw while making judgements about these quantities. This serves the dual purpose of placing the evidence base on record and helping the expert(s) recall all relevant pieces of evidence before making any specific judgements.

5. The Elicitation

   It is at this stage that the iterative process identified by Garthwaite et al. (2005) fits in:

   (a) Elicit specific summaries of the expert's distribution
   (b) Fit a probability distribution to those summaries
   (c) Assess adequacy: if adequate stop, if inadequate repeat the process, asking the expert to make adjustments.

Much of this book deals with the very substantial body of literature concerning step 5, the elicitation itself. However, it is extremely important in practical elicitation to give full consideration to all the other steps.

2.3 Conventions in Chapters 3 to 9

We adopt the following conventions in the following chapters.

1. We suppose that the elicitation is conducted between expert(s) and a facilitator; that is, we do not distinguish between the roles of the facilitator and the statistician, and we suppose that the role of the client has already been fulfilled in planning the elicitation.
2. To avoid repetition of phrases such as 'he or she', 'his or her', we will write as if the expert is a female and the facilitator is a male (However, when referring to real instances of elicitation the true genders of the participants will be used.).
3. In Chapters 3 to 8, we suppose that the elicitation is conducted with a single expert. The case of multiple experts is considered in Chapter 9.

2.4 Conclusions

2.4.1 Elicitation practice

- Elicitation is best conducted as a face-to-face interaction between the expert(s) and the facilitator.
- The elicitation process has several stages, as in Section 2.2.2, all of which are important. The actual elicitation comes at the end, and its success depends on the foundations built in the preceding stages.

2.4.2 Research question

- The value of many aspects of the overall elicitation process has not been formally studied. For instance, does keeping a transcript of the interview help the expert(s) to commit more fully to the exercise or might it inhibit them?

Chapter 3

The Psychology of Judgement Under Uncertainty

3.1 Introduction

3.1.1 Why psychology?

In this chapter, we review some of what is known about the way in which people form judgements about uncertain quantities: how they assess probabilities or make predictions about the future value of some quantity. This involves considering a body of mainly psychological research and theory. In expert elicitation, we rely upon the knowledge and the data that reside in the mind of the expert. Making this knowledge accessible involves processing information. Psychology is often described as the science of the mind or the study of behaviour; and the dominant view within modern psychology conceives behaviour as a series of information-processing tasks. Therefore, the processes at play within elicitation fall fair and square within the remit of psychology.

Our purpose in reviewing this material is practical. We wish to know how best to elicit uncertain quantities. The literature that we review identifies some common pitfalls that people encounter when they make judgements under uncertainty. Knowing about these pitfalls is the first step to avoiding them. However, simply knowing some of the ways in which a judgement can go awry may not be sufficient to avoid difficulties or reduce errors to an acceptable minimum. We think it is important to understand some of the principles and processes that, as evidence suggests, lie behind these potential problems. Armed with an understanding of these processes, anyone engaged in practical elicitation ought to be better equipped to counter the problems associated with them. It should also help those

Uncertain Judgements – Eliciting Experts' Probabilities A. O'Hagan, C. E. Buck, A. Daneshkhah, J. R. Eiser, P. H. Garthwaite, D. J. Jenkinson, J. E. Oakley and T. Rakow

involved in elicitation to foresee and avoid further problems that have the same root cause, even if we have not discussed them in this book. To this end, we also summarise some of the key theoretical ideas from the field of judgement under uncertainty.

For those who wish to pursue this literature in greater detail, there is a large body of work to examine that falls within the sub-discipline known either as *behavioural decision-making* or as *judgement and decision-making*. Work published in this area can be found in journals from several academic fields including psychology, economics, management and consumer sciences, medicine, law and decision analysis. For those interested in more detail concerning judgement under uncertainty, two edited volumes are particularly worthy of mention. Gilovich et al. (2002) provide an overview of the recent developments in research on judgement under uncertainty; and Koehler and Harvey (2004) survey the key topics of judgement and decision-making. Readers preferring an accessible textbook would do well to examine Hastie and Dawes (2001).

3.1.2 Chapter overview

We begin by examining some questions concerning the capabilities of the human brain (Section 3.2). Is it well equipped for the kinds of intuitive statistical tasks that we might require of the expert in elicitation? What is the nature of the processes that people use to form judgements? What are the consequences of this for elicitation? In Section 3.3, we provide some background to the research that we discuss in this chapter. We consider the populations of participants that have been studied in the research on judgement under uncertainty, and discuss the extent to which research with non-expert populations can inform our knowledge of how experts form judgements. We close this section with a brief overview of some early experiments on probability judgement.

In Section 3.4, we examine the body of psychological research and theory that has had the greatest impact on ideas about judgement and uncertainty: the heuristics and biases research programme. Here we describe the 'quick-and-dirty' strategies that often seem to replace more effortful processes when people are asked to assess probabilities or uncertain quantities. Several errors or biases are associated with these strategies, which have considerable practical significance for elicitation exercises. Consequently, these biases are discussed in detail, drawing upon empirical results and theoretical insights.

Because this book focuses on expert elicitation, Section 3.5 provides an overview of the research on experts and the nature of expertise. We end the chapter by summarising the key points of three 'broad-brush' theories that say something about the way in which experts might apply their knowledge when assessing uncertain quantities (Section 3.6). Chapter 4 draws heavily upon the material in this chapter and continues our exploration of some questions of a more psychological nature. How well calibrated are subjective probabilities? In other words, how well do the probabilities that people state for events correspond to the observed relative frequency of those events? How do people interpret the tasks they encounter in elicitation? What specific steps can be taken to reduce bias and error in an elicitation exercise?

3.2 Understanding the task and the expert

3.2.1 Cognitive capabilities: the proper view of human information processing?

How well equipped are people to undertake the kinds of statistical tasks that we require of them when we seek to elicit statistical information from experts? How faithful to what they have observed or experienced is their information processing? Psychological research into these questions took off in the 1950s. It has continued for half a century and has influenced thinking in many areas, including economics, management and medicine. In an influential review of the first decade or so of this research, Peterson and Beach (1967) were generally positive about people's ability to process the data that they had observed. In general, when called upon to estimate descriptive statistics, people were good 'intuitive statisticians'. Estimates of the sample proportions or sample means of recently observed samples are accurate. Estimates of variance are not quite so successful; they increase appropriately with increasing sample variability but they are influenced by the sample mean, as if people judge the coefficient of variation (i.e., the standard deviation divided by the mean). People are often poor in assessing the degree of association between a pair of binary variables. However, they generally fare better in assessing the strength of correlation when many-valued dimensions are involved. Peterson and Beach were also generally positive about human inferences from samples in relation to the normative model: "Inferences made by subjects are influenced by appropriate variables and in appropriate directions ... [though] the degree of sensitivity is often less than optimal." (Peterson and Beach, 1967, pp. 42–43). Within a decade, a very different view prevailed among many psychologists. Typical of this view is Hogarth's (1975) statement that "man is a selective, stepwise information-processing system with limited capacity, ... ill-suited for assessing probability distributions." (p. 273)

Why such starkly differing summaries? Much of this shift was prompted by Tversky and Kahneman through their *heuristics and biases* research programme (e.g., Tversky and Kahneman, 1974). Their focus was on the *process* of judgement under uncertainty. Results obtained in a sequence of clever experiments and observations indicated that when people judge probabilities, assess associations or estimate quantities in uncertain environments, they often use heuristics (short cuts or approximate strategies that use only limited information). These heuristics are efficient – they can operate with limited information in a limited amount of time. Kahneman and Tversky stated that these heuristics could often be relied upon to provide answers that were good (i.e., *sufficiently* accurate for the purpose), but their findings demonstrated that they were error-prone. Errors were not distributed randomly; rather, there were systematic deviations from the optimal solution or predictable violations of probability theory. These *biases* were the markers for the processes that Kahneman and Tversky proposed. They indicated that human judgement can go awry when judgements made using these quick and easy-to-use heuristics are substituted for more complex assessments involving more effort. Errors and biases were sufficiently numerous to leave the reader with the impression that man is a 'cognitive miser' (Taylor, 1981) with insufficient processing capacity or insufficient motivation to make sound judgements under uncertainty. The implications of this for successful elicitation ought to be clear.

All agree that the demonstration that judgement *can be flawed* does not imply that it always *will be flawed*. Nonetheless, there has been considerable difference of opinion expressed over just how good or bad judgement under uncertainty is and the circumstances under which it is better or worse. In the sections that follow, we consider the research that has fuelled this debate, highlighting the findings that are relevant to the practicalities of elicitation. We focus upon processes (the strategies and means by which people make probability judgements) and upon performance (how accurate or appropriate these judgements are).

3.2.2 Constructive processes: the proper view of the process?

From the perspective of cognitive psychology, many would view the term 'elicitation' as something of a misnomer. It implies that there is some information or well-formed opinion already encoded in the mind of the expert and that our task is to extract it or to encourage the expert to 'read it off' in a suitable format. However, psychologists have long viewed memory, judgement and estimation as constructive processes, and do so with strong justification. In contrast to a crude view according to which items of information are accumulated one by one in memory in a kind of 'storage bin', both classic (Bartlett, 1932) and more recent theoretical approaches regard memory as organised in terms of patterns of learnt associations. In other words, different thoughts and feelings will tend to activate (or bring to consciousness) other thoughts and feelings with which they have been associated (e.g., Bower, 1981). Among other things, this means that the probability of a person calling to mind any particular piece of information or knowledge in response to any given question (whether concerning self-reports of behaviour, attitudes, preferences or estimates of probability) will depend very much both on their own personal experience and on situational cues. Such cues can include the way in which a question has been asked and other questions that are asked within the same testing session (Schwartz and Strack, 1991). Therefore, we cannot assume that people always encode and decode information faithfully or that they resist interference from irrelevant sources.

3.3 Understanding research on human judgement

3.3.1 Experts versus the rest: the proper focus of research?

It is a common criticism of academic psychologists that too large a proportion of the conclusions they draw are based on the empirical investigation on undergraduate students. For instance, approximately 75% of published social psychology research has undergraduates at American universities as its subject population. In fact, over 50% of studies use American *psychology* undergraduates (Sears, 1986). The main accusation is that this limits the generalisability of results. The behavioural decision-making literature from which we draw much of this chapter does examine a variety of study populations, but, arguably, it is focused more on undergraduate participants than is desirable.

One standard defence is that when conclusions are based upon the experimental method (in which an independent variable is manipulated and the attendant changes in a dependent variable are observed), our concerns over the precise make-up of study populations should be reduced. If an experimental intervention reliably improves performance in one study population, we should consider it to be quite exceptional that the same manipulation would have the opposite effect in another population (so the defence runs). For instance, we might expect the average psychology undergraduate to be more intelligent than the average citizen of his or her city and less intelligent than the average Nobel Prize winner. However, the learning strategies that enhance the memory for factual material in one individual are likely to be beneficial for other individuals. The magnitude of an effect may vary across populations, but the direction of the effect will vary only infrequently. If so, the broad message of the research will generalise, from the student population to the experts whose judgements we wish to elicit in practice. Nonetheless, this is a moot point and, certainly, it cannot be blindly assumed that the findings obtained with one group of people will necessarily generalise to another. For instance, in medicine it is not uncommon to find drugs that are effective on one group of patients but not on another. In this chapter, and in Chapter 4, we consider many studies with undergraduate populations. However, where possible, we report whether the findings of the research have been seen to generalise to experts.

However, a second defence to the charge against the psychologist is worth mentioning – namely, that we should not confuse the 'professional' with the 'expert'. People can be experts in domains and tasks for which they receive no pay, and professionals may not have the kind of substantive expertise that is desired for a particular elicitation exercise. For instance, many studies of human judgement have examined predictions relating to the outcome of sporting events. This is a domain to which many people devote great amount of time, money and interest to acquire knowledge (arguably sometimes a level of knowledge beyond that which they are required to obtain within their professional life). And, especially for the keen gambler, much of their mental representation of this domain is distinctly probabilistic in nature (Johnson et al., 1991). Similarly, students should be expert in many of the cognitive tasks that they are given in judgement research (e.g., they have had plenty of opportunity to learn about their chances of answering multiple-choice questions correctly, see Section 4.2).

3.3.2 Early research on subjective probability: 'conservatism' in Bayesian probability revision

Edwards is credited by many with introducing psychologists to Bayesian statistics and with prompting the interest in subjective probability among psychologists that spawned the sub-discipline of behavioural decision-making (Kahneman et al., 1982). Edwards pioneered an experimental paradigm for investigating decision-making that became known as the 'book-bag and poker-chip paradigm' (for reviews see Edwards, 1982; Fischhoff and Beyth-Marom, 1983; Rapoport and Wallsten, 1972). A typical experiment would proceed as follows. The participant is faced with two urns (or 'book-bags'): in one urn, the ratio of white to black balls (or 'poker-chips') is 3:2, in the other it is 2:3. One urn is selected at random, and a sample of balls is selected. The participant then estimates the probability that the sample came from the urn with a black-to-white ratio of 3:2. Participants' responses were then compared with the relevant posterior probabilities computed using Bayes' theorem. Variations that were examined included the consideration of *optional stopping* (sampling with replacement ball by ball at a cost until sufficiently confident to bet upon which urn was being sampled from) and *fixed stopping* (specifying in advance the number of 'experiments' one wished to pay to observe before betting which urn was being sampled from). Some studies were of the abstract kind described above, while others involved more context-rich real-life decision scenarios. The behaviour of participants often, though not always, implied *conservatism in the revision of subjective probabilities*. In other words, revisions in the light of new evidence were not as large as Bayes' theorem would prescribe, and participants purchased more than the optimal amount of information before making decisions. Thus, participants in these experiments 'do not extract as much certainty from information as Bayes' theorem would permit' (Edwards, 1965, p. 325).

Following these early experiments, various lines of research have suggested that people do not appreciate the true value of the information that can be used to update or improve their judgement. For instance, the amount that people are willing to pay for information is often insensitive to the factors that should affect it but sensitive to factors that should not affect it (Connolly and Thorn, 1987; Hershman and Levine, 1970). People purchase too little 'good' information and too much 'poor' information.

3.4 The heuristics and biases research programme

Tversky and Kahneman initiated the heuristics and biases research programme, in which they examined the kinds of instinctive processes that people employ when making judgements under uncertainty. Many disciplines have drawn upon this work, and the ideas explored by Tversky and Kahneman have been widely discussed and frequently cited. One key aspect of their programme of research is the notion of a heuristic, which is a strategy that is quick and easy to use. The term can be applied to strategies that are used to solve problems, make decisions or form judgements. It is these heuristics that are used when making judgements that concern us here, for, as we have argued already, a probability assessment is necessarily a judgement (Chapter 1). The speed and ease with which a heuristic can be used to form judgements derives from its requirement for limited information processing. A heuristic may make use of only some of the information that could be used in a task and/or it may require only a shallow level of processing.

A classic paper in the journal *Science* set out three heuristics, namely, *availability, representativeness* and *anchor-and-adjustment* (Tversky and Kahneman, 1974). Judgements are made using the availability heuristic according to how easily instances come to mind. Judgements made using the representativeness heuristic can often be characterised as judgements made according to the similarity between instances and expectations of those instances. When using the anchor-and-adjustment heuristic, quantitative judgements are the result of simple adjustments from an initial starting value. It is assumed that these heuristics are sufficiently instinctive or automatic that it may require conscious effort to stop oneself from employing them and to use more effortful strategies instead. Tversky and Kahneman referred to judgements made using the representativeness and availability heuristics as 'natural assessments', whereby a simple judgement of availability or representativeness often serves as a proxy for a more complex judgement of probability. In this section, we review the research that indicates the problems that can be associated with this.

For detailed coverage of this material, readers may refer to a pair of edited volumes (Kahneman et al., 1982; Gilovich et al., 2002). There are many suitable articles that provide a brief overview (e.g., Tversky and Kahneman, 1974; Arkes, 1991; Fraser and Smith, 1992) and texts that deal with this research programme at rather greater length than is possible in a journal article (e.g., Hastie and Dawes, 2001). We review the key messages of this programme in the following sections. Some material is of direct relevance to how the elicitation process is structured. Other aspects are of interest because they bear more generally upon how people habitually reason about probability and uncertainty.

3.4.1 Availability

The availability heuristic is used to judge the probability of an event or the frequency of class membership (Tversky and Kahneman, 1973, 1974). Estimates are made according to the ease with which events or instances are called to mind. The more easily examples of events are recalled, the more probable this event is judged to be. The more easily instances of a class come to mind, the more sizeable this class is judged to be. Thus, a simple and instinctive judgement of availability (how easily things come to mind) is used in place of the complex processes that might otherwise be used to judge the probability or frequency. The heuristic is effective on many occasions because frequent events and large classes generate more examples that can be called to mind than infrequent events and small classes do. However, some biases are associated with the use of the heuristic. This is because there are factors other than frequency or probability that can influence the *ease* with which examples and instances come to mind.

Probably some of the most compelling examples of the dissociation of availability and event frequency are when very low frequency events receive considerable media coverage. The disproportionate media attention given to particular high-impact or newsworthy events will increase the ease with which examples come to mind. Therefore, the relative frequencies of well-publicised events (e.g., tornados) tend to be overestimated, whereas the relative frequencies of more mundane events (e.g., common influenza) are often underestimated (Slovic et al., 1979).

Events that have occurred recently, or that have particular personal significance will also come to mind more easily. For example, I might judge the probability of having a car accident by recalling instances when friends and associates have had a car accident. If my brother has recently had an accident, my subjective probability of having a car accident will temporarily rise because of the salience of the event.

One bias associated with the availability heuristic that is of particular significance for elicitation is illusory correlation: the erroneous belief that two uncorrelated events have a statistical association. For instance, people sometimes misreport the frequency of co-occurrences of pairs of events that they have observed (Chapman and Chapman, 1971). They report more co-occurrences than they actually saw when prior beliefs led them to expect that a particular pair of events would occur together frequently. Roughly speaking, the assumption is that this misperception arises because co-occurrences of the two events come to mind easily. However, this is not because they have been observed together frequently, but because of other reasons that are strongly associated in the mind. For instance, the belief in a casual mechanism that relates two events can also set up a strong associative bond. Putting it simply, prior beliefs about the reasons why two variables should correlate may blind people to the data that indicate that these variables are uncorrelated (Gilovich, 1991).

The possibility of illusory correlation has several implications for elicitation. Sometimes an expert is asked to estimate the value of a correlation coefficient or to provide judgements from which a correlation coefficient can be calculated (Clemen et al., 2000). In addition, beliefs about the correlation between variables are a part and parcel of the substantive expertise that the expert is expected to bring to the elicitation exercise. Consequently, her beliefs about the relationship between two variables could inform the expert's judgements for all kinds of uncertain quantities (e.g., the assessment of conditional probabilities or the estimation of frequencies). In any of these situations, it is clearly less than ideal if the expert has been misled as to the true strength of the relationship between two variables. Further discussion on the assessment of correlation and explicit conditional probabilities can be found in Chapter 5 (see especially Section 5.3).

Several other biases associated with the use of the availability heuristic have been identified, though their pertinence to elicitation is not so obvious. For those wishing to examine these biases in greater detail, the book edited by Kahneman et al. (1982) includes several of the important early papers on availability. For more recent discussions of the availability heuristic see Gilovich et al. (2002) or Sedlmeier and Betsch (2002).

In many elicitation situations, the potential influence of the availability heuristic will be indirect rather than direct. For instance, when eliciting probabilities pertaining to a unique future event, judgements cannot be influenced by the ease of recall of previous instances if no such instances exist. However, if the expert's reasoning for this event is influenced by her beliefs about classes of similar events, for which availability may be influential, then this can influence the outcome of such an elicitation. For example, when asked to assess the probability that mean life expectancy will exceed 10 years for a new cancer drug, X, there are no instances of 10-year survival for drug X to call to mind. However, such an assessment may reasonably be influenced by the relative frequency of existing cancer drugs that offer mean life expectancies in excess of 10 years, which could be made using the availability heuristic. The ease with which examples of drugs in this category come to mind may well be influenced by things other than the proportion of drugs offering a mean life expectancy of 10 years or more. Such reasons could include media coverage, advertising or other forms of publicity.

3.4.2 Representativeness

The *representativeness* heuristic is used to assess the probability that A belongs to some class B or that A originates from some process B. Specifically, Tversky and Kahneman (1983) define representativeness as "an assessment of the degree of correspondence between a sample and a population, an instance and a category, an act and an actor or, more generally, between an outcome and a model" (Tversky and Kahneman, 1983, p. 295). This natural or instinctive assessment of representativeness can be used as a proxy for an assessment of probability. At this most general level, many kinds of outcome can be assessed in this way, and the 'model' assessed may be one of many kinds of entity (e.g., a person; an object; or a biological, mechanical or economic system). This heuristic is applicable to the assessment of single event probabilities for non-repeatable events. Therefore, we describe the research on the representativeness heuristic in some detail, as it is directly applicable to the kinds of one-off events that are typically the focus of an elicitation exercise.

Representativeness is particularly pertinent to the elicitation of conditional probabilities. Using representativeness, the probability is assessed by how 'representative' outcome A is of model B. In cases where A and B are described in the same terms, representativeness can be reduced to 'similarity' (Tversky and Kahneman, 1983). In natural language, we might speak of the 'degree of resemblance between

A and *B*' or of 'how typical *A* is of *B*'. *A* will be judged to have a high probability of belonging to class *B* if *A* is similar to *B*, or *A* will be judged to have a high probability of originating from process B if it is typical of what can be expected of process *B*. Thus, intuitive evaluations of the conditional probability $P(A \mid B)$ are made using representativeness by assessing the similarity between *A* and *B*. The problem with this kind of natural assessment is that there are factors that should influence probability judgement that do not influence similarity or representativeness. Thus, while representativeness tends to covary with probability, it is not determined by it. Therefore, representativeness can be at odds with the actual relative frequency (e.g., specific outcomes can be representative although unlikely, such as when diagnostic outcomes are representative of a disease, yet are improbable). Furthermore, representativeness is not bounded by class inclusion. An instance can be unrepresentative of a category yet highly representative of a subset of that category (Tversky and Kahneman, 1983; see also Tversky and Kahneman, 1982).

Several biases have been described that have been attributed to the representativeness heuristic, including the conjunction fallacy, base rate neglect, insensitivity to sample size, misperception of chance and randomness, and confusing $P(A \mid B)$ with $P(B \mid A)$ ('confusion of the inverse') (e.g., Kahneman and Tversky, 1972, 1973; Eddy, 1982; Tversky and Kahneman, 1983). As it may be necessary to watch out for these biases in an elicitation exercise, these types of errors are briefly described below.

3.4.2.1 The conjunction fallacy: an example of incoherence in probability assessment

Consider this frequently researched example:

Linda is 31 years old, single, outspoken and very bright. She majored in philosophy. As a student, she was deeply concerned with issues of discrimination and social justice and also participated in anti-nuclear demonstrations. Please tick the most likely alternative:

(1) Linda is a bank teller.

(2) Linda is a bank teller and is active in the feminist movement.

Most people (typically more than 80%) rate the second option as more likely than the first one. However, it cannot be more likely that Linda is a bank teller and a feminist than that she is a bank teller, as the conjunction of two events can never be more probable than either event separately. Hence, this error is referred to as the *conjunction fallacy* (Tversky and Kahneman, 1983). Representativeness accounts for this, as the description of Linda is unrepresentative (untypical) of the stereotypical bank teller, but more representative of the stereotypical feminist. Thus, the addition (in the second alternative) of 'active feminist' to bank teller (in the first alternative) increases the similarity between the description of Linda and the second alternative.

Teigen et al. (1996a, 1996b) also explored conjunction fallacies using two real-life situations: the outcome of the national political referenda and the combinations of football matches in the 1994 World Cup. In the political referenda study, participants assessed probabilities such as *P*(Finland votes to join the European Union), *P*(Norway votes to join the European Union) and *P*(both Finland and Norway vote to join the European Union). In the football study, participants assessed probabilities such as *P*(Brazil beats Norway), *P*(Brazil beats Costa Rica) and *P*(Brazil beats both Norway and Costa Rica). They found that the conjunction fallacy is still committed by a large proportion of people in the referenda study, though conjunction fallacies occurred less frequently in the World Cup study than in previous laboratory studies. In the referenda study, conjunction fallacies occurred least when participants made judgements for referendum outcomes contrary to their own voting preferences.

A number of alternative accounts of why conjunction fallacies occur have been put forward that do not appeal to representativeness (see Fisk, 2004). For instance, Fisk and Pigeon (1998) and Fisk (2002) invoke Shackle's (1969) theory of 'potential surprise', which proposes that subjective probabilities are (inversely) related to the surprise that we would experience if a particular event occurred. The probability assigned to a conjunction of events is then largely determined by the surprise associated with the event with the lower probability (e.g., 'Linda is a bank teller').

3.4.2.2 Base rate neglect (or the base rate effect): ignoring (under-weighting) prior probabilities

The base rate of an event is another term for the relative frequency of an event within some defined class. The base rate is pertinent to assessing the probability of an event conditional upon further information. For instance, it could serve as the a priori probability for the event when Bayes' theorem is used to revise probability judgement in the light of new information. In base rate neglect, people ignore the average frequency, relative frequency or a priori probability of an event. They base probability judgements only upon information about the individual instance currently being assessed, even when this 'individuating information' has limited diagnostic power. Thus, a particular event, known to be rare in the physical world, may be assigned a reasonably high probability on the basis of imperfect indicators if these indicators are judged to be representative of the event. 'Base rate effect' has replaced the term 'base rate neglect' as even the most striking examples of inappropriate use of prior probabilities indicate some (if insufficient) sensitivity to prior probabilities (Koehler, 1996).

The summary of many experiments on this phenomenon is that in many circumstances people are insufficiently sensitive to the base rate occurrence of an event. However, the degree of insensitivity varies considerably with a host of task characteristics, and the manner in which normative principles have been applied in the analysis of this phenomenon is not without controversy (Koehler, 1996). For instance, Gigerenzer and others have shown how frequency representations can improve the reliance upon base rates (see Kurzenhäuser and Lücking, 2004).

For instance, frequency formats (e.g., '8 out of 173 patients') encourage answers to diagnostic reasoning problems that appropriately incorporate disease prevalence. Others are adamant that the investigated problems lend themselves to multiple legitimate interpretations by participants that invalidate the conclusions of experimenters (e.g., Cohen, 1981).

3.4.2.3 Insensitivity to sample size and the 'law of small numbers': inappropriate inference

Sampling theory tells us that the size of a sample greatly affects the probability of obtaining certain results. However, people often ignore the sample size and only use superficial similarity measures when making inferences. For example, people ignore the fact that (or underestimate the extent to which) larger samples are less likely to deviate from the mean than smaller samples.

Tversky and Kahneman (1971) describe several situations in which people seem to apply the '*law of small numbers*' (a deliberate play on words in reference to Bernoulli's law of large numbers). The law of small numbers is an example of erroneous judgement that properties of a population are manifested in small samples. Unfortunately, even experts underestimate the variability in the statistical properties of small samples and are consequently overconfident with regard to what can legitimately be concluded about the population parameters (Tversky and Kahneman, 1971).

3.4.2.4 'Confusion of the inverse': misunderstanding conditional probabilities

A bias or difficulty that has been frequently noted is to confuse a conditional probability with its inverse. In other words, people act as if $P(A \mid B)$ and $P(B \mid A)$ were the same thing (Villejoubert and Mandel, 2002). For instance, doctors are often seen to confuse test sensitivity, P(positive test | disease), with the predictive value of a positive test, P(disease | positive test). They can also confuse test specificity, P(negative test | no disease), with the predictive value of a negative test, P(no disease | negative test) (Eddy, 1982).

This can be argued to be consistent with judgement by representativeness. Consider the case where a positive test is seen as representative of the presence of the disease, and the presence of the disease is seen as representative of a positive test. In this instance, representative thinking offers the opportunity for a perceived symmetry that does not exist in the world, namely, P(positive test | disease) = P(disease | positive test) (Hastie and Dawes, 2001). Representative thinking seduces one into this error in the following manner. From Bayes' theorem, disease prevalence P(disease) is used to calculate P(disease | positive test). However, P(disease) has no bearing upon how representative a positive test is of the presence of the disease or how representative the presence of the disease is of a positive test. Thus, P(disease) is inappropriately excluded from consideration. A more prosaic explanation of the confusion of the inverse is that it arises from semantic confusion reflecting a rather poor conceptual grasp of conditional probabilities that many people suffer from (Koehler, 1996). Whichever explanation one favours, this is an important error to be aware of in elicitation.

3.4.2.5 Insufficiently regressive predictions: translating rather than predicting

When giving predictions for values of one variable conditional upon values of a second variable, people often fail to take proper account of the fact that the two variables are not perfectly correlated (Kahneman and Tversky, 1973). For example, consider the task of predicting the height of males from their weights and assume that the correlation coefficient between height and weight is 0.5. For a male of weight one standard deviation above the mean weight, the best prediction of his height lies 0.5 standard deviations above the mean height. For a male of weight one standard deviation below the mean weight, the best prediction of his height lies 0.5 standard deviations below the mean height. However, the tendency is for people to predict heights far too close to one standard deviation above the mean height or one standard deviation below the mean height, respectively, for the two predictions. The data that Kahneman and Tversky (1973) present suggest that people are inclined to make predictions just as if they were translating one scale into another scale, placing each new value at a point of equivalent extremity to its position on the original scale.

The explanation in terms of the representativeness heuristic is as follows. Knowing that weight and height are correlated, people view being tall as representative of being heavy and being short as representative of being light. Consequently, the extent to which someone is tall largely determines the extent to which they are predicted to be heavy. By thinking of the association between height and weight in this way, people are inclined to ignore the fact that the association is imperfect when they make predictions. It is as if they overlook the fact that, although tall people tend to be heavy, not all are. The result is predictions that are insufficiently regressive to the mean (i.e., there is too much variability in the set of predicted values).

One important implication of this bias is that several methods of elicitation used for assessing dependence or correlation require a series of judgements for one variable conditional upon a series of values for another variable (Clemen et al., 2000). These methods are discussed in Section 5.3. One prediction that can be made on the basis of this bias is that correlation coefficients elicited in this way will tend to imply stronger correlations than is warranted. Equivalently, the magnitude of the regression coefficients elicited according to these procedures will be too large. Of course, this prediction only holds if the expert makes use of the representativeness heuristic. In Subsection 4.5.3, we discuss some specific proposals for overcoming insufficient regressiveness in judgement.

3.4.3 Do frequency representations remove the biases attributed to availability and representativeness?

Perhaps the errors in judgement that we have described occur only because the questions that elicit them are phrased in such a way as to encourage biased reasoning? Can simply expressing the question in a different way remove bias? If we know the 'right' way to phrase the question, will we get a good answer? If yes, perhaps we could be positive about the prospects of eliciting successful assessments for uncertain quantities. Several studies have examined whether changing the way in which uncertainty is described can improve probability judgement.

Several experiments have found that representing probabilities using frequencies rather than single-event probabilities can improve performance on many of the problems described above (e.g., see Koehler, 1996). For instance, Fiedler (1988) asked "Of 100 individuals like Linda, how many would be bank tellers, and how many would be active in the feminist movement and bank tellers?' The incidence of conjunction fallacy errors fell to 20%, whereas variations on the problem that describe a 'unique Linda' almost always observe conjunction fallacy errors in over 50% of participants (Tversky and Kahneman, 1983). Gigerenzer (1996b) observed that frequency formats improve adherence to Bayes' theorem in diagnostic testing problems, thereby greatly reducing the tendency towards confusion of the inverse. Griffin and Buehler (1999) found that relative frequency formats reduced the magnitude of the base rate effect but found no equivalent advantage with two other types of problem that they examined.

Jones et al. (1995) make an important observation from a series of experiments that they conducted. They note that those biases that were discussed under the heading of the representativeness heuristic (Section 3.4.2) are primarily associated with single-case judgements. On the other hand, those biases attributed to the use of the availability heuristic (Section 3.4.1) primarily occur in judgements of relative frequency where reference classes of multiple cases are specified. Jones et al. (1995) examined standard problems that have previously been found to elicit biased responses. In some problems, the biases under consideration had been attributed to representativeness. In other cases, the biases examined were those consistent with the use of the availability heuristic. An interesting dissociation was observed. Frequency formats reduce the incidence of bias in representativeness problems in comparison to single-event formats. However, frequency formats make biased responding *more likely* in availability problems.

Practically speaking, it will sometimes make little sense to employ frequency formats in elicitation. For instance, it is meaningless to estimate how many times out of 100 a new drug's mean efficacy will exceed a given level, since there will only ever be one population mean value for this drug. However, what may still be of value in such instances is to appeal to frequency-based representations of probability in the training or problem structuring that precedes elicitation. One message from this research is that it may help experts to think of probabilities of one-off events as being 'analogous' to frequencies of repeatable events. Indeed, it is common to explain that a probability of, say, 0.7 is such that

70% of events with this probability would, in the long run, occur (a calibration explanation).

3.4.4 Anchoring-and-adjusting

3.4.4.1 The anchor-and-adjustment heuristic

When people are asked to estimate a quantity or assess an uncertainty, they often start with an initial estimate (an 'anchor') and then adjust up or down. The strategy is particularly efficient for repeated judgements for the same kind of situation, where the value from a previous instance can serve as an anchor and then adjustment can be made according to the differences between the previous and the current situation. This can reduce mental processing time by removing the need to re-process information or to re-perform calculations on the basis of information common to both situations. Unfortunately, people tend to stick too closely to the initial value, not adjusting sufficiently. The classic demonstration of this involves asking whether a certain quantity is greater than or less than some arbitrary (anchor) value X, before asking for an estimate of the quantity. If different anchors are provided, answers are seen to be systematically affected by the anchor provided. Answers cluster around the anchor, which is consistent with insufficient adjustment. The effect persists even when anchors are *seen* to be arbitrary and to contain no informational value (e.g., they are generated randomly by a spinner). Furthermore, anchors need not be in the same units as the response to be elicited in order to have an impact. For instance, Wong and Kwong (2000) found that anchoring on a larger numerical value of length (7300 m rather than 7.3 km, by means of the question "Is the length of the runway of the Hong Kong Kai Tak Airport longer or shorter than X?") resulted in higher subsequent estimates of the cost of a bus in tens of thousands of dollars.

As with the availability and representativeness heuristics, anchoring-and-adjusting is seen as an intuitive strategy. Anchors can be adopted relatively spontaneously, and, unless people are prompted to reflect upon their own judgement processes, they may not be particularly conscious of the anchors that influence their judgement. (See also the related discussion of the effects of explicit anchors upon judgement in Section 1.4.1.)

Adverse effects resulting from the use of the anchor-and-adjustment heuristic ('anchoring effects') are of particular concern for elicitation, as they can apply to *any* kind of quantitative judgement. Consequently, *both* assessments of probability *and* judgements of observable quantities can be subject to anchoring effects. For example, the statistician may ask the expert to assess the probability that the working lifetime of a power station exceeds 30 years. Alternatively, the expert may be asked to state the date at which there is a probability of 0.9 that the power station is still operating. Both kinds of judgement are quantitative and, therefore, there are risks associated with the introduction of anchors. This is particularly so when considering a series of related questions such as when assessing the probability that the working lifetime of the power station exceeds 20 years, then 25 years, then 30 years, and so on. There is a danger that the previous assessment acts as an anchor for the assessment currently under consideration. Obviously, the two assessments *should* have some relation to each other. However, the danger is that they are closer together than they should be because of the tendency to adjust insufficiently from an anchor.

3.4.4.2 Anchoring effects

Previously, we described conservatism in the revision of subjective probabilities (Section 3.3.2), which is the tendency to revise probabilities insufficiently upon the receipt of new information. This can be explained as an anchoring effect. If the prior probability provides an anchor from which people adjust insufficiently, posterior probabilities will consequently be insufficiently extreme. Hogarth and Einhorn (1992) consider the sequential updating of probability judgements on the basis of several items of information and describe a model of belief updating that takes account of anchoring effects. Their data show how the order in which items of information are considered can affect the final assessment that is made (even though identical sets of items are considered).

The anchor-and-adjustment heuristic has important implications for the nature and ordering of questions when eliciting any numerical quantity. For instance, Wright and Anderson (1989) observed strong anchoring effects in experiments where participants first judged whether the event probability was above/below 0.25 (low anchor) or 0.75 (high anchor). The tendency for probability judgements to be greater in the high anchor condition persisted across scenarios of differing familiarity for the participants and when financial incentives were offered for good performance (though the effect was reduced when these incentives were high).

More subtle anchoring effects are also worthy of attention. Consider the well-known 'pruning bias' observed with 'fault trees' (e.g., Fischhoff et al., 1978a). A typical experiment is as follows. Mechanics are asked to consider reasons a car might fail to start, and to assign probabilities to different reasons. One group of participants considers seven categories (branches of the fault tree): 'Battery charge insufficient', 'Starting system defective', 'Fuel system defective', 'Ignition system defective', 'Other engine problems', 'Mischievous acts of vandalism' and 'All-other-problems'. A second group of participants considers a 'pruned fault tree' where, for example, the final four categories in the seven-branch tree are subsumed into a single 'All-other-problems' category. Responses across the two conditions are systematically mismatched. Directly equivalent categories (e.g., 'Battery charge insufficient') are rated more probable when four categories (rather than seven) are explicitly named. The differences in probability can be in excess of 0.2 for equivalent categories. The 'All-other-problems' category in the pruned tree condition attracts lower probability estimates than the sum of its logical equivalents in the un-pruned condition ('Ignition system defective + Other engine problems + Mischievous acts of vandalism + All-other-problems'). The size of the effect reduces with the expertise of the participants, but can still be considerable with participants with professional knowledge of the scenario under consideration (Ofir, 2000).

Van Schie and van der Pligt (1994) argue convincingly that anchoring contributes to the pruning effect. Having seven categories provides an implicit anchor for each category of approximately 0.14 (by equiprobability), whereas having four categories initially anchors participants on a probability of 0.25. The results observed are consistent with insufficient adjustment from these anchors and illustrate how relatively subtle changes in the structure of the elicitation task can have considerable impact. Pruning effects can be seen as another manifestation of one aspect of the *range-frequency compromise*: the tendency to use different regions of a response scale with broadly equal frequencies (see Section 1.4.1). Thus, the anchor-and-adjustment heuristic is a judgement strategy, which, at least in some instances, can explain *why* the range-frequency compromise should operate.

Notably, Johnson et al. (1991) report a study in which baseball experts exhibited no pruning effects and provided probabilities very close to the actual relative frequencies for various events that occur in baseball matches. They propose that this was because they considered events for which the experts' knowledge is inherently probabilistic in nature (the relative frequency of hits, strikeouts, and so on, by major league baseball players). Thus, it would seem that having the right kind of expertise permitted these experts to avoid using anchors when making their judgements or, if they did rely upon anchors, to avoid the tendency to adjust insufficiently from them.

Winkler (1967) and Alpert and Raiffa (1982) have addressed the anchor-and-adjustment heuristic and pointed out that when people are asked to estimate an uncertain quantity, they often anchor on what seems to them to be the most likely value of the quantity, then systematically underestimate the variability or uncertainty of their estimate. For example, when asked in experiments to give an interval with 98% of the observed data covered, people tend to anchor their assessments on the median and insufficiently modify outwards. This results in ranges that cover only approximately 70% of the data (see Chapter 5 on eliciting intervals). The pervasiveness of anchoring effects has important implications for the ordering of questions when eliciting several quantiles of a probability distribution or several probabilities covering different ranges of an uncertain quantity (see Chapter 5 on eliciting probability distributions). For some specific approaches to combating this heuristic in elicitation exercises, readers should consult O'Hagan (1998).

3.4.5 Support theory

Support theory represents a key psychological theory of probability assessment (Tversky and Koehler, 1994; Rottenstreich and Tversky, 1997; Brenner and Koehler, 1999; Brenner et al., 2002; Koehler et al., 2003). It has important implications for the presentation of information during the elicitation process and provides a description of the processes that the expert might follow. One key observation of support theory is that "different descriptions of the same event can give rise to different probability judgements" (Tversky and Koehler, 1994, p. 547). Normative models of probability judgement assume description invariance: the probability of an event does not depend on how the event is described. However, numerous experiments indicate that people can be induced to violate this principle of description invariance. For example, participants give lower probability ratings for the packed hypothesis 'Death from homicide, rather than accidental death' than for the more detailed but equivalent unpacked hypothesis 'Death from homicide by an acquaintance or by a stranger, rather than accidental death'.

Support theory uses the terms 'event' and 'hypothesis' in order to emphasise the distinction between an event, per se, and *different possible descriptions* of that event. In the terminology of support theory, a hypothesis is a description of an event. Thus, there may be several different hypotheses for the same event. For example, consider an ice-hockey match in which Sweden defeats Russia. This event can be described by several hypotheses: 'Sweden beats Russia', or 'Russia fails to win or draw against Sweden', or even 'Sweden beats Russia after having been ahead, behind or equal at the end of the first period of play'. Each of these three descriptions (i.e., hypotheses) is a logically equivalent description of the same event. However, they may not be psychologically equivalent and may give rise to different assessments for the probability of the event that they describe.

A key element of support theory is the *unpacking principle*: providing a hypothesis that gives a more detailed description of an event generally increases its judged probability. For instance, the sum of the (three) estimated probabilities that a person's cause of death is heart disease, cancer or some other natural cause usually exceeds the (single) estimated probability that it is any natural cause. In the terminology adopted within support theory, these assessments are *subadditive*: the sum of the three probabilities is *greater* than the single probability to which it should be equal. In this example, 'any natural cause' is the *implicit disjunction* (of the event) or *implicit hypothesis*; and 'heart disease', 'cancer' and 'other natural cause' are the *explicit disjunction* of this event. Increasing the number of events specified within an explicit disjunction further inflates the sum of the judged probabilities. If estimates are obtained from different individuals for different components of the disjunction of some event, then the sum of the estimated probabilities can exceed 1 by a considerable degree. Eliciting responses using frequencies (e.g., 'How many people die from natural causes in a year?') reduces the magnitude of the factor by which the (inferred) probabilities for the explicit disjunction exceed that of the implicit disjunction.

Support theory accounts for these phenomena associated with unpacking by proposing that probability judgements are negotiated by assessments of evidence for and against hypotheses (i.e., descriptions of events). The support for a hypothesis may be reasons for its occurrence or evidence or arguments favouring the hypothesis. Probability assessment for an implicit disjunction relies upon global assessment, typically upon heuristic strategies such as focussing on the most available or representative case. However, assessment can be performed through more effortful processes of evaluating cues (i.e., evaluating those features or variables that predict the occurrence of an event). Unpacking the event into an explicit hypothesis can remind participants of possibilities that they might have overlooked or focus their attention on scenarios that they may have given limited consideration to. Explicitly identifying a possibility usually increases its salience and the support for a hypothesis. This increases the perceived probability of the event and accounts for why an explicit hypothesis often encourages a higher probability assessment than an implicit hypothesis describing the same event.

Support theory has mainly been tested using a student subject pool, though a number of examples of generalisations to expert populations exist. For instance, subadditivity has been observed in doctors' judgements of disease probability (Redelmeier et al., 1995) and for professional option traders' probability judgements concerning future stock prices (Fox et al., 1996). The magnitude of subadditivity is similar to that observed in student populations. Support theory has also been tested on sports fans' judgements for the outcome of sporting events (e.g., how support for complementary hypotheses is weighted), and the data fit well with the theory (e.g., Tversky and Koehler, 1994; Brenner and Koehler, 1999). Some other theories of probability judgement are reviewed in the following chapter when we examine the calibration of subjective probabilities (Section 4.3.2).

3.4.6 The affect heuristic

Slovic et al. (2002) have identified a number of findings, all of which point to the importance of initial *affective evaluations* of stimuli (objects or events) when forming judgements about them. Affective evaluations are those intuitive global judgements or emotional gut reactions through which we assess whether we view objects or events as generally positive or generally negative. The *affect heuristic* is a judgement strategy whereby these affective evaluations serve as proxy for other judgements (or at least heavily influence them by providing an initial anchor for them). Thus, the affect heuristic sits happily within the heuristics and biases tradition, because it is proposed that a simple instinctive judgement (the affective evaluation) stands in the place of a more complex and effortful one. For instance, I might judge both the number of benefits and the probability of serious side effects that I can expect to receive from a surgical procedure according to how good I feel about the operation (Lloyd et al., 2001). Affective evaluations are inevitably influenced by one's mood, which opens up the possibility that irrelevant external factors will inappropriately influence judgement. Consider the advertising industry's careful positioning of scantily clad women in the proximity of cars, stereos or cigarette packets. This can evoke a strong affective response, thereby influencing judgements of product quality or value.

As is the case with the anchor-and-adjustment heuristic, it is proposed that the affect heuristic could be used to make many different kinds of quantitative judgement. Included within this are the probability judgements and assessments for uncertain quantities that are the focus of elicitation. Particularly when dealing with issues that can evoke strong emotional responses (e.g., nuclear power or safety in medicine), we ought to be aware that emotional gut reactions may drive technical assessments more than one might suppose. For instance, Fischhoff et al. (1978b) argue that the levels of risk and benefit that are associated with various hazards are probably positively correlated, as societal regulation ensures that high-risk activities are only permitted when they yield substantial benefits. However, judgements of risk and benefit imply that people perceive a negative relationship between risk and benefit: high-risk activities are judged to be low in benefit, whereas low-risk activities are judged to be high in benefit (Slovic et al., 2002). This is consistent with an influence of affective evaluations. Activities that I feel good about are judged to be low risk and high in benefit, whereas activities that evoke negative affective evaluations are judged to be high risk and low in benefit (Finucane et al., 2000). This is similar to the well-known 'halo effect' whereby global evaluations of value or worth influence judgements about specific features of an object or situation.

Reliance upon the affect heuristic seems to increase under time pressure or other circumstances that reduce the capacity to process information using more effortful strategies. It would follow that structuring an elicitation exercise so as to encourage analysis of the problem under consideration ought to reduce the inappropriate influence of intuitive affective evaluations (see Section 3.6).

3.4.7 Critique of the heuristics and biases approach

Although highly successful and influential, it is important to note that the heuristics and biases research programme is not without its critics. Some of their criticisms have been raised in the preceding sections. These include questioning the normative standards by which supposed biases are judged, insufficient use of expert populations and familiar judgement tasks and imprecise definition of the proposed judgement heuristics. See Gigerenzer (1996a) or Lopes (1991) for further consideration of these issues.

Even if one is dissatisfied with the level of explanation that heuristics provide for the biases that are attributed to their use, it is true that many of these errors are easily replicated. Given the implications that these biases have for the assessments elicited from experts, it does suggest that they ought to be kept in mind when structuring elicitation exercises.

3.5 Experts and expertise

Experts and expertise have been of considerable interest to psychologists, but the comparative ease with which non-expert populations can be examined implies that there are fewer empirical data on the nature and quality of expert judgement than we might wish to draw upon for a book such as this. Some principles underlying research into expertise and its application are discussed in van der Pligt (1988), Shanteau and Stewart (1992) and Bolger and Wright (1994). In the subsections that follow, we summarise some of the key views that have been expressed concerning the nature of expertise.

3.5.1 The heuristics and biases approach

The messages that emerged from the heuristics and biases research programme in the 1970s had obvious implications for practitioners in many disciplines including medicine, law, management and finance. This prompted research into whether professionals in a variety of fields were susceptible to the biases enumerated by Tversky and Kahneman. If so, it would suggest that they might rely upon heuristics in much the same way as the typical undergraduate seems to when forming judgements under uncertainty.

Bias consistent with the use of judgemental heuristics has been observed in a number of professional contexts (e.g., Northcraft and Neale, 1987; Tolcott and Marvin, 1988; Tolcott et al., 1989a, 1989b; Allen, 1982; Arkes, et al., 1981). However, we ought to ask whether biases occur with the *same* frequency in the real world and whether they are reduced by experience. Although experience is no guarantee of accuracy, there are a number of studies that find fewer or smaller biases among professionals with more experience of the specific task under investigation (e.g., Adelman et al., 1993; Adelman and Bresnick, 1992; Anderson and Sunder, 1995). In their extensive review of heuristics and biases research with professional auditors, Smith and Kida (1991) found many examples of biases that could be attributed to the use of heuristics. However, when expert auditors performed familiar tasks, the incidence or magnitude of biases was often reduced. It would appear that both task-specific and more general benefits can result from experience with probabilistic judgement tasks.

3.5.2 The cognitive science approach

This chapter may have given the impression that some psychologists have a rather pessimistic view of the quality of human judgement under uncertainty. They seem to question whether anyone is capable of providing probability assessments that could be described as 'expert' in any sense. While such a summary is an oversimplification, it is probably fair to say that cognitive science has a more optimistic outlook concerning expertise; namely, that skilled human judgement is an attainable goal, though it is one that can take considerable time to realise. A number of authors have characterised the development of expertise as a staged process. Patel and Groen (1991) identify a progression (primarily in terms of knowledge) from novice (layperson or beginner) through intermediate to sub-expert (having adequate generic knowledge of the domain) and finally to expert (having specialised domain knowledge). Under this characterisation, several studies of 'expert' judgement have really only considered 'intermediates' and 'sub-experts' and, as such, might not inform us about those with considerable knowledge and experience. Dreyfus and Dreyfus (1986) identify 'Five Steps from Novice to Expert', focussing on how individuals approach tasks at each stage. The 'novice' learns to recognise relevant facts and features according to the rules that they have been given. The 'advanced beginner' uses similarity to prior examples to solve problems or to make decisions and starts using signs that cannot be described verbally. At the third stage ('competence'), an appreciation of what is important develops, and a more structured approach to decision-making is adopted. The 'proficiency' stage is characterised by intuitive judgement based on past events and known patterns, followed by a more detached and deliberate decision-making process. Finally, the 'expert' has sufficient experience to proceed on the basis of intuitive judgement and knowledge of what normally works. This theme, that expertise consists of intuitive judgement and an automaticity of action deriving from a wealth of knowledge and experience, is a common one in the cognitive science literature (e.g., DeGroot, 1978). A number of authors have identified ten years as the approximate time span required to attain the levels of performance worthy of the labels 'international expert or performer' (Simon and Chase, 1973; Hayes, 1981; Bloom, 1985; Ericsson and Crutcher, 1990). Thus, even if genuine expertise is attainable in a domain, we should not expect to observe it immediately!

3.5.3 'The middle way'

Shanteau provides one of the most extensive accounts of the nature of expertise that is to be found in the psychological literature (Shanteau, 1988, 1992a, 1992b; Abdolmohammadi and Shanteau, 1992; Ettenson et al., 1987). He has argued for a third alternative that sits somewhere between the pessimism of some decision psychologists and the optimism of cognitive scientists. He notes that the literature contains examples of both good and bad professional judgements. Consequently, arguments over the quality of professional judgement, per se, are of little value. Rather, we should ask what distinguishes those professions or professionals who perform well from those who do not.

Firstly, Shanteau identifies that an adequate grasp of domain knowledge is a prerequisite of expertise. Ayton (1992) suggests that the evidence supports the view that expert knowledge is organised in such a way as to affect the way in which experts solve problems. Not only can experts call upon greater knowledge but this greater knowledge (and presumably the manner in which relations between items are stored) also contributes to strategy and reasoning. Anzai (1991) identifies the difference in knowledge representation between novices and experts (concrete/naive vs abstract) as key to explaining the misconceptions to which novices easily fall prey and the more successful problem-solving strategies adopted by experts. However, not all studies that observe differences in knowledge observe differences in reasoning (Elstein et al., 1978).

Shanteau observes that experts exhibit a common set of psychological traits, a distinct 'self-presentation' of the expert role. Experts are creative, outwardly confident and handle adversity well. Shanteau argues that experts are both effective in communicating their expertise to others and (in apparent contradiction) generally inarticulate about decision-making processes (Shanteau, 1988), presumably the result of greater automaticity. If, with increasing expertise, information processing does indeed become more automatic, more intuitive, less analytic and less easily articulated, then this may present particular problems for an elicitation process that requires conscious reflection upon the relations between variables that drive the expert's opinions.

Shanteau identifies a variety of formal and informal decision strategies that experts employ in order to overcome their cognitive limitations. These include continually using feedback to adjust initial judgements, seeking others' advice, using decision aids and employing divide-and-conquer strategies (which split a large problem into more easily solved sub-problems). In relation to the strategies that experts employ, Beach (1992) identifies 'causal reasoning' (focussing on the unique characteristics of individual elements and causal networks) as more typical of expert reasoning than 'aleatory' (probabilistic) reasoning.

The final determinants of expert competence that Shanteau considers are the task characteristics. Quite simply, is it possible to do the task well? This is arguably a much-overlooked determinant of performance (with the exception of a body of research which investigates the impact of the availability of feedback). Shanteau identifies domains (broad areas of expertise) and professions for which low levels of competence in judgement are consistently reported, and those for which good performance is reported. 'Good' performance is noted for weather forecasters, agricultural judges, auditors, accountants, mathematicians, physicists and chess masters. 'Poor' performers include clinical psychologists, psychiatrists, court judges, stockbrokers, parole officers, personnel selectors and those responsible for student admissions. The pattern of competence for physicians is said to be less clear cut. Shanteau speculates that the 'good' is distinguished from the 'bad' because the former set of domains predominantly involve judgements about (fairly constant) objects or things, rather than the 'moving target' of human behaviour. Dawes (1987, cited in Shanteau, 1992b) notes that the domains for which superior performance is common exhibit greater predictability, yet, paradoxically, do not demand such high standards – thus, some mistakes are accepted. Furthermore, the tasks in these domains are more repetitive, allowing more focused practice and the provision of regular and relevant feedback. Gigerenzer (1989, cited in Shanteau, 1992b) asserts that domains shift from 'poor' to 'good' levels of performance over time, decision aids are more readily available in the successful domains and tasks in these domains are perhaps more easily decomposable into manageable sub-tasks.

3.6 Three meta-theories of judgement

Many of the biases that we have discussed in this chapter are somewhat ephemeral. Changing the task characteristics can alter the magnitude of many of these biases and in some cases cause them to disappear entirely (e.g., Tversky and Koehler, 1994; Gigerenzer, 1996a; Koehler, 1996). Also, some individuals are more prone to biases in judgement and reasoning than others: more intelligent people are more likely to follow the normative rules of logic and probability (Stanovich and West, 2000).

This presents a challenge to psychological theory and to those who seek to undertake practical elicitation, as it stretches the ability of any one process theory to account for all experimental findings and field observations. A number of frameworks or meta-theories have been proposed that describe the switch between different judgement strategies. Several of these might be described as 'dual process accounts'. These emphasise a distinction between more effortful 'analytic' processes that lie closer to the prescriptions of normative approaches and quicker heuristic, intuitive and largely automatic processes that are less exact. Two such accounts are briefly described in the following text, followed by a third meta-cognitive account that provides some insight into when human judgement can be expected to succeed or to fall short.

3.6.1 The cognitive continuum

Cognitive continuum theory was developed by Hammond as a unified framework for research and theory in judgement and decision-making (for detailed overviews see Cooksey, 1996; Hamm, 1988). One key element of the framework is that cognition shifts along a continuum from intuitive to analytical thinking. Intuition is typified by rapid information processing, low cognitive effort, inconsistency, the perceptual evaluation of cues, the weighted averaging of information and difficulties in articulating the judgement process. Analysis is typified by the slow and effortful processing of information according to logical rules and quantitative cues. Hammond stresses that it is rarely the case that cognition is *either* strictly intuitive *or* strictly analytical. Rather, most cognition exhibits elements of both poles on this continuum. Thinking and reasoning can more usefully be described as *more intuitive* or *more analytical*. In many situations, cognition can be described as *quasi-rational* – displaying elements of both analytical and intuitive thinking in good measure.

A second keystone of Hammond's cognitive continuum theory is that different tasks induce or encourage thinking that is more intuitive or more analytical. Thus, the position of the decision maker on the cognitive continuum varies as a function of the task characteristics. Analytical thinking is encouraged by well-structured tasks involving few cues, where prior rules or procedures exist, and high accuracy is deemed feasible. Intuitive thinking is induced by tasks with many cues (of which several are redundant) that are linearly related to an outcome, for which minimal feedback is obtained, and high accuracy is not expected.

Under this framework, some have speculated that the elicitation process could be viewed as a 'bridge' between the more intuitive expert, who gleans knowledge from ill-structured multiple-cue probabilistic environments, and the analytical world of formal statistical models.

3.6.2 The inside versus the outside view

Kahneman and Tversky have also identified the tension between intuitive or heuristic reasoning on the one hand and logical or analytical reasoning on the other.

Although they state that "judgements of probability vary in the degree to which they are analytic or intuitive" (Tversky and Kahneman, 1983, p. 30), their writing seems to emphasise a sharper distinction between these two modes of cognition than Hammond does. For instance, in discussing the conjunction fallacy, they do not allude to any quasi-rational 'middle ground':

> "A direct test of the conjunction rule pits an intuitive impression against a basic law of probability. The outcome of the conflict is determined by the nature of the evidence, the formulation of the question, the transparency of the event structure, the appeal of the heuristic, and the sophistication of the respondents. Whether people obey the conjunction rule in any particular direct test depends on the balance of these factors." (Tversky and Kahneman, 1983, p. 30).

In further reflections on two modes of thought, Kahneman and Tversky (1982a) identify two ways in which 'external' (i.e., aleatory) uncertainty can be assessed. In the *distributional mode*, a case is viewed as one of a class of similar cases, for which relative frequencies can be estimated or perhaps obtained. In the *singular mode*, probabilities are assessed according to the features of the individual case under consideration. They conjecture that people typically focus on the singular mode, adopting an 'inside view', in which they concern themselves mainly with the individual features of the current target case. Judgement accuracy is typically improved if people can be encouraged to adopt an 'outside view', viewing the current target case as one of a class of cases, thereby making it easier to consider useful distributional information (see also Kahneman and Lovallo, 1993; Kahneman, 2003; Lagnado and Sloman, 2004). Often, manipulations that reduce biases such as the base rate effect or violations of the conjunction rule can be seen in this light (see Section 3.4.2). For instance, it seems likely that one benefit of frequency formats is to focus attention upon a class of events, thereby reducing the tendency to rely on individuating information.

The inside/outside (singular/distributional) distinction is important for practical elicitation because we often want experts to address questions using all the individual knowledge that they have to distinguish a single case from other cases. Inevitably, experts will tend to refer the individual case to a class of cases that they judge to be similar to this one. So they will use both distributional thinking to evaluate the frequency in the class and singular reasoning to differentiate the individual. In such cases, final judgements will depend upon the particular 'mix' of distributional and singular reasoning that the expert has chosen, or has been encouraged, to employ during the elicitation process. For instance, where reasoning is mostly singular with little reference to distributional reasoning, there will be a tendency to under-represent aleatory uncertainty. Consequently, estimates for the median of an uncertain quantity will tend to be insufficiently regressive to the mean (see Section 3.4.2.5), or the limits for the upper and lower bounds of the quantity will be too narrow (see Chapter 5).

3.6.3 The naive intuitive statistician metaphor

Fiedler and Juslin argue that a dissociation between perception and inference lies at the heart of variability in performance in human judgement under uncertainty (Fiedler, 2000; Fiedler and Juslin, 2006). Specifically, they argue that people are good at logging the frequency with which they observe events occurring. They are also adept at learning the associations between events (i.e., the correlations between variables and the conditional probabilities for different combinations of events). Thus, they are good 'intuitive statisticians', faithfully representing what occurs in the world around them. However, they are naive with respect to how the *samples* of events that they have observed might be biased with respect to the *populations* from which they are drawn. They are therefore easily misled as to how these samples might serve as less-than-perfect guides to inferences and decisions. Thus, people are not good at adjusting their judgement to correct for sample bias; presumably because either they do not realise when a given sample is likely to be biased or they do not appreciate the appropriate correction that is advisable. Alternatively, perhaps they have some appreciation of how to correct for sample bias, but make adjustments that are insufficient (cf anchoring-and-adjusting, Section 3.4.4). This approach is very much 'under development', but appears to have some useful lessons that can contribute profitably to our understanding of how to conduct probability elicitation and of the status of the information that we elicit from experts.

3.7 Conclusions

3.7.1 Elicitation practice

- Research shows that humans cannot be guaranteed to act as rational agents who follow the prescriptions of probability theory and decision theory. In many situations where people are called upon to provide judgements under uncertainty, they are liable to choose from a selection of easy-to-use strategies (heuristics). These strategies can be effective, especially when time or information is scarce, but they are not always optimal.

- Substantive expertise in a specialist area is no guarantee of normative expertise in providing coherent probability assessment. Careful thought needs to be given to the training in probability and statistics that the expert should receive at the beginning of the elicitation exercise. Furthermore, assistance and coaching will often be desirable as the elicitation proceeds.

- The facilitator and, quite often, the expert ought to be aware of the main biases that can occur in judgement under uncertainty and the reasons proposed for their occurrence. The elicitation process ought to be structured with this in mind.

- The facilitator should recognise the potential for inadvertently introducing bias into the elicitation process (e.g., anchoring effects induced by particular orders of questions) and should seek ways to reduce or avoid such problems.
- The level of detail with which an event is described can alter the probability assessments that are provided for that event. Particularly when eliciting assessments for a series of mutually exclusive events, it will be necessary to assess the coherence of the set of assessments provided by the expert.
- Several errors and biases can be attributed to adopting too narrow a focus: focussing too closely on a single instance of some broader class of events, or examining only one hypothesis as opposed to considering competing alternatives. The expert may need to be actively encouraged to avoid this.
- It can be helpful to introduce procedures that explicitly encourage an expert to think analytically. For instance, the facilitator should explore whether aids to judgement are available that help the expert to analyse all the substantive expertise that is relevant to a particular assessment. Or, the facilitator should consider whether and how the expert could be involved in the process of checking her own set of assessments for coherence.
- The facilitator and the expert ought to consider how the expert's particular experience (her access to samples of data) may influence or bias the assessment of uncertain quantities. A full consideration of the uncertainty associated with a quantity ought to take into account how well the expert's past experience informs the specific elicitation tasks.
- Where the 'outputs' of an elicitation exercise become 'inputs' for decision analytic models, the sensitivity analyses on such models can usefully be informed by research on judgement biases. The statistician or decision analyst can use what is known about the likely direction and magnitude of biases to help him set the range of possible parameter values to examine within the model.

3.7.2 Research questions

- Is it possible, or useful, to screen experts for those with sufficient normative expertise to take part in an elicitation exercise?
- In the context of elicitation, what kinds of training and aids to judgement are effective in reducing biases in probability judgement?
- It is known that different methods have a tendency to yield different assessments, possibly ones that are biased in different directions. Does this mean that more accurate or more coherent assessments can be achieved by using multiple methods when an individual expert provides an assessment of an uncertain quantity?

Chapter 4

The Elicitation of Probabilities

4.1 Introduction

Building upon the research and theory that we examined in Chapter 3, this chapter considers the elicitation of probabilities: how it has been done, how it can be done and how it can be improved.

We begin with a discussion of the rather large literature on the calibration of subjective probabilities (Sections 4.2 and 4.3). In calibration research, probability judgements for events in a set are compared to the observed relative frequency of these same events. This provides a means of determining whether probability assessments are in accord with actual outcomes. Testing probability assessments in this manner is rarely, if ever, practical in elicitation. This could be because the relative frequencies, of events for which we seek expert assessments are not known (i.e., insufficient data exist) or because they are not knowable, as in the case of one-off events. Why, then, consider this literature? Again our purpose is practical. Our assumption is that by examining situations in which it is possible to make statements about the accuracy or appropriateness of probability judgements, it is possible to establish some principles that ought to apply to situations in which it is difficult to make such statements. These principles relate to issues such as how well people usually assess probabilities, whether there are common problems that one ought to be aware of and what steps can be taken to help experts when they assess probabilities. In examining these issues, both empirical results and theoretical considerations are discussed. Again, we assume that developing an understanding of the theory behind the findings gives the best chance of successfully applying this literature to the practicalities of elicitation. Our assumption that the messages from calibration research apply to assessments made in different circumstances may be challenged, but we feel that there is sufficient merit in this assumption to warrant

Uncertain Judgements – Eliciting Experts' Probabilities A. O'Hagan, C. E. Buck, A. Daneshkhah, J. R. Eiser, P. H. Garthwaite, D. J. Jenkinson, J. E. Oakley and T. Rakow

a detailed examination of this body of work. The extent to which calibration can be used to assess the accuracy of elicitation is revisited in Chapter 8.

In Section 4.4, we review the various means by which probabilities have been elicited, describing the different tools and techniques that have been employed for this purpose. We discuss the pros and cons of different ways of presenting the elicitation problem to the expert and the various means by which the expert can report her judgement.

We close the chapter by examining the literature on debiasing, considering specific strategies for improving judgements for uncertain quantities (Section 4.5). This leads naturally to the presentation of a number of practical recommendations for the best practice in elicitation.

Much of our discussion in this chapter develops themes and ideas that were introduced in Chapter 3. In particular, we consider how probability assessments might be influenced by the particular information-processing strategies that experts use instinctively or that they might be encouraged to adopt. In Chapter 3, we saw that many judgements under uncertainty seem to be made using *heuristics*: efficient and relatively effortless short cut strategies that allow judgements to be made quickly (see Section 3.4). However, judgements that are made in this way are often found to be suboptimal, as heuristics can generate incoherent probability assessments or biased quantitative judgements. Understanding the heuristics that experts might use when assessing probabilities is important for understanding the data on the calibration of subjective probabilities (Sections 4.2 and 4.3) and for appreciating how to improve judgements for uncertain quantities (Section 4.5).

4.2 The calibration of subjective probabilities

'There is a 0.3 probability of rain tomorrow.' How is it possible to empirically determine the accuracy, appropriateness or correctness of this statement? Whether it rains or not does not determine the veracity of this single prediction. Only statements of absolute certainty (probability of 0 or 1) can be proved incorrect from a subsequent observation.

This problem is most commonly dealt with by considering a number of probability judgements provided by the same 'source'. A source may be an individual, a group, a collection of individuals or a quantitative method such as a statistical prediction model. Accuracy or appropriateness for this *set* of judgements can then be assessed. Specifically, calibration refers to the match between subsets of probability judgements grouped by magnitude and the observed relative frequency with which the target event occurs for each subset. For perfectly calibrated judgement, for all sets of judgements of identical subjective probability, p_i, the observed relative frequency of the specified events should be p_i. For instance, when all judgements of 0.1 probability are grouped, 10% of these events should occur; for all judgements of 0.2 probability, 20% of these events should occur, and so on.

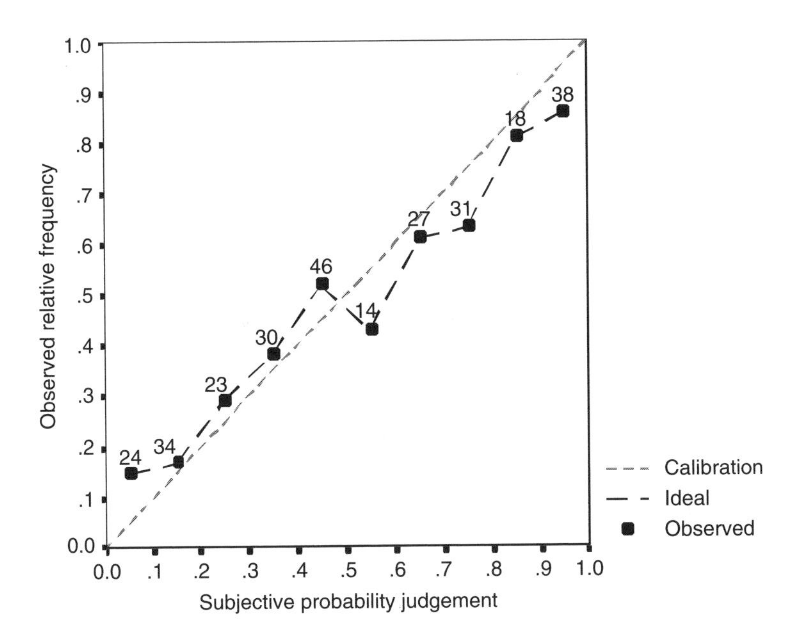

Figure 4.1: Example of the calibration curve for a reasonably well-calibrated set of probability assessments (full-range format). Integers show the number of judgements provided within each class interval.

The calibration of subjective probabilities can be represented graphically and assessed qualitatively by means of a *calibration curve* (see Figure 4.1). In most studies (including those that we review here), calibration is assessed quantitatively by means of formal scoring rules. However, calibration curves often serve to provide readers with an easy grasp of the general pattern of results in a study. A technical discussion of *scoring rules*, including definitions, can be found in Chapter 8.

On a calibration curve, perfect calibration is represented by points falling on the identity line when the judged probability and the observed relative frequency are plotted against each other. In studies where participants are free to give any response over a specified range, judgements are often grouped. For instance, probability judgements between 0 and 0.2 may be grouped into a 0.1 category to provide sufficient judgements in the same subset. It is commonly observed good practice to indicate how many judgements lie in each category. Sometimes, judgements are grouped into categories of unequal width so that each category contains the same number of assessments.

When assessing the calibration of a set of subjective probabilities, there are some partially related features that are often commented upon that readers ought to be aware of: *over/under-confidence*, *over-extremity* and *discrimination*.

If a set of probability judgements is perfectly calibrated, then the mean probability judgement will equal the relative frequency of the target event in the set.

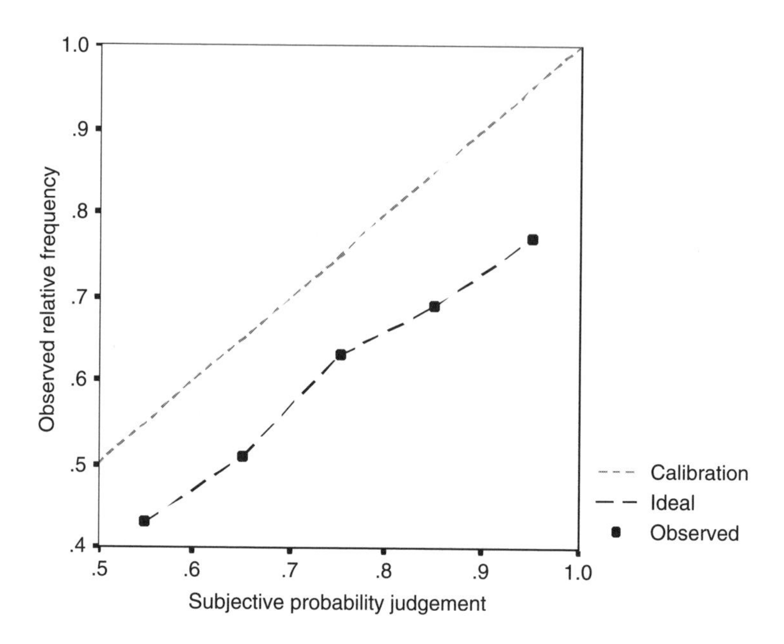

Figure 4.2: Example of the calibration curve for an *overconfident* set of probability assessments (half-range format). For all subsets of judgements, grouped according to subjective probability, the observed relative frequency is lower than the subjective probability.

The reverse of this conditional statement is not true. Therefore, a perfect match between the mean probability judgement and relative frequency of events is a necessary rather than a sufficient condition for perfect calibration. If the mean judgement exceeds the relative frequency of the event, then the assessor is said to be *overconfident* and her set of judgements is said to exhibit *overconfidence*. The terminology reflects the notion that by overestimating how likely events are (in general), the assessor is too confident in her belief that the event will occur. When plotted on a calibration curve, a set of judgements that exhibits overconfidence will typically be represented by a calibration curve that lies below the identity line for most or all of the plotted points (see Figure 4.2). The larger the difference between the mean judgement and the relative frequency of the event, the greater the degree of overconfidence. When the mean probability judgement is less than the relative frequency of the event, the set of judgements is said to exhibit *under-confidence*. Examining whether and to what extent a set of judgements exhibits overconfidence or under-confidence is described variously as assessing confidence, assessing over/under-confidence, or assessing calibration-in-the-large. We will see that overconfidence occurs frequently in calibration research. Under-confidence does also occur, but is less common.

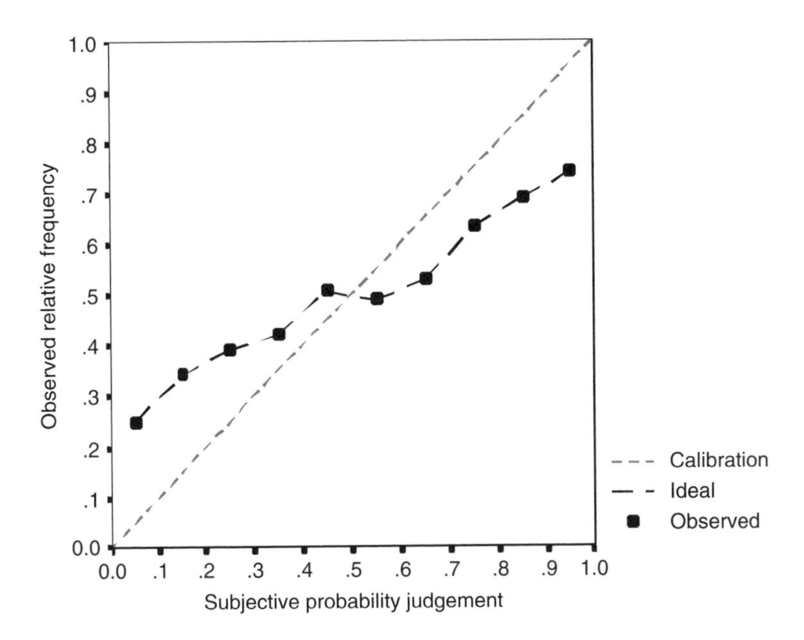

Figure 4.3: Example of the calibration curve for a set of probability assessments that exhibit *over-extremity* (full-range format). For low subjective probabilities, the relative frequency exceeds the probability judgement, whereas for high subjective probabilities, the reverse is true.

An entire set of judgements may be neither over- nor under-confident but may still be less then perfectly calibrated. One of the common patterns of results that fall in this category is best described as *over-extremity*. Subsets of events judged to have low probabilities (e.g., 0.1) have a relative frequency that is somewhat higher than their judged probability, whereas those events judged to have high probabilities (e.g., 0.9) occur with relative frequency less than their judged value. When a set of judgements exhibit over-extremity, the calibration curve will be more shallow than the identity line and will cross it (see Figure 4.3). The term over-extremity is used to indicate that miscalibration would presumably not have occurred had the extremes of the scales not been used or if they had been used less frequently (e.g., had the assessor quoted a probability 0.2 on every occasion that they had previously quoted 0.1). Many authors prefer to use the term overconfidence when describing the pattern of results that we have referred to here as over-extremity. This choice of terminology reflects the notion that the assessor is deemed too confident in her ability to 'pick' either when an event is highly unlikely or when it is highly likely to occur, which she signals with inappropriately low or high probabilities respectively.

The term *discrimination* relates to this ability to correctly identify when an event is more, or less, likely to occur. The following example will serve to illustrate the concept. Imagine that I make eight probability judgements for the event 'heads when a coin is tossed' (H). On four occasions I quote $P(H) = 0.25$ and on four occasions I quote $P(H) = 0.75$. For perfect calibration in this small data set, we would require that heads were observed once out of four times when $P(H) = 0.25$ and three out of four times when $P(H) = 0.75$. However, imagine that heads comes up twice when I assess $P(H) = 0.25$ and twice when I assess $P(H) = 0.75$. Then my set of judgements is neither over- nor under-confident: my mean judgement is 0.5 and this matches the relative frequency of H (4 out of 8). However, my set of judgements did not discriminate according to how likely I was to observe H: the relative frequency of H was 0.5 irrespective of whether I assessed $P(H) = 0.25$ or $P(H) = 0.75$. In order to have a set of probability judgements that discriminate appropriately, we would require fewer observations of H when I quote $P(H) = 0.25$ than when I say $P(H) = 0.75$. In crude terms, the better the discrimination exhibited by a set of probability judgements, the steeper the calibration curve. When probability judgements discriminate poorly, the calibration curve is close to horizontal (see Figure 4.4), and a perfectly horizontal calibration curve denotes a complete absence of discrimination. Discrimination is often assessed independently of other features when calibration is examined (see Section 8.2 for a discussion of the scoring rules used for this purpose).

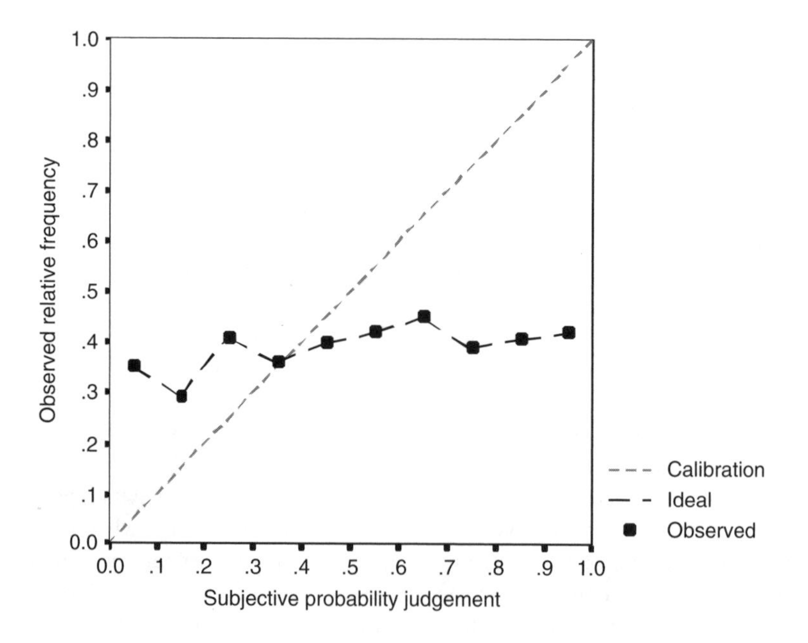

Figure 4.4: Example of the calibration curve for a set of probability assessments that exhibit *poor discrimination* (full-range format). Irrespective of the subjective probability that was quoted, the relative frequency with which the event occurs is approximately 0.3 to 0.4.

4.2.1 Research methods in calibration research

Calibration research by psychologists and those in related fields is very much a mixed bag. It focuses on judgements for different types of events, employs different modes of response and examines a variety of study populations.

4.2.1.1 Response modes

Typically, one of two types of response is required, though other response alternatives are used (see Ferrell and McGoey, 1980; Lichtenstein et al., 1982). 'Half-range tasks' employ a two-alternative forced choice (2AFC) paradigm: participants select one of two answers, then state the probability that they are correct on a scale from 0.5 to 1. Here, the question to the participant is often phrased in terms of 'confidence': 'How confident are you that your chosen answer is correct?' This kind of question would seem to focus on epistemic uncertainty. It implies that the assessor ought to think about how successfully she can translate her knowledge into a probability judgement. 'Full-range tasks' ask participants to state the probability that a given statement is correct or that a specified event will occur and employ the full probability range (from 0 to 1). Here, the phrasing of the question is more likely to be 'How certain is it?' or 'How likely is it?' Questions like this seem to be more about aleatory uncertainty. Uncertainty seems to be a property of the event, and the assessor may not be prompted to think about the uncertainty that lies within her head.

When comparing results from studies that employ different response modes, the formal equivalence of the two methods is usually assumed (e.g., that 80% confidence that a statement is false implies a 0.2 probability that it is true and *vice versa*). However, the two modes may not be psychologically equivalent. Combining a full-range response mode with a 2AFC paradigm, Gigerenzer et al. (1991) found that participants sometimes assigned confidence ratings below 50%. Thus, having said that one of two mutually exclusive and exhaustive options is more likely to occur, participants sometimes assign a 40% chance to the 'more likely' option. It would seem that in, or after, the act of choosing, participants sometimes decide that they have chosen poorly. Response modes that prohibit participants from correcting their own judgement can therefore lead to systematic inaccuracies in probability judgement.

Ronis and Yates (1987) had previously found a similar effect of response mode. Two topics were examined: the outcomes of basketball games and general knowledge questions. On each question, participants made a forced choice as to the more likely of two options. Then they indicated their probability judgement for their chosen option using either a full-range scale (the 'Choice-100 method') or using a half-range scale (the 'Choice-50 method'). Calibration was found to be superior for the Choice-100 method. Ronis and Yates (1987) were less enthusiastic than Gigerenzer et al. (1991) about the improvements in calibration that accrue when participants can state that they are 40% certain that the more likely of two options will occur. Not unreasonably, they point out that such responses are difficult to interpret. In the context of interactions between facilitators and experts in elicitation, incoherent responses of this nature would obviously warrant clarification.

4.2.1.2 Event types

Perhaps the largest body of calibration studies has used almanac questions (typically with student populations). For example, 'Which is further north? (a) New York or (b) Rome. How confident are you?' (half-range format); or, 'What is the probability that absinthe is a precious stone?' Such studies are solely concerned with epistemic or internal uncertainty. Rome is definitely to the north of New York (always!), and absinthe is not a precious stone (and has no prospects of becoming one!) – any uncertainty is epistemic uncertainty and reflects doubt as to the correct answer. Studies of the calibration of experts have typically used forecasting tasks of one kind or another: 'What is the probability of hail within the next hour?', 'What is the probability that the lab test will indicate pneumonia?' or 'What is the probability of this patient dying within six months?' Answers here presumably reflect *both* aleatory and epistemic uncertainties. For example, when assessing the probability that a patient dies within six months: if two patients have identical characteristics, one may survive and one may die; and, also, my knowledge of how cues such as patient characteristics and symptoms predict survival will inevitably be imperfect.

4.2.2 Calibration research: general findings

4.2.2.1 How well calibrated are probability judgements?

Several reviews of this literature have been performed that reveal some regularities in the empirical findings (Lichtenstein et al., 1977; Lichtenstein et al., 1982; Beach and Braun, 1994).

For most student populations and many expert populations, subjective probabilities elicited in calibration studies typically exhibit over-extremity in full-range tasks (being extreme in the sense that they too often lie close to 1 or 0). When very low probabilities are estimated, the relative frequency of events tends to be greater, and when very high probabilities are assigned to events, the relative frequency of events tends to be lower. In half-range tasks, subjective probabilities tend to be higher than the observed relative frequency of occurrence, markedly so for higher assigned probabilities. Hence, overconfidence is the norm when half-range formats are used.

The practice of using 'overconfidence' to describe what we have earlier described as 'over-extremity' can be seen to make good sense when one translates the typical pattern of (over-extreme) results for full-range tasks into the half-range format. Imagine reflecting probabilities from the lower half of the full-range into the upper half. This is equivalent to translating an assessor's judgement of a 0.3 probability of rain into a 0.7 probability of no rain. This is presumably what the assessor would have said had her assessment been preceded by the typical half-range-format question: 'Which is more likely, rain or no rain?' When transformed, low probabilities such as 0.1 and 0.2 become high probabilities of 0.9 and 0.8, and, being frequent, serve to inflate the mean probability judgement to a value greater than the relative frequency of the target event. The result is a set of judgements (on a half-range scale) that exhibits overconfidence.

A small number of studies have examined lay probability forecasts for future events (e.g., current affairs). Overconfidence is typically not observed in these studies (Wright, 1982; Wright and Ayton, 1986; Wright and Wisudha, 1982). Carlson (1993) found that calibration was better for future forecasts than for probability judgements of past events (current affairs in Experiment 1, and college football matches in Experiment 2), though calibration for future forecasts was still less than perfect. One possibility is that when uncertainty is epistemic (past events) rather than *both* epistemic *and* aleatory (future events), participants sometimes confuse inferences for memories. Thus, *mistakenly* thinking that they remember that something has occurred, they are *inappropriately* prepared to give probability judgements at or close to 1 for past events. This would not happen for future events, where apparent memories cannot mislead in the same way. Consistent with this, at higher probability judgements, overconfidence is less severe for future events than for past ones in Carlson's (1993) data.

This reduction in overconfidence for events where uncertainty is both epistemic and aleatory represents an improvement in calibration. This is encouraging for expert elicitation, as this is usually the kind of event for which elicitation is required. For determining the best practice in elicitation, we might therefore be justified in taking a greater note of calibration research that focuses on events of this kind (as opposed to studies which focus on almanac questions where uncertainty is purely epistemic).

Over/under-confidence in half-range tasks varies reliably with the proportion of correct responses. Overconfidence decreases as the proportion of correct responses in a 2AFC paradigm increases, such that under-confidence may be observed for sets of questions yielding a large proportion of correct responses (the 'hard–easy effect'). For instance, for sets of true–false questions where students get 60% correct, overconfidence is the norm; whereas, for sets of true–false questions where students get 90% correct, under-confidence is the norm. Seemingly, participants in calibration studies make insufficient adjustment to their subjective probabilities in response to gross level changes in the relative frequency of an event. This could be considered to be another example of anchoring bias (see Section 3.4.4).

Calibration is typically no better for those who perform better at a task (i.e., getting more answers correct), though there are exceptions to this (Wright et al., 1994). While performance might increase with knowledge, practice, effort or the amount of information available, the corresponding increase in confidence is typically at least as great (Paese and Sniezek, 1991; Ronis and Yates, 1987; Oskamp, 1965). Kahneman et al. (1982) have suggested that *substantive expertise*, the expert's capability within her specialised knowledge domain, has no necessary relationship with *normative expertise* (the ability to provide coherent and unbiased probability assessments). However, at least some evidence with experts assessing probabilities about professional subjects with which they are familiar suggests that they may be better calibrated than non-experts (see Winkler and Murphy, 1984, for an example of substantive experts who exhibit normative expertise). Outcome feedback (finding out whether the judged event subsequently occurred or not) appears to be particularly important for helping to improve calibration. This presumably helps the expert to learn how predictable events are, allowing her to map her level of subjective certainty appropriately onto the probability scale. The absence of outcome feedback can have serious adverse consequences for calibration.

4.2.2.2 How effective are corrective procedures?

Calibration can improve with training, though typically extensive intervention involving outcome and calibration performance feedback is required. Subbotin (1996) found that outcome feedback reduced overconfidence for non-independent items (i.e., related items of a similar nature), but had no effect upon underconfidence when almanac items were sufficiently easy that a high proportion of answers were correct. However, typically, outcome feedback alone is relatively ineffective in improving performance unless provided together with feedback on task structure (e.g., cue-criterion relationships) or information that summarises the probability judge's performance over a set of judgements (e.g., Remus et al., 1996). Stone and Opel (2000) note an interesting dissociation between the effects of performance feedback for the overall accuracy of a set of judgements and environmental information about the cue-criterion relationships relevant to the judgement. Calibration was improved through performance feedback (showing participants their own calibration graph from a pre-training session), whereas discrimination was improved by environmental information (a lecture on relevant cues for the probability judgement in question).

One reason why providing assessors with formal feedback on their calibration performance may be important is that people seem to be less than perfect at remembering the probability judgements that they have previously made. They are prone to *hindsight bias*: the tendency to misremember or misreport what one previously predicted such that it is more consistent with what subsequently occurred. For instance, Fischhoff and Beyth-Marom (1975) asked participants for probability judgements predicting events in the arena of current affairs. They asked participants to recall their previous judgement when the outcome of these events was known. When an event had occurred, participants often remembered giving a higher judgement than they had actually provided some weeks earlier. They often recalled giving a lower judgement than they had previously quoted on those occasions when the event in question had failed to occur. A consequence of this is that an expert may genuinely believe that her probability judgement is more successful than it actually is. Overcoming this may be one of the important functions of training.

Simple manipulations such as warning people about overconfidence, offering financial incentives or requiring that they explain their judgements typically do not improve probability judgement (Fischhoff et al., 1977; Siegel-Jacobs and Yates,

1996; Hammersley et al., 1997). Making participants accountable for the *process* by which they reach their judgement (i.e., requiring them to justify why they used the information as they did) does increase the amount of information used (Siegel-Jacobs and Yates, 1996). However, as it increases the use of *both* relevant *and* irrelevant information, the effects of this are not uniformly beneficial.

One of the few simple manipulations to succeed in reducing overconfidence is reported by Koriat et al. (1980), who asked their participants to list reasons why they might be wrong before assigning probabilities to their choices. However, close reading of this paper shows that the effect is rather smaller than academic folklore would typically have one believe. In general, a failure to give appropriate attention to *alternatives* to the outcome being judged is a theme that recurs frequently in empirical work on judgements of many kinds. Further discussion of this theme can be found in Sections 3.4.5 and 3.6.2 and in McKenzie (1998, 1999).

Browne et al. (1999) describe an interesting technique, which may merit further empirical investigation. Participants (undergraduates) created their own categories for which they subsequently indicated what the appropriate probability label was. The authors focussed on the effect of identifying differing numbers of categories – with increasing numbers of categories, the calibration worsened but discrimination improved. Further comparisons between a problem-structuring approach of this kind and other more-established methods could be of value.

Some studies have found that recomposing two or more conditional probability assessments can provide better-calibrated judgements. Kleinmuntz et al. (1996) found that the calibration for general knowledge items was better when probability assessments were recomposed from three component assessments: a pair of mutually exclusive conditional probabilities and a background assessment (e.g., P(IBM 1988 export sales > \$ 5 billion | IBM 1988 total sales > \$ 60 billion), P(IBM 1988 export sales > \$ 5 billion | IBM 1988 total sales < \$ 60 billion) and P(IBM 1988 total sales > \$ 60 billion)). Wright et al. (1994) found a similar advantage for recomposed assessments over direct assessments for forecasting outcomes in the World Snooker Championship. The probabilities of any given player reaching a given stage in the competition were better assessed by estimating and combining the probabilities of winning against all possible opponents that they could meet (as defined by the tournament draw). For example, the probability of a player reaching the final was better assessed by multiplying subjective probabilities for P(wins quarter final) and P(wins semi-final | wins quarter final) than by directly eliciting P(wins semi-final). The improvements attributable to recomposition are not necessarily large, but are, nonetheless, of some practical importance.

O'Hagan (1988) discusses the recomposition of probability judgements, preferring to use the term 'elaboration' to refer to any process of constructing a measurement of a probability from measurements of other quantities that are composed using any theorem of probability. He proposes elaboration as the principal way to improve probability measurements, arguing that the essential role of probability theory is to provide us with elaborative measurement devices.

4.2.3 Calibration research in applied settings

It is natural, now, to ask whether calibration research with experts finds the disappointing pattern of results that is typically found with non-expert populations. Christensen-Szalanski and Beach (1984) performed a ten-year search of *Psychological Abstracts*, covering judgement and decision-making (i.e., taking a broader remit than that solely focussing on probability judgement). They found 37 articles detailing 'good' performance and 47 that identified 'poor' performance. All the examples of poor performance were laboratory studies, typically with student participants. Fifty-eight percent of the examples of good performance were in applied settings. The overall pattern of better performance in applied settings seems to be reflected in the probability judgement literature.

In the field of probability judgement, meteorological forecasting represents one of the most keenly investigated domains. Probabilistic weather forecasts have been commonplace in the United States since the 1960s. Calibration was initially moderate but improved over time (Murphy and Brown, 1984); the progression was attributed by some to the regularity of outcome and performance feedback (Bolger and Wright, 1994). Examples of excellent calibration exist (Murphy and Winkler, 1977), and forecasters have been able to outperform predictive models constructed from analyses of past instances under a range of circumstances (Root, 1962; Sanders, 1963) or to improve upon these 'objective forecasts' when provided with them (Murphy and Brown, 1984). Calibration is better for shorter than for longer forecast periods (Winkler and Murphy, 1968). Meteorologists' good calibration is task specific; in other words, it is not the case that they are well-calibrated individuals who are necessarily well calibrated for other tasks (Keren, 1985).

Examples of good calibration are also found in the realms of sports and games. Keren (1987) found that experienced tournament bridge players were almost perfectly calibrated for predictions of the probability that a contract would be made. In contrast, keen club players were on average overconfident and showed the common pattern of using the extremes of the probability scale too frequently. Vertinsky et al. (1986) reported generally good calibration among college field hockey players for their probability judgements of the outcomes of tournament games in which they were involved. Bookmakers have been found to be well calibrated for predictions of the outcome of sports events (Dowie, 1976; Yates and Curley, 1985). These findings support the assertion of Johnson et al. (1991), discussed in Section 3.3.1, that an expert's ability to express her knowledge in probabilistic form ought to be better when key aspects of her field of knowledge reflect aleatory uncertainty (e.g., predicting the outcome of sports events).

Calibration in the business domain is somewhat variable. Braun and Yaniv (1992) found that economists' forecasts of economic downturn were more accurate than a base-rate model for short time frames, with the reverse being true for longer time frames. Wilkie and Pollack (1996) found that expert probability judgements for currency exchange forecasts outperformed statistical prediction models.

However, several studies of probabilistic forecasting for stock market values report quite modest calibration performance and suggest a rather complex relationship between expertise and performance. Staël von Holstein (1972) investigated probabilistic forecasts of stock prices by a variety of professional groups with relevant interests and experience. Few individuals performed better than a uniform judge (always stating the same probability derived from past event frequencies). The order of performance from best to worst was as follows: stock market experts, statisticians, business administration students, university business teachers and, finally, investment bankers. This suggests that, in terms of the accuracy of probability assessment, there are benefits that accrue with high-level substantive expertise (viz stock market experts) and high-level normative expertise (viz statisticians). However, at lower levels of expertise, additional knowledge or information will not always ensure greater accuracy, as shown by the decreasing levels of performance across students, business teachers and bankers.

Two further studies of probabilistic stock forecasting seem to confirm the adage that 'a little knowledge is a dangerous thing'. Yates et al. (1991) compared the forecasting performance of undergraduate business students ('novices') and graduate finance students ('semi-experts'). The novices' performance was superior, predominantly because the probability assessments of the semi-experts were too extreme. The semi-experts assessed large changes in stock price to be more likely than was warranted by the actual movements in stock prices. In contrast, the novices judged large changes in stock value to be relatively improbable. Thus, they responded to the task as if they knew the limits of their ability to predict which stocks would rise quickly, or fall rapidly, in value. Önkal and Muradoğlu (1994) compared the probabilistic stock forecasts of business students who had no experience of stock trading (novices) with business students who did have trading experience (semi-experts). As in the previously described study, the probability assessments provided by the novices were better, though in this case their superiority was primarily attributable to better discrimination.

Yates et al. (1991) propose that the inappropriately large variance in the assessments of the semi-experts in their study was most probably because these graduate students were using unpredictive cues to guide their judgement. Such behaviour is consistent with the use of heuristics (short cut strategies), which can produce biased judgements (see Section 3.4). For instance, a reliance on cues that fail to predict is consistent with illusory correlation, which is the persistence of a belief in a relationship between variables despite data to the contrary. This is attributed to the use of the availability heuristic (which involves making judgements by what comes easily to mind; see Section 3.4.1). Another possibility is that these semi-experts were responding too strongly to cues that imperfectly predict movements in stock prices. This is consistent with relying upon the representativeness heuristic, which results in predictions that are insufficiently regressive (see Section 3.4.2, especially Section 3.4.2.5). For instance, an increase in company stock value is representative of recent reports of high profits. Therefore, large increases in stock value might be expected to follow reports of substantial profits. However, profit levels imperfectly predict changes in stock value. A failure to recognise this would result in inappropriately extreme judgements.

A further study of probabilistic stock forecasting by Önkal and Muradoğlu (1996) suggests the possibility of a curvilinear relationship between the level of expertise and the accuracy of probability assessments. Experts (portfolio managers) were more accurate than semi-experts (other banking professionals) and were not less accurate than novices (students).

Clearly, examples of both good and poor calibration can be found in applied settings. However, the general level of calibration performance is better than that observed in studies of students who provide probability assessments for the truth of almanac items. It may be important to note that the distinction between these types of studies is not restricted to the differences between the participants. In studies that use almanac questions, all the uncertainty is epistemic. In these applied studies, the uncertainty was *both* epistemic *and* aleatory. For additional discussion of this distinction with respect to calibration research, see Sections 4.2.1.2 and 4.2.2.1.

4.2.4 A case study in probability judgement: calibration research in medicine

4.2.4.1 The accuracy of physicians' probability judgements

Physicians have received considerable attention in the broader literature on judgement and decision-making under uncertainty and within the more narrowly focussed literature on probability judgement. No doubt some of this attention has been prompted by the importance of their work, the potential impact of their judgement and the difficulties associated with the decisions that they take. For the same reasons, physicians are common participants in elicitation exercises, where their expertise is used to provide inputs into decision analytic models such as those used in cost-effectiveness analyses. Given this focus, and the range of research available, we examine the research on the calibration of subjective probabilities in medicine as a separate case study in expert probability assessment.

Investigations of doctors' probability judgement have yielded a range of results with respect to their calibration. Two broad categories of judgement have been investigated: diagnosis and prognosis. In diagnostic tasks, physicians estimate the probability that a patient has a given disease or condition; typically, subsequent testing (e.g., laboratory tests or angiography) is used to provide definitive diagnosis. In prognostic tasks, estimates of survival or functional status at a specified future point in time are made, the accuracy of which can be determined by careful follow-up.

Christensen-Szalanski and Bushyhead (1981) found doctors to be imperfectly calibrated for probability judgements for pneumonia diagnosis. Although sensitive to relative differences in the predictive value of symptoms, the probability of pneumonia was overestimated and the accuracy was worse than would be achieved by a uniform or base-rate judge (see Yates, 1990). Dawson et al. (1993) observed modest discrimination and calibration among physicians for confidence ratings that they provided for estimated values of several measures commonly used by cardiologists (pulmonary capillary wedge pressure, cardiac index and systemic vascular resistance index). No relation was observed between confidence and accuracy (the mean absolute error in judgement, that is, |actual value – estimated value|). Experienced physicians were no more accurate, but were more confident. Other studies that reveal inaccurate judgements of uncertainty include Poses et al. (1985), Thornbury et al. (1975), de Smet et al. (1979), Eisenberg et al. (1982), Tierney et al. (1986) and Dolan et al. (1986).

A number of examples of good judgement by physicians are found in the literature. McClish and Powell (1989) examined physicians' judgements of the probability of survival for patients in an intensive care unit (ICU). The probabilities supplied by the physicians working in the ICU were compared with probabilities generated by a logistic model derived from the APACHE II illness severity index. The physicians displayed greater discrimination than the model and could more readily identify those who were likely to die. Overall, the model was better calibrated, as a result of better calibration in the mid-range of the probability scale. Poses et al. (1990) obtained prognostic judgements for critically ill patients on admission to ICU and 48 hours after admission. There was considerable variation between doctors, but discrimination and calibration were good for both sets of judgements. In a study of the diagnosis of throat infection (streptococcal pharyngitis or 'strep throat'), physicians had better discrimination than a statistical model, though the overall accuracy was not greater (Centor et al., 1984). A pattern that seems to emerge from these studies is that physicians' discrimination is often better than that of a statistical model, but greater bias and/or random variability means that overall accuracy is better for the models. Bias could be attributable to insufficient knowledge of base rates or their under-utilisation. Increased variability typically represents unreliability in judgement.

4.2.4.2 Factors affecting physicians' probability judgement

A number of studies have considered features that might determine the quality of physicians' judgement.

Several studies indicate that personal disposition or individual experience can bias judgement, and that, in some cases, this may be linked to the use of heuristics (i.e., relatively quick and effortless strategies for judgement, Kahneman and Tversky, 1974). For instance, Christensen-Szalanski et al. (1983) suggest that physicians sometimes use the availability heuristic (assessing probability according to how easily instances or examples come to mind; see Section 3.4.1). They asked physicians to estimate the mortality rates of various diseases. The physicians' estimates were unrelated to the incidence of reporting in journals, but estimates for a disease were biased by personal encounters with people with that condition.

Poses and Anthony (1991) found physicians to be poorly calibrated for probabilistic judgements for bacteremia, generally providing probabilities that were too high in relation to the rate at which the disease actually presented in their patients. Patients seen at times when doctors recalled that they had frequently cared for patients with bacteremia (which is non-contagious) were assigned higher probabilities of having the disease. This is consistent with the influence of the availability heuristic, which would predict that recent experience of a disease would make it easier to recall examples of patients with the disease. Furthermore, lower probability estimates were assigned to patients who they thought would be at the greatest risk of death if they were to have bacteremia. This is consistent with a 'value bias', which the authors characterise as 'wishful thinking'. Value biasing occurs when the value or utility of an event influences the probability assigned to the event. One possible interpretation of Poses and Anthony's (1991) findings is that when a doctor is particularly concerned about the consequences of a disease for one of their patients, it reduces his/her strength of belief that the patient does actually have this disease.

The possibility of value biasing is also suggested by other studies. In a bold study, Arkes et al. (1995) obtained probabilistic survival estimates for seriously ill patients from physicians, the patients and from a 'designated surrogate' (typically a relative). As might be hoped, doctors displayed better discrimination than the other participants. The doctors' judgements reflected which patients were more, or less, likely to die. However, the physicians had a pessimistic bias, whereas the lay participants were essentially unbiased. One interpretation suggested by the authors is that doctors may deliberately overstate the risk of death, as outcomes better than anticipated are favourably received, whereas outcomes worse than expected increase the chance of complaint or litigation. (Note that this explanation for the bias runs in the opposite direction to that proposed by Poses and Anthony (1991). This may suggest that caution ought to be exercised in interpreting these, arguably post hoc, explanations too strongly.)

We often expect our physicians to be analytical scientists *and* caring individuals with good knowledge of our personal circumstances. Is it possible that these roles might sometimes work against each other? A study by Bobbio et al. (1992) arguably highlights the importance of objectivity in interpreting data. They found that cardiologists were more accurate in predicting angiographic coronary artery disease from 15 patient variables for anonymous patients than with personal knowledge of the patient.

Burgus et al. (1998) demonstrated that the order of clinical information can influence judgement, with the information received last appearing to receive greater weight. Order effects are a common finding in judgement research (Hogarth and Einhorn, 1992), which are often explained in terms of the process of anchoring-and-adjusting that we discussed in Section 3.4.4. However, it is important that, even when experts are performing a familiar task, the order in which they consider the relevant information can have a substantial bearing on the final judgement that is formed.

Arguably, too few studies have considered the role of the environment in medical judgement. Tape et al. (1991) investigated regional variation (between three US states) in the accuracy of (probabilistic) pneumonia diagnosis. Using the lens-modelling approach (see Glossary), where the relationship between judgements and multiple cues is modelled by examining judgements for many cases, they found that the diagnostic accuracy varied between regions. However, this could be accounted for by regional differences in the reliability of cues. In general, physicians' 'policies' (i.e., linear models representing the relationship between patient variables and physician judgement) did reflect the relative importance of clinical factors in their own region. Rakow et al. (2005) observed that doctors' assessments of the probability of early mortality following paediatric heart surgery were appropriately influenced by the risk factors for surgery that had been reported in medical journals. However, few of these published risk factors predicted the outcomes of surgery in the hospital for which probabilities were assessed. Consequently, probability assessments failed to discriminate between patients according to the outcome of their surgery, irrespective of whether the assessments were provided by doctors or were generated by logistic regression models that had been reported in a medical journal.

Some have suggested that the poor calibration observed in some studies may be attributed to a lack of experience with probability estimates. While physicians may consider uncertainties in their professional decision-making, explicit statements of numerical probabilities are not always a part of their everyday practice (Keren, 1987). One could say that they do not have sufficient normative expertise (in probability) to transfer their substantive expertise (in medicine) into sound probability judgements. Formal interventions to improve the appropriateness and accuracy of judgement have been shown to be successful (Poses et al., 1992). However, a classic paper (Poses et al., 1995) demonstrates that decision-making may remain unchanged even when judgement has improved. Poses and colleagues found that prescribing patterns remained unchanged even though they had successfully improved physicians' assessment of the probability of throat infection (streptococcal pharyngitis). It would appear that doctors simply lowered the threshold at which they would prescribe in parallel with the lowering in their probability judgements achieved by educational intervention. Or, more likely, prescription decisions made no recourse to formal probability assessments, but were based purely upon unaltered feelings of subjective certainty that doctors had simply learned to 're-label'.

4.3 The calibration of subjective probabilities: theories and explanations

4.3.1 Explanations of probability judgement in calibration tasks

A number of proposals have been put forward to account for the pattern of results in laboratory and applied settings described in the preceding text (see Griffin and Brenner, 2004).

As already indicated, several authors identify the importance of feedback for improving judgement. This is true of judgement in general (Fischhoff, 1989) and of probability judgement specifically (Ferrell, 1994; Keren, 1987). In particular, it has been argued that both outcome and performance feedback are required and that this, coupled with a direct dependency between good calibration and 'success', accounts for the accurate judgement observed in applied settings such as meteorology or bridge playing (Lichtenstein et al., 1982; Keren, 1987). Outcome feedback alone (Benson and Önkal, 1992) or simple cumulative accuracy feedback alone (Staël von Holstein, 1971, 1972) typically prove unsuccessful at improving calibration.

In the context of elicitation, one cannot, of course, provide feedback for a one-off event or for a series of events that have yet to occur. However, one could use feedback in the initial training in probability that is given to the expert. The evidence above suggests that this should give her a better 'feel' of the appropriateness of her probability assessments.

The results of a number of studies are consistent with 'value biasing'. A classic example of this is provided by Rothbart (1970) who observed correlations between attitudes and probability judgements for questions on Quebec separatism. (The availability heuristic provides an alternative explanation for this. A belief in the likely success of one's own political aspirations would be encouraged if one's peers consisted of a disproportionate number of like-minded individuals – a not unreasonable conjecture.) Value biasing, or the tendency to allow the utility of an event to influence the assessment of its probability, has also been suggested as an explanation for suboptimal probability assessment in medicine (as discussed in Section 4.2.4.2). Wallsten (1981) found that radiologists' judgements of the probability of a tumour were influenced by how important it was to rule out the possibility of a tumour. Poses et al. (1985) proposed that factors important for the management of sore throat were incorrectly associated with the relative frequency of the disease, contributing to doctors' overestimation of the frequency of streptococcal infection in their patients. Perhaps unsurprisingly, value biasing has also been observed in the sports fans' prediction of the outcomes of sporting events (Babad, 1987). Such findings relate to a broader literature on the so-called 'optimistic biases' whereby people tend to rate their own positive chances of success as better than average, but their own chances of failure, disease or mishap as lower than average (Alicke et al., 1995; Eiser et al., 2001; Weinstein, 1980). A related proposal is that such overconfidence is a function of individuals having exaggerated beliefs in their own ability to control events (Armor and Taylor, 1998; Harris and Middleton, 1994; Langer, 1975).

It will often be difficult to identify when value judgements have biased an expert's probability judgements in an elicitation exercise. Knowing how to work with the expert to address this may be even harder. The problem may be moderated when there are multiple experts with different (perhaps competing) points of view. Combining judgements will, more often than not, yield an aggregate assessment that is better than an assessment from an individual expert chosen at random (see Section 4.5.2 and Chapter 9 on multiple experts). However, this approach only partially addresses this problem.

In recent years, some psychologists have argued that much of poor calibration is an artefact of the experimental setting or the response scale people are required to use. Gigerenzer (1994) argues that frequencies rather than probabilities are the response format to which we are best adapted. There are some experimental examples of the reduction of errors in probabilistic reasoning or improvements in judgement accuracy when frequency formats are used (e.g., Gigerenzer et al., 1991; Price, 1998). Aggregate (frequency) judgements such as 'How many of the last 20 items did you get correct?' reliably exhibit less overconfidence than the mean subjective probability calculated from the same items (Treadwell and Nelson, 1996). Juslin (1994) and others convincingly argue that the unrepresentativeness of item selection helps create an illusion of overconfidence by selecting sets of questions that are 'harder than average'. (Rome being warmer than New York would usually imply that it is further south – the fact that this is not so makes this a deceptively tricky question.) Price (1998) discusses two possible effects of relative-frequency elicitation. Firstly, this approach may reduce the random response error in probability judgements (see Section 4.5.2). Secondly, it may encourage the use of frequency information and simpler, more effective, strategies for making probability judgements (see Section 4.2.2.2). Analyses of studies that employ representative design reveal lower (sometimes insignificant) levels of overconfidence (Juslin et al., 2000). Gigerenzer et al. (1996) argue that aggregate judgements focus judges on the appropriate reference class of events – 'typical general knowledge tests' – and for this event, the cues available to participants are not misleading. Treadwell and Nelson (1996) favour a slightly different 'dual source' account, under which aggregate judgements are simply different kinds of judgement to single-case probability judgements, which therefore rely upon different kinds of evidence for their production.

As we have discussed elsewhere (Section 3.4.3), when one-off events are the focus of an elicitation exercise, it will rarely make sense to speak of frequencies. However, results such as these raise the possibility that it may sometimes help the expert if she is prompted to think about frequencies of events within a class. Asking 'Of 100 power stations like this one, how many would still be operational in 20 years' time?' may be an odd question, but it may help the expert consider the role of aleatory uncertainty. Asking the expert 'Of the 20 events that I have asked about, how many do you think will occur?' may help her consider just how predictable or unpredictable the events under consideration are, perhaps encouraging reflection upon the success of past predictions for which the outcomes are now known.

4.3.2 Theories of the calibration of subjective probabilities

The proposals discussed above do not constitute theories of how people make subjective probability judgements. Such theories of subjective probability are reviewed in detail by McClelland and Bolger (1994), more briefly by Harvey (1997) and more selectively by Bolger and Wright (1993). We outline some of the key theories here, as they have something to say about what is going on in the head of the expert when she provides a probability assessment and how different features of an elicitation problem can have an impact upon her judgement.

4.3.2.1 Stage models

Some theorists propose models involving three stages of processing (Koriat et al., 1980; May, 1986a, 1986b). Information processing precedes the emergence of subjective certainty, and this is then quantified as a numerical response. Ferrell and McGoey (1980) propose a model on the basis of the signal detection theory that identifies two stages: detection and assigning a probability value. Under such 'staged' models, miscalibration can derive from any one stage or the transition between any two adjacent stages.

The 'strength and weight model' is proposed by Griffin and Tversky (1992). 'Strength' refers to the extremeness of evidence, 'weight' to its predictive validity. They argue that people focus on the strength of evidence and adjust (possibly insufficiently) according to its weight. Overconfidence is then expected (in a half-range task) where evidence has high strength but low weight. Support theory (e.g., Tversky and Koehler, 1994) also describes the process of probability judgement in terms of the assessment of evidence. Support theory was discussed in the previous chapter (Section 3.4.5).

Early research in calibration appealed to biased information processing as the source of overconfidence and over-extremity in probability judgement (e.g., a tendency to seek support for a hypothesis to exclude contradictory possibilities). However, more recent theoretical developments have raised the possibility that at least some of the key findings in calibration research are artefacts of the experimental setting or of the common methods of data analysis associated with calibration research. More recent theories of probability judgement are briefly reviewed in the following three subsections.

4.3.2.2 Ecological theories

Ecological theorists argue that individuals are well adapted to extracting and storing information regarding the frequency of events and of the co-occurrence of events. Gigerenzer et al. (1991) proposed that people construct a probabilistic mental model (PMM) when making judgements under uncertainty. Relevant cues are searched for in memory, and the relative frequency with which a given cue predicts the target event (as previously observed by the assessor) defines the probability that she estimates. For instance, imagine that I base my level of confidence that New York is north of Rome on the observation that it has a colder climate. If I believe that for pairs of northern hemisphere cities, the one that is colder is also the one that is further north on 90% of occasions, I will assign a probability of 0.9 to the proposition 'New York is north of Rome'. As indicated above, poor calibration is argued to be a function of the experimental setting in which question sets are constructed wherein cue-criterion relationships that people have learned outside the experimental setting are not preserved within it. This class of models draws support from the rather better results reported in applied settings and in studies where representative sets of items are used (e.g., the same judgement is made for each pair of a finite set of cases).

4.3.2.3 Error theories

Erev et al. (1994) undertook an important examination of the phenomenon of overconfidence. In particular, they considered the apparent contradiction between overconfidence or overly extreme judgement in calibration research, and conservatism (insufficiently extreme adjustment) in the Bayesian paradigm. They conclude that some overconfidence can be considered a manifestation of 'regression to the mean', if subjective probability estimates are considered as 'true' judgements perturbed by random error. It is easy to argue that a *response error* of this kind is likely to be present. The same subjective feeling of certainty will not always be 'labelled' with the same subjective probability; and events that engender different subjective feelings of certainty may well be labelled with the same subjective probability (not least because of the limited precision with which assessors are prepared or permitted to respond). Crucial to the argument is the observation that the response scale is bounded (between 0 and 1 in full-range tasks and between 0.5 and 1 in half-range tasks). Consequently, at the extremes of the response scale the response error cannot be symmetrically distributed, making it less likely that probability estimates at the extremes of the probability range are unbiased estimates. Erev et al. (1994) ably demonstrate how such an assumption predicts the characteristic pattern of the apparent over-utilisation of extreme probabilities (i.e., very low and very high probabilities).

4.3.2.4 Ecological plus error theories

Juslin (1993) and Soll (1996) describe theories that combine and develop the ideas described in the previous two subsections. For more recent work, see Juslin et al. (1997, 2003). These authors presume that our cognitive architecture provides us with a reasonably accurate impression of the cue-criterion relationships (that we have observed), and that these cues can be used to guide our probability judgement. However, unlike Gigerenzer et al. (1991), they are explicit that error in judgement means that probability judgements will not always be perfectly calibrated, even for representative sets of judgements. In contrast to Erev et al. (1994), they identify two sources of error. Like Erev et al. (1994) they argue that response error, the error attributable to human inconsistency, plays a part in miscalibration. In addition, they point to the role of sampling error (or aleatory uncertainty), which is the error attributable to variability in the environment. The samples of events that we observe provide us with an imperfect guide to the cue-criterion relationships that are used to guide probability assessment. This results in errors in probability judgements, which are intended to apply to populations of events (i.e., are population parameters). For instance, at the population level ('in the real world'), two variables may be correlated with $r = 0.7$. However, for samples we have observed ('in our experience'), the correlation between these two variables may take a range of values. Consequently, the samples that we have observed cannot represent 'perfect information' about the world around us. See Section 3.6.3 for the discussion of the particular difficulties that can arise when our experience exposes us to biased samples.

4.3.2.5 Theories of calibration: anything left to explain?

Although important, and convincingly presented, arguments concerning error and unrepresentative task design may not be sufficient to explain all miscalibration. Brenner, et al. (1996) and Pfeifer (1994) note that error, or regression effects, can account for over-extremity in probability judgements, but not for all overconfidence in full-range tasks (i.e., when the mean probability judgement exceeds the mean proportion that are correct). Brenner et al. (1996) report a prediction task using a representative design where gross overconfidence was observed for judgements of both frequency and confidence (i.e., probability): where either judgement was highly correlated with independent assessments of item representativeness. On the other hand, Juslin et al. (2000) argue vehemently that, when all potential artefacts associated with calibration tasks and their analysis are taken into account, there is little evidence of any meaningful cognitive overconfidence bias. Even if one accepts their view, an important concern remains. In calibration studies with items such as almanac items, one is dealing only with epistemic uncertainty and can determine accuracy simply and immediately (because the 'true' answers are known). With real-world elicitation tasks, accuracy may not be easily assessed (and when it can be, it may involve considerable delay). The research we have reviewed implies suggestions regarding how tasks should be structured. Yet, when the characteristics of the assessments are ill understood, it is not always clear exactly what action should be taken to improve the assessments provided in an elicitation exercise.

4.3.2.6 Key lessons from the calibration literature

Can people learn to be well calibrated? Yes, but the process is far from straightforward, and results of training may often not generalise well to new tasks. A number of authors have set out necessary conditions for good probability judgement or guidelines and prescriptions for improving calibration. For instance, Murphy and Winkler (1984) identified a number of factors that could contribute to the generally high quality of probability forecasts in meteorology, including (1) practice (forecasts are made many times every day), (2) prompt feedback (outcomes become known quickly), (3) evaluation (quantitative scoring rules are used to assess performance), (4) importance (effort and technological resources are devoted to making forecasts) and (5) objectivity (unlike some areas, the forecaster has no influence upon the outcome). Fischhoff (1989) specifies four conditions for improving judgement: (1) abundant practice with a reasonably homogeneous set of tasks, (2) well-defined criterion events, (3) reinforcement specific to the judgement task and (4) an explicit admission of the need for learning.

4.4 Representations and methods

In this section, we focus on some of the technical details associated with the practicalities of elicitation. We begin by considering the language, both mathematical and natural, that can be used when assessing uncertain quantities; and then discuss the various means by which the expert can commit herself to an assessment.

4.4.1 Different modes for representing uncertainty

4.4.1.1 Numerical expression of probability

Statements about chance and uncertainty can be expressed in a number of different ways: as probabilities, percentages, relative frequencies, odds or natural frequencies.

Probability: 'There was a 0.08 probability of being a victim of crime in 2002.'
Percentage: 'There was an 8% chance of being a victim of crime in 2002.'
Relative frequency: '1 in 12 people were victims of crime in 2002', or '80 in every 1000 people were victims of crime in 2002.'
Odds: 'The odds against being a victim of crime in 2002 were 11 to 1.'
Natural frequency: 'From a population of 56 million people, 4.5 million were victims of crime in 2002.'

These different methods of expressing chance are mathematically equivalent (on the assumption that all statements refer unambiguously to the same event), but a growing literature indicates that they are not always psychologically equivalent.

In a number of places in this chapter and in the preceding one, we have identified biases in reasoning with probabilities or in estimating proportions and relative frequencies, that are reduced in magnitude when information is given or elicited using frequencies (e.g., Section 3.4.3). Two views seem to be prevalent in accounting for this. Firstly, as discussed earlier, it is proposed that frequencies focus attention upon classes of similar events and the distributional properties of these classes such as the base rate of event occurrence (Kahneman and Tversky, 1982a). For instance, Koehler (2001a, 2001b) compared potential jurors' reactions to different ways that the random match probability for a DNA test could be specified in a courtroom. Specifying that 'one in one thousand people in Houston would match the blood sample' or that 'the probability that the suspect would match the blood sample is 0.1%' ought to be equivalent statements. However, participants were less likely to attribute guilt when presented with the former statement, presumably because it leads them to call to mind the several hundreds of people in a large city (containing many thousands) who, in addition to the defendant, could be the source of the blood. Secondly, Gigerenzer (2002) proposes that frequencies remove ambiguity. For example, imagine that a patient is told that a drug carries a 0.3 chance of a side effect of sexual impotence. The patient might legitimately interpret this as 3 in 10 people are expected to encounter (some or consistent) impotence or that all patients can expect to experience impotence in 3 out every 10 sexual encounters. Gigerenzer argues that the ambiguity arises because the reference class, patients or sexual encounters, is unclear. In contrast, expressions using frequencies tend to be explicit about the reference class. Consequently, they are less likely to be open to misinterpretation. It may not be deemed practical to use frequencies as part of an elicitation exercise. Nonetheless, this point does highlight the importance of ensuring that the expert and the facilitator are not unknowingly at odds in their interpretation of exactly what event or outcome is under consideration.

A further benefit attributed to the use of frequencies will only rarely be of practical significance in elicitation, as it relates more to reasoning with probabilities that have already been stated than to making probability judgements. It is proposed that frequency formats simplify the mathematical operations that must be performed (e.g., avoiding decimal arithmetic). They reduce errors by reducing the frequency of miscalculation and increasing the chance that implausible/impossible answers resulting from arithmetic slips are identified and rectified (Gigerenzer, 1996b). This issue would only be relevant to elicitation where the expert is expected to combine probabilities in her head without any assistance from the facilitator. It would be remiss of the facilitator not to assist the expert in this way, as this is part of the normative expertise that we expect the facilitator to bring to the elicitation process.

Results from Slovic et al. (2000) highlight the differences in expert elicitation that can arise between frequency and percentage response scales – but suggest that a strict 'frequencies good, probabilities bad' interpretation may be too simplistic. The task performed by forensic psychologists and psychiatrists was to assess the chance that a (hypothetical) psychiatric patient described to them would harm someone within six months of hospital discharge. Frequency formats led to consistently lower judgements. However, in a subsequent study, when asked to assess the danger posed by a patient with a 0.2 probability of committing violence, those described as having a 20% chance were deemed less dangerous and more suitable for discharge than those with a 20 in 100 chance. Presumably, the effects of discharging 100 such patients become transparent and somewhat frightening (see Section 3.4.6 on the affect heuristic). Thus, when action depends upon the interplay between several judgements and assessments, effects in one direction on one task (e.g., risk assessment judgements) encouraged by frequency formats may be cancelled by effects in the opposite direction on another task (e.g., judgements of suitability for discharge).

Yamagishi (1997a, 1997b) cautions that responses using relative frequency are affected by the choice of the number that the assessment is made 'out of'. He compared eliciting estimates of relative frequency using two alternative formats: "X out of 100" and "X out of 10,000". His experimental results show that frequency estimations, expressed as percentages, are greater in the "Out of 100" format than in the "Out of 10,000" format when frequencies of low-probability events are estimated. When frequencies of high-probability events are estimated, the reverse is true: probabilities are greater for the 'Out of 10,000' than for the 'Out of 100' format. Yamagishi (1997a, 1997b) elegantly accounts for his findings by appealing to anchoring effects, where insufficient adjustments from initial anchors would predict these effects (see Section 3.4.4). Thus, it appears that care needs to be taken when determining the values used for the elicitation of relative frequency.

Experts may find it cumbersome to express highly improbable events using probabilities and percentages. We might suppose that we have a better feel for the difference between a 1 in 10,000 chance and a one in a million chance than we do for the equivalent difference between a probability 0.0001 and that of 0.000001. Interestingly, Wright (1984) asserts that estimates of uncertainty expressed in odds tend to be more extreme than those expressed using probabilities, possibly because an odds scale is not bounded (in the sense that one cannot specify what the longest, or shortest, odds possible are). Rightly or wrongly, using odds seems to make people more likely to provide probability assessments close to 0 or 1. This discrepancy between odds and probabilities also indicates that the psychological inequality of mathematically equivalent expressions of uncertainty is not restricted to comparisons that include frequencies or relative frequencies.

We might suppose that something similar would be true for assessments of highly unlikely events made using relative frequency scales. At its lower end, a relative frequency scale is unbounded in the same sense as an odds scale, (i.e., one cannot specify the largest number with which to complete the phrase: 'A 1 in – chance'). This may well make it easier to report very low probabilities. See Section 4.4.2.1 for more details on the practicalities of elicitation for events with very low probabilities.

4.4.1.2 Verbal expressions of uncertainty

Descriptions that include terms such as 'possible', 'probable' or 'unlikely' express belief in how likely an event is. As such, they could be considered to be estimates of frequency or proportion or to be imprecise probability judgements. However, since such terms are relative rather than absolute, they need to be interpreted differently according to the context. A 'rare' disease is only rare compared with other more 'common' diseases: nothing in the use of such terms implies that a person is any more or less likely to have a 'common' disease than to have a 'common' name. For instance, Teigen (2001) observed that a probability of (or close to) $\frac{1}{6}$ can be variously described as 'a good chance' (e.g., when it applies to one of six runners in a race) yet also as 'improbable' (e.g., when it applies to one athlete in a two-man race). Thus, for 'good chance' read 'as good as anyone else's chances', or for 'likely' read 'as likely as any other event'. Nonetheless, attempts have been made to map verbal expressions onto a numerical scale of probability such that we could use them as a means of measuring probability assessment (Wallsten et al., 1986; Clark, 1990; Windschitl and Wells, 1996; van der Gaag et al., 2002).

Several studies have demonstrated considerable interpersonal variation in the assignment of numerical equivalents to verbal terms (e.g., Beyth-Marom, 1982; Budescu and Wallsten, 1985; Lichtenstein and Newman, 1967; Reagan et al., 1989; Sutherland et al., 1991). For instance, each of the expressions 'possible', 'probable' and 'good chance' elicits highly variable interpretations (Wallsten et al., 1986). My numerical interpretation of 'probable' will often be different from yours. In the context of communicating to cancer patients, Sutherland et al. (1991) report one study in which the standard deviation of percentage chance estimates assigned to the terms 'possible', 'unlikely' and 'rare' all exceeded 30. The standard deviation of estimates exceeded 20 for several other terms such as 'frequently', 'likely' and 'occasionally'.

The interpretation of verbal terms can be shown to be context dependent, that is, influenced by the kinds of events or scenarios being considered. For instance, Wallsten et al. (1988) found that lower numerical equivalents were assigned to the *same* verbal expressions when applied to events assumed to happen only rarely, as opposed to more frequently. This sensitivity to context, however, is not exclusively an aspect of verbal probability descriptions but can be observed with effectively *any* kind of rating, including numerically expressed probability estimates and predictions (e.g., Eiser and Hoepfner, 1991; Fischhoff et al., 1978a; van der Pligt et al., 1987).

Taken at face value, the term 'fifty-fifty chance' would seem to be a verbal term with such clear numerical equivalence (a probability of 0.5) that it would not be associated with the variability in interpretation associated with other verbal expressions of uncertainty. However, studies suggest that people often use it as a proxy for a 'Don't know' response when there are two logical possibilities rather than as a genuine expression of a 0.5 subjective probability (Bruine de Bruin et al., 2002; Fischhoff and Bruine de Bruin, 1999). Excessive (and arguably inappropriate) use of this 'fifty-fifty' response falls away when people are also provided with a 'Don't know' category.

A number of authors have noted that verbal expressions of uncertainty elicit different patterns of response to numerical expressions of probability, suggesting that the former encourages more intuitive (and less analytic) reasoning. For instance, while binary complementarity is usually well preserved for numerical assessment – for example,, if $P(A)$ is assessed to be 0.65, then $P(\text{not A})$ is explicitly assessed to be 0.35 – the equivalent does not always hold for verbally expressed uncertainty. One can feel that it is 'more likely' that event A will occur but fail to feel that its complement 'not A' is less likely to occur (Windschitl, 2000).

It has been suggested that using verbal expressions in the early stages of the elicitation process is more intuitive than using numbers. If one takes the view that sometimes intuition is more appropriate than analysis, this may be of merit (see Section 3.6.1). For instance, Phillips and Edwards (1966) claimed that the rate of systematic error or bias in directly assigning numerical probabilities is too high. Perhaps for this reason, communicators tend to prefer to communicate probability estimates in verbal rather than numerical form in contexts as diverse as sporting gambles (Erev and Cohen, 1990) and (hypothetical) treatment prognoses (Brun and Teigen, 1988). This preference for verbal forms was also found when doctors were asked to think aloud when solving a decision task (Kuipers et al., 1988). This suggests that verbal forms of expressions may help decision makers represent a problem so that a choice can be made, even though they are less precise than numbers. However, this same research indicates that *recipients* of such information prefer it to be given in numerical form.

Numbers have a compelling attraction over verbal expressions for statisticians and their clients because they are more precise, allow us to implement calculations, including combining them across experts, and have a unique rank order. In contrast, verbal expressions are more naive and more variably expressible (see Wallsten et al., 1993). These drawbacks of verbal expressions can explicitly be found in the substantial empirical studies of numerical versus verbal expressions and studying the influence of context (e.g., Budescu and Wallsten, 1985; Budescu et al., 1988). The key messages about verbal expressions of uncertainty are that they mean different things to different people and sometimes mean different things to the same person in different contexts. Therefore, they will rarely serve any formal function within elicitation, as it will always be necessary to clarify with more precise methods exactly what the expert means by a particular expression. Nonetheless, as will be seen in the following section, some of the precision attributed to numerical assessments of probability may be illusory (see also Chapter 7).

4.4.2 Different formats for eliciting responses

A common distinction in cognitive psychology is between explicit or direct tests of performance (where the purpose of the test is clear to the participant) and implicit or indirect tests (from which performance levels are inferred without the participant necessarily having express knowledge of what is being tested). We can apply a similar distinction to the assessment of probabilities.

4.4.2.1 Direct methods

(a) *Direct estimation.* This simply involves asking the expert to state her response.

(b) *Response scales.* Response scales provide a visual representation of the feasible range of responses. This may simply be a list of ordered categories where the chosen category is circled, ticked or underlined to indicate the expert's response. Alternatively, a visual analogue scale (a line on which the expert indicates her response by a line or cross) provides a continuous response scale. Such scales will typically be labelled at each end (e.g., '0' and '1' or 'impossible' and 'certain') and may or may not be labelled at intermediate points such as the mid-point. Values are obtained by measuring the distance from one end, and the measured lengths may be subject to transformation to convert them into probabilities. Values extracted from visual analogue scales are typically treated as having the properties of an interval scale (which one might challenge).

The labelling of response scales has been shown to have important effects upon judgement. For instance, Slovic et al. (2000) found substantial differences in expert probabilities (risk assessments) between scales with an upper label of 'greater than 40%' and those with an upper label of '100%'. Differences extended to the use of those parts of the scale that did not differ between conditions. Response scale effects can be quite considerable in magnitude and are reviewed by Poulton (1989) and Schwartz (1999). Participants, perhaps not unreasonably, seem to be 'cued' by the particular categories on a frequency (or other rating) scale as to what might be 'reasonable' or 'average'. If there is any uncertainty in the participant's judgement of absolute event frequency, the particular choice of candidate responses can have considerable influence upon responses. These effects are prime examples of the constructive processes that we described in Section 3.2.2 and in Chapter 1.

Eliciting probabilities for sets of events containing some very rare events presents a particular challenge to the elicitation process. Woloshin et al. (2000) report the use of a visual analogue scale where probabilities below 1 in 100 are presented on a 'magnified' logarithmic scale (1 in 100, 1 in 1000, 1 in 10,000 and 1 in 100,000). This 'magnified' scale and a traditional visual analogue scale outperformed a '1 in X' direct assessment scale on inter-rater correlation, intra-rater test–retest reliability and usability. In addition, the magnified scale successfully facilitated the expression of small probabilities (e.g., 10^{-5}) that could not be easily expressed with the traditional scale.

(c) *Probability wheels*. Probabilities may be presented (radially) around the circumference of a disk. The expert or facilitator then manipulates two radii to divide the disk into two segments (typically of two different colours) – one representing the probability of the event and the other, its complement. One therefore has three equivalent ways of representing probability: (1) via the scale on the circumference, (2) via the proportion of the circumference swept out by one segment and (3) via the proportion of the area of the circle shaded in one colour. Of course, we have already encountered several instances where what is mathematically equivalent is not psychologically equivalent. However, we do not know whether this would be the case in this instance.

4.4.2.2 Indirect methods

(a) *Bets*. Bets can be used to infer an expert's subjective probabilities. For example, 'With the opportunity to win 10, would you rather bet on event A or on throwing a '6' on a fair die?' Preference for betting on event A implies that it has a subjective probability greater than $\frac{1}{6}$. A sequence of bets can be used to specify subjective probabilities more precisely, by iterating until an indifference point is reached. The 'Ellsberg Paradox' (Ellsberg, 1961) suggests that this logic of the preference or indifference between gambles may not be reflected in people's choices if one or more of the events have ambiguous or ill-defined probabilities. For instance, when asked to *bet on black*, most people prefer to pick from an urn containing 30 black balls and

30 red balls rather than from an urn with 60 black and red balls in unknown proportions. This implies belief that the probability of black is at least as high in the '30-30 urn' (and no higher in the other urn). However, when asked to *bet against black*, most people also prefer to pick from the urn containing 30 black balls and 30 red balls rather than from an urn with 60 black and red balls in unknown proportions. This is typically interpreted as showing aversion to ambiguity: a preference for well-defined (precise) probabilities over ill-defined (ambiguous or uncertain) probabilities.

(b) *Probability wheels.* Probability wheels are also quite commonly employed with the betting methodology. This may involve removing any numerical labels, and simply asking the expert to manipulate the wheel until they are indifferent between a bet that wins if the event occurs and a bet that wins if a 'spinner' stops in the indicated section of the wheel.

4.4.3 Key lessons

The key lesson of this section is that what is mathematically equivalent may not be psychologically equivalent. Equivalent representations of the same information will not be interpreted as equivalent – and this presumably applies to representations for and by the expert. Consequently, probabilities elicited using frequency formats can be expected to deviate systematically from those elicited using probabilities. Similarly, equivalent processes that, in principle, 'tap' the same beliefs in different ways can be expected to yield different responses from the same expert.

4.5 Debiasing

We have already considered a number of specific proposals for improving the probability assessment for individual events. In this section, we open out the discussion to consider the literature on debiasing, which not only focuses on the elicitation of probabilities but also sets out general principles for addressing biases in any kind of judgement under uncertainty. A number of biases were discussed in Chapter 3, where these systematic errors in judgement under uncertainty were often attributable to the use of short cut strategies (heuristics). Heuristics and biases were discussed more specifically in Sections 4.2 and 4.3, in relation to the elicitation of probabilities. Thus, this section on debiasing is tied closely to much of the material covered in the last two chapters. However, as we consider some general principles, this discussion of debiasing is also relevant to several topics that will be covered in subsequent chapters. For instance, the debiasing literature includes a number of recommendations for good practice that could apply to many elicitation tasks, including the elicitation of probability distributions, correlation coefficients and linear models, either from individual experts or from groups (see Chapters 5, 6 and 9).

4.5.1 General principles for debiasing judgement

Fischhoff (1982) identifies several different debiasing strategies that have been proposed or tested, each of which is based on different observations or assumptions about the reasons behind these biases (see also Arkes, 1991). His analysis neatly partitions these strategies and is reproduced below (Table 4.1). Put simply, the options are 'fix the task', 'fix the expert' or 'match the expert to the task' (or *vice versa*). Fischhoff (1982) describes attempts to debias overconfidence in probability judgements (see Section 4.2.2.2). We have discussed many of these in the preceding text, and reported, like Fischhoff, that most of these attempts have had little impact on reducing overconfidence in probability judgements. However, there is no reason

Table 4.1: Debiasing methods according to underlying assumption (Fischhoff, 1982, p. 424).

Assumption	Strategies
Faulty tasks	
Unfair tasks	Raise stakes
	Clarify instructions/stimuli
	Discourage second-guessing
	Use better response modes
	Ask fewer questions
Misunderstood tasks	Demonstrate alternative goal
	Demonstrate semantic disagreement
	Demonstrate impossibility of task
	Demonstrate overlooked distinction
Faulty judges	
Perfectible individuals	Warn of problem
	Describe problem
	Provide personalised feedback
	Train extensively
Incorrigible individuals	Replace them
	Recalibrate their responses
	Plan on error
Mismatch between judges and the task	
Restructuring	Make knowledge explicit
	Search for discrepant information
	Decompose problem
	Consider alternative formulations
	Offer alternative formulations
Education	Rely on substantive experts
	Educate from childhood

to suppose that debiasing procedures will not succeed in other classes of task. Therefore, Fischhoff's framework provides a useful checklist for how one might approach this.

Suggestions for removing or reducing some of the biases discussed earlier have been put forward on the basis of (a) observation of the apparent sources of these biases and the factors that affect their magnitude and (b) evaluations of the effectiveness of specific debiasing procedures and techniques. Broadly speaking, there seem to be two classes of approach to debiasing: (1) preventative measures that seek to act upon the expert and/or the structure of their task to reduce/eliminate bias or error 'at source' or (2) post hoc measures which adjust judgements in an attempt to reduce error and bias after the judgements have been provided.

With respect to task structuring, MacGregor et al. (1988) examined several decision aids for quantitative judgement (which encouraged their student participants to decompose the problem in a number of different ways). Participants' accuracy and consistency generally increased as the decision aids became more structured. However, one practical problem with using decision aids is that people can be resistant to using them (Kleinmuntz, 1990; Larrick, 2004).

Irrespective of the point in the elicitation process where they might act, the principles that lie behind apparently different debiasing strategies are often similar. These principles are discussed below.

4.5.2 Managing noise

Quantitative judgements under uncertainty are demonstrably subject to random variation or 'noise'. These random variations in judgement represent another source of aleatory uncertainty. This can be particularly problematic in judgements provided by people (rather than models), where psychologists often allude to the 'hand tremor' analogy. Just as fine motor skills are executed imperfectly because of the difficulty in controlling the body's physical apparatus, performance in cognitive tasks (such as probability judgement) suffers because of a lack of 'cognitive control' of our mental apparatus (Hammond and Summers, 1972). Consequently, information is not used with perfect consistency, and variability in judgement ensues. A variety of techniques are often effective in reducing the random variation in a set of judgements, thereby increasing the accuracy of the set of judgements.

Several approaches involve averaging or aggregating judgements from more than one source. These 'sources' may be multiple experts, where aggregation may be achieved 'behaviourally' (via interaction) or 'mechanically' (mathematically after individuals have supplied their own judgements) (Hill, 1982; Rowe, 1992). Alternatively, different sources may reflect different strategies, such as when the judgements of an expert are averaged together with those supplied by a prediction model derived from the statistical analysis of past cases (Blattberg and Hoch, 1990). Aggregation produces a set of judgements with less noise than the *average* amount of noise exhibited by the original sources. Most of the possible benefit from combining a large number of expert judgements can be obtained by aggregating the judgements of relatively few experts (e.g., only three or four). The greatest benefits of averaging are found when the sets of assessments provided by different sources do not correlate highly with each other (Johnson et al., 2001). Therefore, meaningful improvements in the assessment of uncertain quantities can be achieved without the requirement for recruiting impractically large numbers of additional experts. See Chapter 9 on multiple experts for further discussion of these and related issues.

Aggregation is not guaranteed to (and, in fact, typically will not) produce a set of judgements that has less noise than *every one* of the contributing sources. What aggregation across experts does offer is a set of judgements that is likely to be better than the set provided by a randomly chosen expert. Relative to those individual experts with small amounts of noise, the aggregate judgements will exhibit greater random variation. Relative to those individual experts with higher-than-average levels of noise, the aggregate judgements will exhibit less random variation.

When several experts have provided judgements, we probably do not have the means to determine directly which one or few of the experts provide the 'best' judgements – otherwise we probably need not have elicited the others' opinions in the first place. However, Sniezek (1990) showed that getting members of a group to decide which one of them will provide a judgement on behalf of the group can be highly effective. This approach outperformed averaging judgement (mechanical aggregation) and consensus judgement (behavioural aggregation). This suggests that this approach is superior to averaging as a method of removing noise, and/or, better still, successfully identifies an individual whose judgements are better than average on other components of accuracy (e.g., discrimination).

Noise can also be reduced and accuracy enhanced by creating a model of the expert's opinion. For instance, in multiple cue judgement, a linear model of the expert (who has supplied a series of case-by-case judgements) derived using multiple regression is usually more accurate than the expert herself (Slovic and Lichtenstein, 1971; Hastie and Dawes, 2001). Thus, in addition to providing convenient inputs into more complex decision analytic models, models that 'smooth' the expressed opinions of the expert may actually be more valuable.

4.5.3 Redressing insufficient regressiveness in prediction

In Section 3.6.2, we discussed two modes of thought: the *distributional* mode, where a case is viewed as one of several similar cases, and the *singular* mode, where probabilities are assessed on the basis of individual properties of the case (Kahneman and Tversky, 1982a). Kahneman and Tversky (1982b) propose that corrective procedures can usefully be organised according to the consideration of two types of information that reflect these two modes of thought: *distributional* data and *singular* data. Distributional data could include base-rate data (data about long-run proportions of events). *Singular* data might also be described as *case* or *individuating* data (data about individual instances). Expert elicitation is most likely to be required in those instances where distributional data alone are insufficient for the purpose. For instance, there may be no database from which one can extract precisely the information one needs, and indeed such a database may not be conceivable. This is necessarily the case when past experience does not provide the answers one seeks. Consequently, there may be considerable value in considering what has been suggested in respect of *both* types of information.

Predictions in uncertain environments are typically insufficiently regressive; in other words, the variance in a set of judgements is typically too great (see Section 3.4.2.5). For instance, a doctor might have unreasonable confidence that healthy patients will enjoy longer-than-average life expectancies or that unhealthy patients will enjoy shorter-than-average life expectancies (Kahneman and Tversky, 1973; Tversky and Kahneman, 1974). Some would consider this to be another example of 'overconfidence', which one might characterise as exhibiting excessive confidence in one's ability to pick which cases will deviate most from the mean (cf Section 4.2.2.1). This is not only a problem for intuitive judgement; statistical models can over-fit the data in a training set at the expense of accuracy in a test set. For instance, shrinking the regression coefficients when deriving a prognostic clinical rule can improve the calibration of a set of probabilities (Steyerberg et al., 2001). In an attempt to rectify insufficient regressiveness, Kahneman and Tversky (1982b) propose a five-step procedure as outlined below.

1. Specify a reference class into which the case under consideration can be meaningfully placed. For example, identify the class of drugs to which Drug X can be considered to belong.

2. Assess the distribution of outcomes for the reference class, considering both central tendency and variability. For example, estimate the average life expectancy for patients treated using this class of drugs and provide a formal estimate of how variable different drugs are in their effectiveness.

3. Make an intuitive assessment of the individual case based on its particular features (i.e., using the singular information). For example, state the anticipated life expectancy for patients who are given Drug X. This intuitive assessment is then subjected to the corrective procedures outlined in steps 4 and 5.

4. Assess the extent to which the information available to the expert permits accurate prediction. For instance, use records of previous predictions or employ expert assessment (perhaps requiring further elicitation questions) to estimate the correlation between predictions and outcomes.

5. Use the assessment from step 4 to correct the intuitive assessment made in step 3. In the case of judgements that are deemed to be insufficiently regressive, this will involve moving the intuitive assessment closer to the mean of the reference class. The magnitude of adjustment varies as a function of the correlation between predictions and outcomes estimated in step 4. Less predictable situations will entail greater adjustment. The procedure should be made explicit to the expert, and the expert and the facilitator should discuss

whether to adopt this new assessment, to retain the original one or to continue the elicitation to generate a further assessment.

Such an approach focuses on how one might correct an expert's 'median' prediction for an uncertain quantity and would need to be adapted if applied to probability distributions that one elicits for the same uncertain quantities (see Chapter 5). For instance, applying corrections to upper or lower tertiles or quartiles may involve correction away from the mean, rather than towards it – as, while 'best estimates' are too often too distant from the mean, best- and worst-case scenarios are often too close to it.

4.5.4 A caveat concerning post hoc corrections

We have identified some circumstances in which biases commonly occur and have discussed some of the statistical manoeuvres that can often improve the accuracy of a set of judgements. If we know what is wrong and how to fix it, it is tempting to argue that it should be common practice to adjust an expert's assessments once they have been placed into the hands of the facilitator. However, there are some problems with this approach.

The first problem is that predicting exactly when a bias *of a particular kind* will occur is not always straightforward. It is true that many biases are quite reliable, in the sense that given the same set of circumstances, they seem to be found again and again. However, they can be ephemeral: under different task conditions they may disappear or there may even be a bias in the opposite direction. For instance, consider how one might make post hoc corrections to reduce overconfidence in a set of probability judgements (see Section 4.2). One might adopt some straightforward strategy for systematically reducing the value of some or all of the probability assessments that an expert has provided. However, overconfidence is not always observed. Over/under-confidence varies as a function of the overall relative frequency of events, such that under-confidence is the norm in some circumstances. This is the hard–easy effect discussed in Section 4.2.2.1. In the absence of the data that are necessary in order to be able to assess calibration formally, it may well be difficult to predict whether we should expect over- and under-confidence. If we get this wrong, our post hoc corrections will have the effect of *increasing* bias, rather than serving to decrease it.

There is a second reason why corrective procedures applied after an elicitation may backfire. Consider how an expert might act if she suspects that her assessments will be corrected after she has made them. Experts are invited to take part in elicitation precisely because their beliefs are of interest to the client. If the expert suspects that the final end point of the exercise (after correction) will no longer give a true reflection of her beliefs, she may be inclined to provide different assessments than she otherwise would have. Anticipating likely corrections, she may provide assessments that are more extreme in the opposite direction to the proposed correction. Given the stated purpose of the exercise (to elicit her beliefs), such an action could be considered to be 'in good faith'. By anticipating any likely correction, the expert is doing her best to ensure that her beliefs are indeed captured by the elicitation exercise. This highlights the importance of ensuring that an elicitation exercise is a genuine collaboration between all involved. It is clear that, for all kinds of reasons, the process of debiasing assessments of uncertain quantities is far from straightforward.

In general, where the objective of elicitation is to obtain an honest representation of the expert's knowledge and beliefs (see Section 8.1), post hoc correction may be indefensible. If the expert accepts that adjustment may correct inaccurate processing by her of her knowledge and experience or if she feels that it reflects some learning that she would wish to do on the basis of past experience (e.g., of any miscalibration of her assessments), then (but perhaps only then) such adjustments might be acceptable.

4.6 Conclusions

4.6.1 Elicitation practice

- It is important that feedback is given on training exercises to help the expert learn how to make better probability assessments.

- Subjective probabilities can be well calibrated, but often they are not. This highlights the importance of training and ought to be taken account of in sensitivity analyses (if expert assessments are subsequently used in decision analysis).

- The interpretation of verbal expressions of uncertainty varies considerably across individuals and situations. Attempts to impute specific values to them that do not take account of this are fraught with danger.

- Alternative ways of describing uncertain events or eliciting a response using different response scales can prompt experts to change their information-processing strategies. These can result in alternative assessments being made. Facilitators ought to be aware of how sensitive probability judgements can be to changes in language and response formats.

- Aids and procedures for debiasing judgement ought to be considered. However, people can be resistant to their implementation, and facilitators ought to be aware of the dangers of imposing debiasing measures upon the expert without consultation.

4.6.2 Research questions

- What are the best ways to provide the substantive expert with the normative expertise that they need to contribute fully to an elicitation exercise?

- The calibration of subjective probabilities is generally better for predictions of future events than it is for the truth of almanac items. Is this because the uncertainty associated with future events is *both* epistemic *and* aleatory, rather than solely epistemic? Or, is this because the typical study populations for these kinds of study differ in the degree of substantive expertise that they can bring to bear on the task? Determining the implications that calibration research has for practical elicitation partly depends upon the answers to these questions.

- How should the *expert's* preferences for carrying out the elicitation be taken into account? Sometimes there are different ways in which an elicitation task can be structured, for instance, when different numbers of response categories or different types of response scale are feasible. In such cases, how helpful is it to ask the expert how they would like to proceed?

- Sometimes the effects of different forms of language or response formats are known, but one cannot say which alternative is best. How does one decide which forms of language or response formats are best in a given situation? Could it be useful to use multiple methods with the same expert as a means of checking for coherence?

- When it is possible to obtain probability assessments from different sources (multiple experts and/or methods), how should one proceed? Should one average across several sources or should one seek to identify the single best method that one believes will be most accurate?

- How can the facilitator best work with the expert to determine what debiasing methods can be applied and to decide how they should be employed?

Chapter 5

Eliciting Distributions – General

5.1 From probabilities to distributions

In practice, we invariably wish to elicit an expert's knowledge about uncertain quantities, such as the volume of oil that can be extracted from an oilfield, the proportion of patients who will experience intestinal bleeding when treated with a new anti-inflammatory drug or the mean time for the first failure of a new type of sewer pipe. To express uncertainty about an unknown quantity, we need to elicit the expert's probability *distribution* for that quantity. A probability distribution is made up of a large, usually infinite, number of probabilities.

For example, consider the proportion of patients having intestinal bleeding, and let this proportion be X. The expert's knowledge about this uncertain quantity is represented as her probability distribution for X, which specifies her probability that X should be less than any given proportion x. Thus, it includes her probability that X is less than 0.1 (a bleeding rate less than 10%), her probability that X is less than 0.2, and so on. It also includes her probability that X is less than 0.10001. Although this is surely almost the same as her probability that X is less than 0.1, and we could not expect, in practice, the expert to be able to distinguish between the two probabilities, they are strictly different. Both must be specified in order to obtain the expert's probability distribution, and indeed we need her probability for every possible x value from 0 to 1. The number of probabilities we need is strictly infinite.

It is clearly impractical to elicit all these probabilities from an expert, and this is one factor that makes the problem of eliciting probability distributions challenging.

Uncertain Judgements – Eliciting Experts' Probabilities A. O'Hagan, C. E. Buck, A. Daneshkhah, J. R. Eiser, P. H. Garthwaite, D. J. Jenkinson, J. E. Oakley and T. Rakow

Since distributions are almost invariably what we need to elicit, it is essential to consider how we can make the transition from eliciting individual probabilities, which has been discussed thoroughly in the preceding chapter, to eliciting distributions.

5.1.1 From a few to infinity

Fortunately, an accurate, if not perfect, representation of the expert's beliefs about an uncertain quantity can be obtained from knowing just a small number of individual probabilities.

In the example of the proportion X of patients with intestinal bleeding, suppose the expert has specified $P(X < 0.05) = 0.2$, $P(X < 0.08) = 0.4$, $P(X < 0.15) = 0.75$ and $P(X < 0.25) = 0.95$. These elicited values are shown as four points on the cumulative distribution function (CDF) in Figure 5.1. We also know that $P(X < 0) = 0$ and $P(X < 1) = 1$, so effectively we also have these two points. The expert's CDF is a curve that passes through these points. One such curve is shown in Figure 5.1. The corresponding density function is shown in Figure 5.2.

One could, of course, draw many other (non-decreasing) curves through the points in Figure 5.1, any one of which might be the expert's real CDF for X. However, if we were to deviate even a little from the line in Figure 5.1 the resulting density function is likely to appear much stranger than Figure 5.2. It might have several modes, plateaus or steep rises or falls. If we believe that the expert's beliefs would not be represented by a density function with such features, then it is hard to deviate from the curve in Figure 5.1 or the density function in Figure 5.2 in more than very small measures.

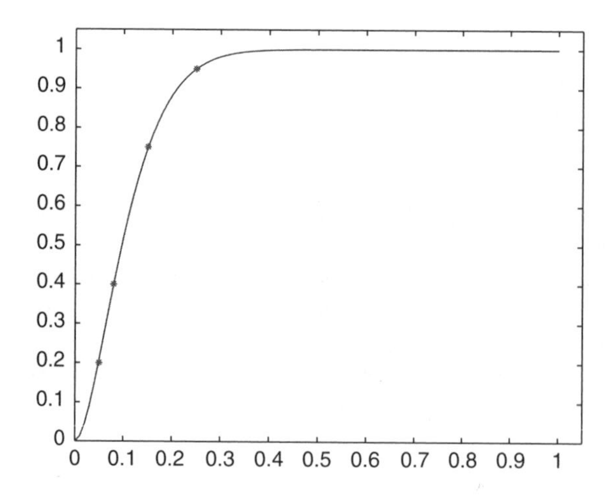

Figure 5.1: Fitting a CDF to four elicited probabilities.

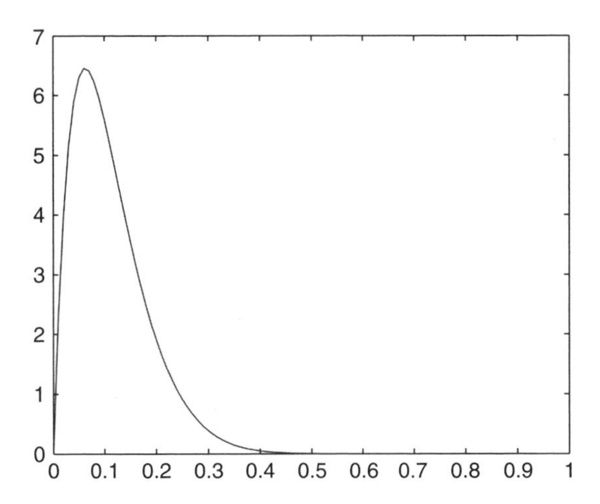

Figure 5.2: The fitted density function.

Thus, a few elicited probabilities may, in practice, be enough to specify the expert's probability distribution with some degree of accuracy. However, this argument rests on two assumptions. The first assumption is that the expert's stated probabilities are themselves accurate. We examine this assumption in Chapter 8. The second assumption is that the expert's probability distribution has the smooth, unimodal, reasonably symmetric form of Figure 5.2. As a partial justification of this second assumption, experience indicates that people's knowledge about uncertain quantities is usually well represented by smooth unimodal densities. In practice, we can confirm this or identify whether the expert has more unusual beliefs (such as a bimodal distribution), by eliciting aspects of the shape of her probability density function.

5.1.2 Summaries

Eliciting shape takes us away from the realm of eliciting probabilities and has not been researched by psychologists in the same way. The point is that a distribution is not simply a collection of probabilities making up a CDF. The probabilities are ordered and structured, and statisticians have other ways to describe a distribution. To a statistician, distributions may also be described and specified using measures of location such as the mean or median, measures of scale such as the variance or standard deviation, and so on (see Section 1.2.3). All these descriptive elements – probabilities, shape features, moments, and so on – are known as *summaries*. The basic technique for eliciting a distribution entails eliciting a (usually quite small) number of summaries of the expert's distribution. If well chosen, these summaries should be enough to identify the expert's distribution with a high degree of precision.

5.1.3 Fitting

Having elicited the desired summaries, the elicitation is completed by fitting a simple and convenient distribution to match the elicited summaries. This is what was done in Figure 5.1 – the line has been drawn to fit the stated summaries. In Figure 5.1, the line is the Beta(2,16) CDF (and Figure 5.2 is the Beta(2,16) density function). The beta distributions are standard distributions in statistics that are well known and relatively easy to work with. Given that the Beta(2,16) distribution fits the elicited summaries and that any other curve that fitted them and had the same (elicited) general shape would be almost identical to the chosen distribution, it is sensible to use this well-understood and convenient distribution.

5.1.4 Overview

The principal questions that need to be addressed in eliciting distributions are discussed in this chapter. They concern what and how many summaries to elicit and how to deal with the imprecision that results from choosing just one of many distributions that might fit the elicited summaries, as well as the inevitable imprecision in the elicited summaries themselves.

5.2 Eliciting univariate distributions

In designing an elicitation method, there is usually some choice as to which summaries the expert is asked to assess and, if possible, summaries should be chosen that are usually assessed reasonably competently. As we found in Chapters 3 and 4, the psychological literature contains many relevant studies.

5.2.1 Summaries based on probabilities

We have already seen that individual probabilities such as $P(X < x)$ can be valuable summaries, and relatively few of these can provide accurate elicitation of a probability distribution in conjunction with some knowledge or assumptions about shape. In eliciting such probabilities, the facilitator specifies the values x at which the probabilities are requested. The choice can be difficult to make without some initial idea of what values the expert thinks are plausible for X. Poor choices will lead to the expert giving probabilities of 0 (if too low) or 1 (if too high), and will provide rather little information about the expert's distribution.

The probabilities are therefore often reversed, in the sense that the facilitator specifies a probability p and asks the expert for a value x such that $P(X < x) = p$. This is the elicitation of *quantiles* of the expert's distribution. The most widely used quantile is the median, which is the case when $p = 0.5$. Starting with the median, a method of bisection is often used to elicit a number of quantiles, which entails a sequence of questions of the following form:

Q1. Can you determine a value (your median) such that X is equally likely to be less than or greater than this point?

Q2. Suppose you were told that X is below your assessed median. Can you now determine a new value (the lower quartile) such that it is equally likely that X is less than or greater than this value?

Q3. Suppose you were told that X is above your assessed median. Can you now determine a new value (the upper quartile) such that it is equally likely that X is less than or greater than this value?

Question Q1 elicits the median; Q2, the lower quartile and Q3, the upper quartile. This can continue with more bisections, although it becomes increasingly difficult for the expert to make the judgements reliably and meaningfully.

To illustrate this process, consider the example presented in Section 5.1.1, where X is the proportion of patients who will experience intestinal bleeding when taking a new anti-inflammatory drug.

The expert is first asked to state a proportion $x_{0.5}$, such that she feels X is equally likely to be above or below $x_{0.5}$. She gives the response $x_{0.5} = 0.1$. Having elicited the median in this way, the facilitator asks the expert to specify a value $x_{0.25}$ such that, on the assumption that X is less than or equal to 0.1, she feels X is equally likely to be above or below $x_{0.25}$; that is, she gives equal probability to X being below $x_{0.25}$ or to being between $x_{0.25}$ and $x_{0.5} = 0.1$. She gives the response $x_{0.25} = 0.06$. Finally, she is asked to specify $x_{0.75}$ by a similar question but with the assumption now that X is *above* 0.1. Her response is $x_{0.75} = 0.15$.

The facilitator would then use the three probability statements $P(X < 0.06) = 0.25$, $P(X < 0.1) = 0.5$ and $P(X < 0.15) = 0.75$ to fit a suitable probability distribution in the same way as the four probabilities were used in Section 5.1.1. Figure 5.3 shows the three points plotted with the Be(2,16) CDF. The points lie almost as close to the CDF as the four points in Figure 5.1.

Another form of quantile elicitation is to elicit *credible intervals*. A credible interval with probability q is a range of values (x_1, x_2) such that $P(x_1 < X < x_2) = q$. Usually, the expert is advised that the range should be a central interval. This means that there should be equal probability that X should lie above the upper limit x_2 or that it should be below the lower limit x_1, that is, a probability of $(1 - q)/2$ in each case. For example, the upper and lower quartiles form a 50% credible interval, $q = 0.5$.

Many experiments have examined people's performance at assessing credible intervals. If the credible intervals calibrated well with reality, then the proportion of credible intervals with coverage probability q that contain the true value of X should be about q. (This notion of calibration is discussed more fully in Section 4.2) The experiments show that assessing credible intervals is a task that people perform reasonably well, but there is a clear tendency for central credible intervals to be too short, so that this proportion is less than q, when q is large. Thus, 90%, 95% and 98% credible intervals are typically found to contain the

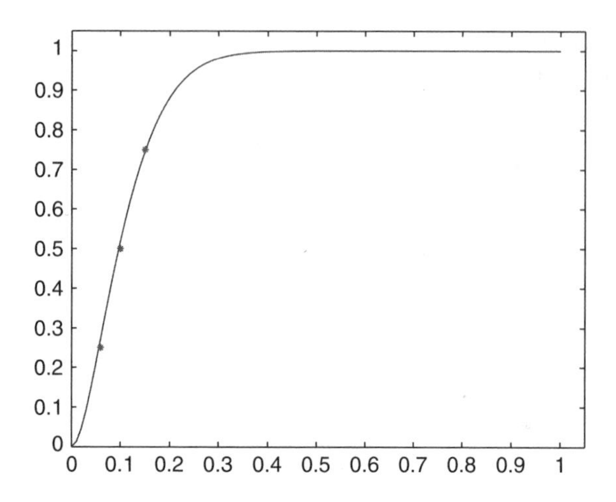

Figure 5.3: Presentation of the three probability statements described above on the CDF of Be(2,16).

correct answer between 40% and 70% of the time (Alpert and Raiffa, 1982; Lichtenstein et al., 1982). The implication is that people (even experts) rarely assess their uncertainty to be as large as they should (Alpert and Raiffa, 1982). See also Section 4.2.2.

This inappropriate narrowness of subjective credible intervals also attracts the label 'overconfidence' (Lichtenstein et al., 1982) – people are too confident that the true value lies between the limits they have specified. Slovic (1972) identifies this as a manifestation of anchoring-and-(insufficient)-adjustment. The assumption is that people anchor on their best estimate and adjust (insufficiently) up and down to derive their upper and lower limits. Yaniv and Foster (1995, 1997) interpret this finding as an outcome of peoples' preferences with respect to the trade-off between accuracy and informativeness. People are reluctant to specify very wide intervals, for fear that they may be unhelpfully uninformative. For instance, I may express a 90% probability that Martin Luther King's age at death was between 40 and 60 years, and although this overstates my confidence (in fact, he died at age 39), it would be of little use for me to say that 'I am at least 99% sure that Martin Luther King died aged between 1 and 119 years old'. The balance of this trade-off seems to result in a strong preference for specifying limits that are equivalent to an approximate 50% subjective probability interval. The strong implication is that people can provide well-calibrated credible intervals, so long as they are not pushed to provide them for high levels of certainty.

The evidence is conflicting as to whether eliciting probabilities or quantiles leads to better calibration. For instance, in addition to eliciting a credible interval directly, by specifying q and eliciting the interval, a credible interval can be obtained by fixing the interval and eliciting the probability q; Seaver et al. (1978) found that the fixed interval method performed better, while Murphy and Winkler

(1974) found the converse. More recently, the Winman et al. (2004) report improved calibration using an adaptive fixed interval method, while Soll and Klayman (2004) found that including other assessments such as a median improved the calibration.

When using quantile elicitation, the facilitator needs to choose the probabilities p or the credible interval coverage probability q. The median and quartiles are the most commonly assessed quantiles (using the method of bisection) and while the resulting 50% interval has sometimes given good results (e.g., Murphy and Winkler, 1974; Peterson et al., 1972), other empirical work has found that overconfidence is less if the 33rd and 67th percentiles are assessed (Barclay and Peterson, 1973; Garthwaite and O'Hagan, 2000).

Context almost certainly influences the finding of overconfidence. For example, Garthwaite (1989) elicited 50% central predictive intervals for the dependent variable in a simple linear regression model and found that far more than 50% of the intervals contained correct values. The subjects had drawn graphs to help make their assessments and this probably improved the accuracy of their median assessments, a fact which they did not appreciate when assessing quartiles, so the quartiles tended to be too far apart.

Much empirical research has investigated respondents' ability to assess the extreme 'tails' of a distribution. For example, Alpert and Raiffa (1969) elicited 98% central credible intervals. They asked 'almanac' questions of the following kind:

> How many foreign cars were imported into the United States in 1968?
>
> (a) Make a high estimate such that you feel there is only a 1% probability that the true answer would exceed your estimate.
>
> (b) Make a low estimate such that you feel there is only a 1% probability that the true answer would be below this estimate.
>
> (Alpert and Raiffa, 1969, pp. 16–17)

It should have been somewhat of a surprise to a subject to find the true value of a quantity falling outside such an interval, but 43% of all assessments produced such surprises. This information was given as feedback before a second session. In this second session, 23% of the assessments produced surprises, which is still very high. One reason for the large number of surprises is that assessing tails of distributions is a difficult task, mainly because it requires the consideration of events that are unlikely, and so comparisons do not come readily to mind. (It is unfortunate that assessing probabilities for rare events is difficult, as expert opinion is of paramount importance when empirical evidence is scarce.)

As noted earlier, a good elicitation method should yield a probability distribution that accurately reflects the expert's opinion, but this is hard to check and a pragmatic alternative is to compare assessed distributions with true values when these are known (see the discussion of calibration in Section 4.2). Several experiments have attempted to train subjects in order to improve the calibration/objective accuracy of their assessments. These have typically found that objective accuracy is improved substantially by training, as with the feedback given by Alpert and Raiffa (1969), but confirm that biases such as overconfidence are tempered rather than eliminated (Schaefer and Borcherding, 1973; Lichtenstein and Fischhoff, 1980). In these experiments, training has usually taken the form of feedback on each assessment; subjects are told the correct values after making the assessments and the trainer stresses the direction of biases and how the expert might reduce them. The benefits of effective feedback can be seen in the performances of weather forecasters, who make regular predictions for the same quantities each day (such as temperatures and the probability of precipitation), and thus soon learn the accuracy of their forecasts. Experiments have generally found them to be quite well calibrated. For example, Peterson et al. (1972) conducted an experiment with two meteorologists in which they gave forecasts of the maximum and minimum temperatures on the following day. Using the method of bisection, the meteorologists expressed their forecasts in terms of 50% central credible intervals. They showed good calibration: out of 55 forecasts, 28 fell inside the 50% credible intervals, 18 fell outside and 9 fell on the boundaries.

An alternative based on eliciting ratios of probabilities is termed the analytic hierarchy process (AHP) method (Saaty, 1977, 1980; Hughes, 1993; Cagno et al., 2000; Monti and Carenini, 2000). Although simply formulated as a way to elicit a discrete probability distribution, it can be used to elicit probabilities for a discretisation of a continuous quantity, which then serve as summaries of the continuous distribution. To elicit a set of probabilities $p_1, p_2, \ldots, p_n$, the expert assesses pairwise relative probabilities of the form $\frac{p_i}{p_j}$ for $i \neq j$. Not all of the $n(n-1)/2$ pairwise ratios need to be assessed, since all the p_is can be inferred from a suitable set of $(n-1)$ ratios (e.g., $\frac{p_2}{p_1}, \frac{p_3}{p_1}, \ldots, \frac{p_n}{p_1}$). However, Saaty proposes that each ratio should be assessed using a scale from 1 to 9, where the points on the scale have simple verbal descriptions: for instance, $\frac{P(A)}{P(B)} = 5$ may correspond to the judgement that A is "strongly more probable than" B. More than $n-1$ assessments may then be made and probabilities fitted by matrix algebra (Saaty, 1980) or by statistical inference (e.g., Basak, 1998). AHP forms part of a relatively large literature on eliciting preferences (not necessarily probabilities) by pairwise comparison, but the use of verbal descriptions suggests (see Subsection 4.4.1.2) that it may be of limited value in the practical elicitation of probability distributions.

5.2.2 Proportions

Psychologists have studied another aspect of assessing probabilities, which is a subject's ability to judge proportions when shown samples. In these experiments (Erlick, 1964; Nash, 1964; Pitz, 1965, 1966; Shuford, 1961; Simpson and Voss, 1961; Stevens and Galanter, 1957), binary data were displayed to subjects for a limited period of time and they were then asked to estimate one of the sample proportions. For example, Shuford (1961) projected 20×20 matrices onto a screen, one at a time. The elements of each matrix were red squares and blue squares, and subjects observed a matrix for 1 second in some trials and for 10 seconds in others. After each trial, subjects had to estimate the proportion of squares that had been, say, red. In this and similar experiments, subjects generally assessed the sample proportion very accurately, with the mean of subjects' estimates differing from the true sample proportion by less than 0.05 in most cases.

The distinction between estimating a sample proportion and assessing a probability is fundamental to understanding the relevance of much of the psychological literature to practical elicitation. It is related to the distinction between aleatory and epistemic uncertainties that was discussed in Section 1.3.1. The number of foreign cars that were imported into the United States in 1968 is a one-off (i.e., unique and unrepeatable) value, and a subject's uncertainty about it is essentially epistemic. In contrast, the task of estimating the proportion of red squares in a 20×20 array is clearly related to judging a frequency probability. Its relevance to the elicitation task lies in the way that frequencies of occurrence in random, repeatable events may guide an expert in assessing a personal probability for a one-off proposition, as discussed in Section 1.3.4.

From this perspective, the evidence of people's ability to judge proportions is reassuring. It suggests that when experts use frequencies of occurrence of quasi-repetitions of related propositions, they will judge such frequencies reasonably well (subject to the caveat that in the experiments all the occurrences are equally available). Having said that this evidence is reassuring, the evidence reported in the preceding section, concerning subjects' ability to judge probabilities, is much more directly relevant and less comforting. The probabilities that were used in those experiments were, in many cases, epistemic, of the kind that, in practice, we wish experts to judge (like the number of foreign cars imported in a given year). The findings are that subjects often judge such probabilities less well, with a particular tendency towards overconfidence. It seems that their errors are not in assessing frequencies of equally available quasi-repetitions, but lie in the more complex tasks of identifying appropriate information, judging relevance to the proposition in question, and so on.

5.2.3 Other summaries

In addition to probabilities, statisticians use various other summaries to describe probability distributions, and it is therefore of interest to consider their use in elicitation. How well are experts likely to perform in assessing, for instance, the mean and variance of their probability distributions for uncertain quantities? As with experiments on proportions, psychologists have tended to look at this via experiments involving random samples.

For instance, experiments have been used to investigate people's ability to estimate measures of location (Beach and Swenson, 1966; Peterson and Miller, 1964; Spencer, 1961, 1963). Typically, a sample of numbers is displayed to subjects who are asked to estimate the mode, median or mean of the sample. When the sample distribution is approximately symmetric so that these three measures are numerically similar, subjects' estimates have shown a high degree of accuracy (Beach and Swenson, 1966; Spencer, 1961). However, an experiment conducted by Peterson and Miller (1964) used a sample drawn from a population whose distribution was highly skewed; subjects' assessments of the median and mode were again reasonably accurate, but assessments of the mean were biased towards the median.

Regrettably, it seems that people are poor both at interpreting the meaning of 'variance' and at assigning numerical values to it. When estimating relative variability, empirical evidence indicates that people are influenced by the mean of the stimuli and estimate the coefficient of variation (i.e., the standard deviation divided by the mean), rather than the variance. For example, Hofstatter (1939) obtained assessments of the variability in the lengths of sticks tied in bundles. He found that the assessments of variance increased with the sample variance, as they should, but as the means increased, the assessments decreased. Lathrop (1967) has replicated this latter result. Even allowing for the effect of means, systematic differences still arise between intuitive judgements of sample variance and the objective values. If large deviations from the mean predominate as, for example, when the sample is drawn from a population whose distribution is bimodal, then the variance is overestimated. On the other hand, if small deviations from the mean predominate, as, for example, when the population distribution is normal, then the variance is underestimated (Beach and Scopp, 1967).

In the main, research concerned with descriptive statistics from samples of a single variable has produced relatively clear-cut conclusions. People are capable of estimating proportions, modes and medians of samples. We are slightly less proficient at assessing sample means if the sample distribution is highly skewed and we often have serious misconceptions about variances. All of these results, while strictly not pertinent to assessing a probability distribution for epistemic uncertainty about parameters, which is the task that experts are almost invariably required to do, may be indicative of how well an expert can be expected to perform that task.

Asking experts for probabilities and credible intervals is a reasonable approach. We are generally competent at quantifying our opinions as probabilities, quantiles or credible intervals. However, there is a general tendency for the assessed distributions to imply a greater degree of confidence than is justifiable. Practice, coupled with feedback, will reduce this bias, but assessing the extreme tails of distributions is difficult (e.g., 98% credible intervals) and while training should reduce bias, it will not eradicate it.

There is essentially no direct psychological research into how accurately an expert might be able to assess the mode, mean or variance of her probability distribution for an uncertain parameter. There is apparently no research into whether experts can meaningfully assert shape properties such as unimodality or skewness. The available research suggests that it is unlikely that an expert will be able to specify a mean or variance reliably, since they are not good at the simpler task of estimating a mean or variance of a sample. The weight that the extreme sample members should carry is typically misjudged, and we can expect that an expert will have even more difficulty in quantifying these summaries in the presence of skewness or heavy tails. But the absence of any psychological research into this aspect of elicitation means that there is no empirical confirmation of these suppositions.

5.3 Eliciting multivariate distributions

5.3.1 Structuring

When the expert's opinion is sought on two or more unknown variables, the output of the elicitation should be the expert's *joint* probability distribution for those variables. The task is now more complex than when eliciting a distribution for a single variable, and the facilitator must inevitably ask more complex questions.

An important special case is when the variables are independent, meaning that if the expert were to obtain new information about some of the variables, it would not change her beliefs about the others. The concept of independence is straightforward to explain, and independence between variables is a relatively simple judgement for the expert to make. It is also a very convenient judgement, because when all the variables are independent, their joint distribution is just the product of their marginals. The elicitation exercise then reduces to eliciting the expert's beliefs about each variable separately, so only univariate elicitation techniques are required. Utilising independence to decompose a multivariate elicitation task into simpler univariate tasks is consistent with the idea of disaggregation.

Psychological research also indicates that, when events are independent, joint probabilities should be assessed via univariate probabilities, as people exhibit systematic bias when making joint probability assessments. In particular, people tend to overestimate the probability of conjunctive events and underestimate the probability of disjunctive events. For example, Bar-Hillel (1973) found that people tended to overestimate the probability of drawing a red marble seven times in succession from a bag containing 90% red marbles and 10% white marbles, and underestimate the probability of drawing a red marble at least once in seven successive draws from a bag containing 10% red marbles and 90% white marbles. These errors can be explained as the result of anchoring: the probability of an elementary event provides an obvious starting point for estimating the probability of both conjunctive and disjunctive events. For conjunctive events, the probability of the elementary event must be reduced, which is done insufficiently, and for disjunctive events it must be increased, which is again done insufficiently.

With many elicitation methods, it is transparent as to what assessments would correspond to independence and, in application of these methods, subjective independence between some pairs of parameters is often observed; see examples in Garthwaite and Dickey (1991, 1992). It should be noted, though, that assumptions of independence often make assessment tasks easier for an expert. For example, if an expert has assessed the marginal distribution of X, and X and Y are independent, then the conditional distribution of X given Y is easily specified as 'no change'. It may be that experts are too willing to accept independence where it does not strictly apply.

Even where variables are dependent, it may be possible to restructure the problem by expressing it in terms of independent variables. An example might be where we seek a medical expert's opinion on the effectiveness of two treatments that are to be compared in a clinical trial. Letting X and Y denote the relevant measures of effectiveness of the two treatments, we would not typically have independence between X and Y. If the expert learnt that X, the effectiveness of the first treatment, was higher than she originally expected, then this would generally lead her to have an increased expectation of the effectiveness Y of the second treatment. This may be because the expert believes that the treatments act in similar ways, but it may also be because of uncertainty in patient recruitment; that is, if X is smaller than expected, say, this may be because the trial has recruited patients who are more ill, and will thereby suggest a smaller value for Y. However, the expert might be willing to accept independence between two *functions* of X and Y. For instance, it may be reasonable to suppose independence between X and $Z = Y/X$. Here, Z is the relative effectiveness of treatment 2 over treatment 1. Such a structure is often appropriate where treatment 1 is 'standard care' or a placebo and treatment 2 is a new or active treatment. Where both treatments are new, the asymmetry of the preceding structure may not be appealing, but the expert might be happy to express independence between $(X + Y)/2$, the mean effectiveness, and the difference $Y - X$. Bayesian hierarchical models are natural examples of structuring dependent variables in terms of conditional independence. O'Hagan (1998) emphasises the role of structuring as an aid to elicitation. Kadane and Schum (1996) provide an extended example of complex structuring of beliefs.

5.3.2 Eliciting association

Where variables are dependent, and cannot obviously be reduced to independence in this way, we cannot escape from the complexity of multivariate elicitation. We can (and generally should) elicit summaries of the expert's marginal distributions, but these no longer characterise the joint distribution completely. The question then arises as to which summaries of the expert's joint distributions are most effective and reliable to elicit.

To introduce the ideas of eliciting association, consider Figure 5.4, which shows the space of possible values of two uncertain quantities, X and Y. Their medians are marked as $X_{0.5}$ and $Y_{0.5}$ and the upper quartile of X is marked as $X_{0.75}$.

So there is a probability 0.5 that X will exceed $X_{0.5}$, which is the region to the right of the dashed line passing through $X_{0.5}$, and similarly there is a probability 0.5 in the region above the dashed line through $Y_{0.5}$, which is the region for which Y exceeds $Y_{0.5}$. The intersection of these two regions is the shaded area in Figure 5.4, and corresponds to the event that both X and Y exceed their median values. If X

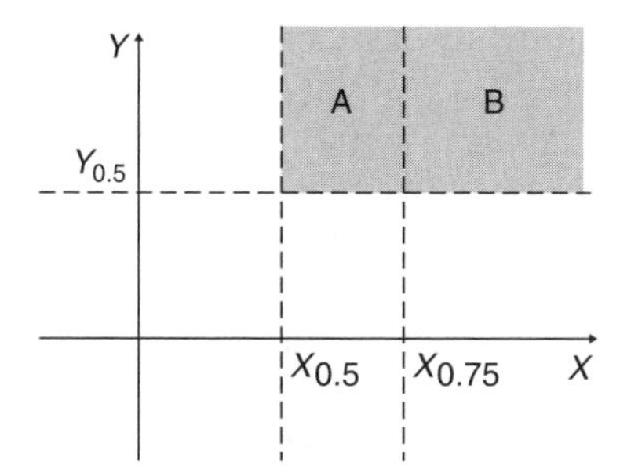

Figure 5.4: Elicitation association.

and Y are independent, then the probability in this shaded region is 0.25. If they are not independent, this probability can take any value from 0 to 0.5. (It cannot exceed 0.5 because the shaded region is a subset of each of the two areas that we have already identified as having probabilities 0.5.)

One way to ask the expert about the association between X and Y would be to ask for her probability in the shaded quadrant. If she gives a value greater than 0.25, then her beliefs are that larger X values will tend to be associated with larger Y values, which we call *positive* association. If she gives a value below 0.25, then this corresponds to *negative* association (larger X values are expected to be associated with smaller Y values, and vice versa). This quadrant probability is a *joint* probability, $P(X > X_{0.5}, Y > Y_{0.5})$. We might also elicit it via the *conditional* probability $P(Y > Y_{0.5} \mid X > X_{0.5})$, that is, the probability that Y exceeds its median *given that* X exceeds its median. This probability can be between 0 and 1, and in fact $P(Y > Y_{0.5} \mid X > X_{0.5}) = 2 \times P(X > X_{0.5}, Y > Y_{0.5})$.

The expert might also be asked for more detailed information, such as the probability in the region marked A or that marked B. Their sum is the quadrant probability. However, these are already becoming difficult for an expert to consider, in practice. For instance, if there is positive association, we would generally expect the probability in B to exceed that in A, yet this might not be immediately apparent to the expert. The reason the probability in B should be larger is that if there is positive association, we would generally think that the conditional probability $P(Y > Y_{0.5} \mid X > X_{0.75})$, in which we condition on X exceeding its upper quartile, would be larger than $P(Y > Y_{0.5} \mid X > X_{0.5})$. The probability in B is $P(X > X_{0.75}, Y > Y_{0.5}) = P(Y > Y_{0.5} \mid X > X_{0.75})/4$, and if we assume that $P(Y > Y_{0.5} \mid X > X_{0.75}) > P(Y > Y_{0.5} \mid X > X_{0.5})$, then $P(X > X_{0.75}, Y > Y_{0.5}) > P(Y > Y_{0.5} \mid X > X_{0.5})/4 = P(X > X_{0.5}, Y > Y_{0.5})/2$. This argument suggests that it would generally be better to ask for the conditional probability $P(Y > Y_{0.5} \mid X > X_{0.75})$ than the joint probability $P(X > X_{0.75}, Y > Y_{0.5})$.

Statisticians usually describe association by measures of correlation, but these are mathematically complex, and directly eliciting correlations may be expected to encounter problems similar to those found in asking experts to specify means or variances. Eliciting conditional or joint probabilities may give more accurate assessments of association. If, for instance, X and Y have a normal joint distribution, then it can be shown that $r = \sin(2\pi(p - 0.25))$, where p is the quadrant probability and r is the (Pearson) correlation coefficient. So an elicited value of the quadrant probability can be turned into a value for the correlation coefficient (if we can assume that the expert's joint distribution is normal).

Several studies in psychology literature have looked at eliciting correlation or association, although these have primarily elicited correlation between variables that might be considered to be drawn from some population. Hence, the uncertainty involved is primarily aleatory. For instance, the seminal work of Clemen et al. (2000) studied the task of eliciting a correlation between the height and weight in a population of male MBA students (amongst others). They examined the following six methods.

1. Ask the respondent for a verbal (non-numeric) description of the strength of a correlation on a 7-point scale ranging from 'very strong negative relationship' to 'very strong positive relationship'. (Clemen et al. made strong assumptions to convert the verbal assessments to correlations.)

2. Ask the respondent to assess the correlation directly by specifying a value between -1 and 1.

3. Ask the respondent to imagine that a person has been picked at random from the population. Give the person's percentile for one variable and ask the respondent to assess the person's percentile for the second variable.

4. Ask the respondent to imagine that two people, A and B, have been picked at random from the population. Conditional on A being greater than B for one variable, ask the respondent to assess the probability that A is also bigger than B for the other variable.

5. Ask the respondent to imagine that a person has been picked at random from the population. The respondent is asked to assess the probability that for both variables the person is below a specified percentile.

6. Ask the respondent to imagine that a person has been picked at random from the population. Conditional on the person being below a specified percentile for one variable, ask the respondent to assess the probability that the person is also below that percentile for the second variable.

Clemen et al. found that method 2 performed best. This is surprising since others have suggested that the direct assessment of moments is a poor method of quantifying opinion (Morgan and Henrion, 1990; Kadane and Wolfson, 1998; Gokhale and Press, 1982). Method 4 asks for a concordance probability to be assessed, which can be equated to a value of Kendall's τ. Assumptions of normality are then made so as to relate Kendall's τ to the Pearson correlation coefficient. Assessment of concordance probabilities to examine correlation has been examined by Gokhale and Press (1982), who found it preferable to the alternative methods they considered, and by Kunda and Nisbett (1986), who concluded that reasonably accurate correlation estimates are obtained provided (a) subjects are very familiar with observations from the population in question and (b) the data relate naturally to a numeric scale. In several experiments, respondents have been shown samples from a bivariate population and then asked to judge the 'degree of relatedness' between the two variables. In these experiments, it has been found that respondents make use of only a limited portion of the available data, sometimes basing their judgements on just the proportion of time the positive outcome for one of the binary variables occurred with a positive outcome for the other (Smedslund, 1963; Inhelder and Piaget, 1958; Jenkins and Ward, 1965; Ward and Jenkins, 1965). Broniarczyk (1994) suggests that respondents may assess covariance more accurately between continuous variables than the binary ones.

Statistically, it is important to distinguish between eliciting an expert's *beliefs about* a population correlation coefficient and eliciting *the value of* the correlation in the expert's beliefs about two variables. In the first case, the correlation coefficient is the variable whose probability distribution we wish to elicit. This task is addressed, for instance, by Gokhale and Press (1982) and Clemen et al. (2000). The second case is the situation that arises in multivariate elicitation, where the correlation (or some other measure of association) is to be elicited as one summary of the expert's joint distribution for two variables. When the two variables can be considered as single draws from a population, some of the methods described above may be appropriate. However, many cases of multivariate elicitation do not fit this situation, particularly when we are eliciting beliefs about two parameters for which the uncertainty is epistemic. Consider, for example, eliciting someone's beliefs about the fuel economy and the acceleration of a new car. This car is not some random draw from any population. Methods 4 to 6 in the study of Clemen et al. are no longer appropriate as phrased, and the concordance probability can only be meaningful for repeatable variables. We could, however, consider analogues of methods 5 and 6 such as the following:

5*. The respondent is asked to assess the probability that both the fuel economy and acceleration are below a specified percentile of each marginal distribution.

6*. Conditional on the fuel economy being below a specified percentile of its marginal distribution, ask the respondent to assess the probability that the acceleration is also below that percentile of its marginal distribution.

5.3.3 Joint and conditional probabilities

Methods 5* and 6* clearly elicit joint and conditional probabilities, respectively. Just as individual probabilities and quantiles are widely used in univariate elicitation, they are a natural choice for multivariate problems. Notice that both 5* and 6* rely on first having identified marginal quantiles, because the marginal distributions referred to here are the respondent's own subjective distributions. Again, this is realistic because multivariate elicitation will quite naturally begin by asking for univariate summaries, which serve to identify suitable joint or marginal probabilities to ask about. However, this makes the elicitation prone to anchoring biases. Thus, with 5*, the expert is likely to be anchored to the marginal probability, from which she needs to adjust the probability downwards to account for the conjunction of the second event. This is the same effect as already noted with eliciting conjunctive propositions.

With 6*, the expert is again anchored to the marginal probability, but must now adjust up or down according to whether the association is positive or negative. It is likely that she will not adjust far enough. Note also that, although we have referred to 6* as eliciting a conditional probability, conditional probabilities are usually elicited in the form of $P(Y \leq y \mid X = x)$, and conditioning on $X \leq x$ rather than on $X = x$ may be cognitively more complex. On the other hand, we might also expect joint probabilities such as 5* to be subject to a kind of representativeness bias, and hence be positively/negatively biased if the association between the two variables is positive/negative.

Many of these suppositions regarding the elicitation of joint and conditional probabilities or quantiles have not been confirmed experimentally. There is a need for more exhaustive cognitive studies on eliciting such summaries, particularly in the context of epistemic uncertainty.

Note that the original Methods 5 and 6 of Clemen et al. do not explicitly require prior elicitation of marginal beliefs, because they refer to underlying aleatory marginal distributions. These distributions will, in practice, be unknown, and although they do not need to be known in order to answer the questions of Clemen et al. Understanding this makes the questions cognitively more complex than 5* and 6*.

5.3.4 Regression

None of the methods examined by Clemen et al. use graphs in any way. It seems likely that graphical methods could help with bivariate elicitation, especially as it is very natural to plot a graph to describe the relationship between two variables such as height and weight. This approach would represent an association between variables in terms of regression, which is related to correlation. For two variables X and Y, for instance, we might try to elicit the regression function $m(x) = E(Y \mid X = x)$. If the expert accepts the proposition that this function is linear, we might simply elicit $m(x_1)$ and $m(x_2)$ for any $x_1 \neq x_2$. Eliciting more than two points on the function ('overfitting', see Section 5.4.3) would allow the assumption of linearity to be checked or a more accurate fitting of a straight line. Here as elsewhere, it may be preferable to elicit medians rather than means.

A body of psychological research has examined multiple regression. In this research, the x-variables are generally referred to as cues, Y is termed the criterion and the regression coefficients are referred to as cue-weights. Subjects predict the value of the criterion, basing their predictions on the known values of the cues. It has been found, in a wide variety of situations, that a subject's responses can be represented quite well by a linear model that relates the criterion to the cues. The correlations between subjects' responses and those predicted by linear models (fitted to the same responses that determined the model) have generally taken values in the 0.70s when the judgemental task is from a 'real world' situation and in the 0.80s and 0.90s for less complex artificial tasks. In some studies, the model derived from one sample of predictions was used to forecast a second one. The forecasts produced in this manner were only slightly less accurate than those produced by a model actually based on the second sample of predictions (Einhorn, 1971; Slovic and Lichtenstein, 1968; Wiggins and Hoffman, 1968). Experiments also show that, provided cues are monotonically related to the predicted variable, a simple linear combination of the main effects will do a remarkably good job at forecasting a subject's assessments, even if subjects know that interactions exist. One implication is that, when eliciting the dependence of one variable on one or more other variables, it is reasonable to constrain an expert's assessments to fit a linear model and to ignore interactions, unless it becomes clear that some interactions are important. An extensive review of cue-weighting experiments is given in Slovic and Lichtenstein (1971).

This kind of elicitation task is particularly relevant to a situation known as *extending the argument*, which may arise when trying to structure an elicitation problem. Suppose a physician is asked for her beliefs about the mean improvement in lung function that asthma patients will achieve using a new inhaler. She may reply that it would depend on how severe the patient's asthma is. The facilitator might point out that the physician was not being asked about an individual patient but the mean effect, averaged over patients with different severities of asthma. The physician's reluctance is nevertheless important, because it shows that the answer will depend on the mix of severities for patients who take the new inhaler. It is right then to 'extend the argument' to include severity and to elicit the physician's beliefs conditional on severity. Then, the facilitator will be led into the kind of regression structure discussed above.

A joint distribution involves more than modelling conditional means or medians. Hence, the task of eliciting a joint distribution is more complex than determining an expert's cue-weights. Generally, conditional probabilities are a natural way to augment marginal probabilities when trying to specify a joint probability distribution and, in particular, allow conditional dispersion to be elicited and modelled. Conditional medians and other quantiles are extensively exploited, for instance, by Kadane et al. (1980), Dickey et al. (1986) and Kadane et al. (1996).

5.3.5 Many variables

When eliciting beliefs about many variables, a high-dimensional joint probability distribution is required. Given the difficulties of eliciting joint and marginal probabilities and other summaries for even two variables, this is a daunting task.

In practice, it is usual to make quite strong assumptions about the form of the joint distribution.

For instance, the facilitator may assume that a multivariate normal distribution would fit the expert's beliefs and concentrate on eliciting summaries that allow him to identify means, variances and covariances for this distribution. Because of the special characteristics of the normal distribution, this will not require the consideration of more than two variables at a time. There are several examples of this kind of elicitation, where a particular parametric form is assumed for the joint distribution, in Chapter 6.

More generally, it is possible to construct a joint distribution using the method of copulas (DallAglio et al., 1991). This approach elicits marginal distributions for the individual random variables and measures of correlation between each pair of variables. Then, the choice of a copula function yields a full joint distribution for the variables. Copulas and several more complex ways of building joint probability distributions from realistic numbers of elicited summaries are developed in Kurowicka and Cooke (2006).

5.4 Uncertainty and imprecision

Elicitation is clearly not a precise process. Although a very large part of the statistical literature relating to elicitation has been concerned with fitting a single distribution to the expert's judgements (see Chapter 6) and behaving as if this distribution perfectly represents the expert's beliefs, some formal and informal methods of accounting for uncertainty and imprecision do exist.

There are two ways in which a distribution, that is chosen by the above method of fitting to elicited summaries, may not exactly represent the expert's beliefs. First, there is inevitable imprecision in the expert's stated summaries. Second, even if we accept those summaries to be precise, there are many alternative distributions that would fit them precisely, any one of which might be the expert's real distribution.

5.4.1 Quantifying elicitation error

One obvious reaction to the possibility of elicitation imprecision is to acknowledge that there is uncertainty about the expert's true distribution and the proper way to quantify uncertainty is through probabilities. So can we formalise this in terms of a probability distribution for the expert's true probability distribution? (The reader should note here the discussion in Section 1.5.3 about the existence of 'true' distributions. The term here may be interpreted, if desired, as referring to whatever 'good' distribution might, in principle, be elicited.)

To do so would seem to require a kind of meta-probability, a probability distribution for a probability distribution. This raises conceptual questions of what such a distribution would mean and practical questions of how it could be determined.

However, in reality, no new kinds of probability need to be invoked. The expert's true probability distribution can be thought of as the accurate representation of her knowledge about the parameter of interest, and, in principle, it might be elicited precisely if the expert were (a) able to specify an indefinite number of summaries and (b) able to specify those summaries perfectly. We know that the expert can do neither of these things in practice, but this simply gives the expert's true distribution the same status as a parameter. To determine the true value of an unknown parameter always requires either an infinite number of observations or infinitely precise observations, or both. The uncertainty we have about the expert's distribution is also epistemic.

There are several formulations in statistical literature in which the expert's stated summaries are treated as data, and a statistical analysis is introduced to use the data to learn about the unknown parameter, using formal (Bayesian) statistical analysis. It is usual in these methods to distinguish the statistician or decision maker, who receives the expert's summaries and who makes inferences about the expert's distribution, from the expert. However, like the distinction we make between the expert and facilitator, this separation is convenient for exposition but, in practice, the expert and the decision maker may be the same person. Lindley et al. (1979) identified two different ways in which the decision maker might use the expert's summaries as data.

The expert's summaries relate to her distribution $f(X)$ for an uncertain quantity X, and the decision maker may use her assessments to improve his own knowledge about X. Lindley et al. call this the *external approach*, but note that the objective in this case is not really elicitation. The decision maker is not interested in the expert's $f(X)$ *per se*, and there is no attempt to convert the elicited summaries into a fitted distribution for the expert. If, in contrast, the decision maker uses the expert's summaries to learn about $f(X)$, then this is what Lindley et al. call the *internal approach*. The resulting statistical inference naturally expresses uncertainty about the expert's distribution. In addition to an estimate of the distribution, which may, for instance, be the decision maker's posterior mean, uncertainty can be expressed through, for instance, the posterior variance or posterior probability intervals. Further development of the external approach is given by French (1980), Lindley (1982) and Lindley and Singpurwalla (1986), and of the internal approach by Oakley and O'Hagan (2005); see Chapter 7 for more detail.

5.4.2 Sensitivity analysis

A much less formal approach to recognising uncertainty in the elicited distribution is simply to present a number of alternative distributions that fit the elicited summaries. This is usually done in the context of a sensitivity analysis which explores the consequences of the single elicited distribution not being the expert's true distribution (and/or of there being more than one 'good' distribution that appropriately represents the expert's beliefs – see Section 1.5.3). The consequences obviously depend on the purpose for which the elicitation is being performed. One particular area where this kind of sensitivity analysis has been quite widely used is when the expert's distribution will form the prior distribution for a Bayesian statistical analysis of some data. Then, changing the prior distribution will lead to different posterior inferences from the same data. The sensitivity analysis seeks to identify how much the key posterior inferences might be influenced by the choice of elicited distribution; would another distribution that also fits the expert's stated summaries lead to materially different inferences?

A body of research in Bayesian statistics has taken this a step further by specifying a class G of prior distributions as some sort of band around the elicited distribution and then mathematically deriving the range of posterior inferences that may be obtained by varying the prior distribution over G. An example of the classes of distributions considered is the *epsilon-contamination* class (Berger and Berliner, 1986), which takes the form

$$G = \{\pi : \pi = (1 - \varepsilon)\pi_0 + \varepsilon\phi, \phi \in F\},$$

where π_0 is the elicited distribution, ε is some small probability and F is a large class of distributions (perhaps even the class of all possible distributions). The class G includes all distributions that differ from π_0 by taking a probability ε and re-spreading it according to the distribution ϕ instead of π_0. (Provided F includes π_0, then so does G.) This technique is known as global sensitivity analysis; see Berger (1994) for an extensive review.

Although these ideas have been explored particularly in this area of eliciting a prior distribution for Bayesian analysis, they can clearly be applied to whatever purpose the elicited distribution is required for, to see whether the elicitation is precise enough for that purpose. Further discussion of the adequacy of elicitation can be found in Chapter 8.

5.4.3 Feedback and overfitting

Two common devices for improving elicitation are feedback and overfitting.

Feedback is the process of checking the adequacy of a distribution that has been fitted to an expert's stated summaries, by telling her some implications of that fitted distribution and asking her to confirm that these are reasonable representations of her beliefs. Suppose, for example, that we wished to elicit an expert's beliefs about a parameter θ that is a proportion, so that it must lie between 0 and 1. As discussed in Section 6.3, a beta distribution would be a natural choice to represent an expert's beliefs about such a parameter. If a beta distribution is to be fitted, then only two summaries need be elicited. For instance, if the expert specifies that her median for θ is 0.3 and her upper quartile is 0.55, then the unique beta distribution that fits these summaries is $Beta(0.77, 1.43)$ (to two decimal places accuracy in the beta parameters). Having fitted this distribution to the expert's statements, the facilitator might now *feed back* to the expert the fact that the lower quartile of $Beta(0.77, 1.43)$ is 0.116. The expert would be asked to confirm whether this is an accurate representation of her beliefs. (If she said that it was not, and if she was not willing to change either of her original probabilities, then the facilitator would need to consider fitting a more complex distribution.)

Overfitting is a related process. Rather confusingly, in the statistical model–fitting literature, 'overfitting' has the negative connotation of fitting an unnecessarily complex model to the data, whereas in the context of elicitation it has the positive sense that one can improve elicitation by eliciting more summaries than needed to fit a simple parametric distribution. Continuing the above example, instead of feeding back the lower quartile of his fitted distribution, the facilitator might elicit the expert's lower quartile. In practice, we would not expect the expert to specify three values that perfectly fit a beta distribution and, more realistically, the expert might set the lower quartile at 0.1. Having the three quantiles $P(\theta \leq 0.1) = 0.25$, $P(\theta \leq 0.3) = 0.5$ and $P(\theta \leq 0.55) = 0.75$, where the facilitator expects to use a beta distribution, is a case of overfitting. In this case, the $Beta(0.77, 1.43)$ distribution fits quite well to the three summaries, but so does $Beta(0.75, 1.4)$ and various others with similar parameter values.

These methods, and this example in particular, highlight a number of issues regarding imprecision, uncertainty and fitting distributions. In the overfitting context, it is clear that no single beta distribution will fit the three elicited summaries exactly. There are some who would argue that to fit a beta distribution then distorts the expert's stated beliefs. For instance, to claim that the expert's distribution can be represented by $Beta(0.77, 1.43)$ asserts that $P(\theta \leq 0.1) = 0.224$, whereas the expert has given 0.25 for this probability. Fitting a beta distribution forces on the expert the facilitator's view that the expert's beliefs should be representable as a beta distribution. In a sense this is true, but if we recognise that the expert's summaries are imprecise, and so we interpret her as saying that $P(\theta \leq 0.1)$ is 'approximately 0.25', it is certainly not clear that the beta distribution is a distortion of the expert's true beliefs. Indeed, in the feedback context, where the expert is not first asked to specify the lower quartile, it is likely in practice that she will express satisfaction with the value 0.116, and thereby confirm the facilitator's choice of the $Beta(0.77, 1.43)$ distribution.

Both feedback and overfitting recognise that, in practice, the expert cannot specify summaries precisely. In the above example, feedback is clearly easier for the facilitator. A unique distribution is fitted to the first two summaries and it is the expert who decides whether $P(\theta \leq 0.116) = 0.25$ is close enough to her true probability. If the facilitator instead uses overfitting, it is more complex to fit a distribution to three summaries, and to decide on a best fit requires a judgement of what metric to use. For instance, he could minimise the sum of the squared differences between the elicited summaries and those implied by the fitted distribution, the sum of absolute differences, the sum of squared percentage differences or some other criterion. Once a distribution has been fitted, it is necessary to decide whether the differences are within the bounds of imprecision in the expert's judgements. For this purpose, it would be appropriate to feed back to the expert the fitted distribution and implied summaries, and to ask the expert to say whether she is satisfied with the latter.

Although overfitting in this manner would naturally be followed by feedback, there are good reasons to suppose that the extra complexity of this process is preferable to simple feedback. Simply asking the expert whether $P(\theta \leq 0.116) = 0.25$ is close enough to her true probability is a leading question. The expert is not invited first to consider what her true value is. At the very least, if she considers it now, we may expect her response to be anchored to the facilitator's suggested value. But there is also a serious potential for bias if the expert feels pressurised to conform to the facilitator's suggestions. Feeding back fitted values *after* overfitting avoids the problem of anchoring and arguably reduces that of the pressure to conform.

Another problem with simple feedback is that it treats the initial summaries as precise and invites the expert to acknowledge imprecision only in the implied summaries that are fed back. Overfitting allows for imprecision in all the elicited summaries, and by finding a compromise 'best fit' distribution may be expected in practice to yield a more faithful representation of the expert's underlying beliefs. After all, no statistician would be comfortable fitting a regression line to two data points: having more data should lead to a more accurate estimate of the regression line, even though the fitted line does not exactly pass through any of the data points.

O'Hagan (1998) describes an example of overfitting, where the median, 33rd and 67th percentiles are elicited with the expectation of fitting a normal distribution, but a lognormal distribution is then used if the median is sufficiently below the midpoint of the other two percentiles. In the same example, feedback of other percentiles is then used to check that the fitted distribution is acceptable to the expert. Overfitting is also systematically used by Al-Awadhi and Garthwaite (1998) when fitting a multivariate t distribution, and by Kadane et al. (1980).

5.5 Conclusions

5.5.1 Elicitation practice

- Eliciting a probability distribution in practice entails eliciting a (relatively small) number of summaries from the expert and then fitting a suitable distribution that conforms to those elicited judgements. It is important to recognise that the fitted distribution implies many more statements that the expert did not make, and feedback may be used to check that these implications are acceptable to the expert.

- To elicit a univariate distribution, summaries based on probabilities are most widely used, including individual probabilities, quantiles, credible intervals and (in some cases) ratios of probabilities. There is conflicting evidence about which summaries are elicited most accurately, but, in general, it seems that eliciting credible intervals with high probabilities (e.g., 90% or higher) should be avoided.

- Although there is some evidence that people find it easier to think of proportions than probabilities, it is impractical to elicit epistemic uncertainties in this manner. However, training in the assessment of probability can use proportions to give an interpretation of the probability scale.
- The task of eliciting a multivariate distribution is substantially more complex, and it is important to begin by structuring the task, for instance, in terms of independent variables (whether the direct variables of interest, functions of them or latent variables).
- Approaches to eliciting association in a joint distribution include joint and conditional probabilities and regression relationships.
- Any fitted distribution is an approximation to the expert's true underlying beliefs and will differ from the ideal distribution because (a) there are many possible distributions that would fit the elicited summaries and (b) the expert cannot, in practice, specify the requested summaries accurately and reliably. It is important to recognise in some way the resulting uncertainty or imprecision in the fitted distribution.

5.5.2 Research questions

- The distinction between eliciting epistemic and aleatory uncertainty has not been clear in much of the research literature. There is a need for study into the extent to which research on people's ability to assess proportions, means, variances and correlations in random samples can provide guidance on eliciting epistemic uncertainty.
- Although univariate summaries have been well studied, there has been little research on eliciting multivariate distributions. It is not clear what kinds of probabilities or other summaries are most useful or how many summaries are needed to elicit a joint distribution effectively in two or more dimensions. In particular, the conjectures in Section 5.3.3 concerning possible biases in the assessment of joint and conditional probabilities should be tested empirically.

Chapter 6

Eliciting and Fitting a Parametric Distribution

6.1 Introduction

It is impractical to quantify an expert's opinion as a probability distribution without first imposing some structure on the distribution. This was noted in Chapter 5. A minimal structure is to assume that the cumulative distribution function is continuous and smooth, so that a few quantile assessments can identify its shape reasonably well. This can be a workable approach for simple univariate problems, but may not be for more complex multivariate problems, since the number of (marginal and conditional) quantile assessments that are required grows rapidly as the dimension of the probability distribution increases.

In practice, it is usual to impose further structure on the probability distribution that is used to represent the expert's opinion. The most common assumption is that opinion can be well represented by a member of some specified family of parametric distributions. This family might, for example, be the set of normal distributions; a member of the family would then be a normal distribution with a particular mean and variance. Modelling opinion by a specified family of distributions has the advantages that (a) it can give a representation of the expert's opinion that is useful and easy to use, (b) fewer assessments need to be elicited, (c) if more than the minimum number of assessments are elicited then assessments that are out of line can be identified and flagged and (d) implications of the expert's assessments can be determined and given to the expert as feedback.

The family of probability distributions that is used to model an expert's opinion should be selected on the basis of whether it is likely to contain a distribution that represents the expert's opinion well – a new family should be chosen if the initial

Uncertain Judgements – Eliciting Experts' Probabilities A. O'Hagan, C. E. Buck, A. Daneshkhah, J. R. Eiser, P. H. Garthwaite, D. J. Jenkinson, J. E. Oakley and T. Rakow

choice seems poor. It must be remembered that the facilitator makes pragmatic simplifying assumptions *on behalf of* the expert, not in defiance of her, so any modelling assumptions should, as far as possible, be checked with her. A suitable family of distributions will also depend upon the purpose for which the subjective distribution is required. If the elicitation is to formulate uncertainty about input to a decision project or a mathematical model, such as in the risk assessment of a complex engineering project, then many different forms of distribution would be suitable. In such cases, relatively simple distributions are commonly used, such as the uniform and triangular distributions (see Section 6.4). In contrast, if the elicitation is to obtain a prior distribution which will be updated in a Bayesian analysis of some additional data, it is common to assume that the subjective distribution comes from the family of conjugate prior distributions. A conjugate prior distribution is generally in the form of a prior distribution that is most easily updated when sample data become available. Different sampling models have different conjugate distributions. Of course, advances in computational methods make it far less imperative to choose prior distributions that are easily updated, but choosing some structure is necessary and a conjugate distribution can typically model a wide variety of opinion, so it is usually a good idea to explore whether a conjugate distribution can adequately represent the expert's opinion.

Probability distributions within a family of parametric distributions are fully determined by specifying the values of parameters. Hence, when an expert's opinion is to be modelled by a distribution from a specified family, the elicitation problem reduces to the task of determining parameter values that correspond to her expressed views. For instance, if it is assumed that the expert's opinion can be well represented by a member of the family of normal distributions, then the elicitation problem is to determine suitable values for the mean and variance parameters. Typically, the parameters that must be identified include a measure of location (mean, median or mode) and a measure of spread (usually a variance or standard deviation). If prior opinion about two or more quantities is to be quantified, then their joint distribution is likely to also include covariances or correlations as parameters. The other common quantity in probability distributions is a 'degrees of freedom' parameter. In the same way that uncertainty about a mean is given by its standard error, the level of uncertainty about a variance is often determined by the number of degrees of freedom associated with it.

6.2 Outline of this chapter

Although similar assessment tasks can be useful in a variety of situations, there is no single elicitation method that can be applied to all problems. Rather, a suitable elicitation method depends on the nature of the situation to which it relates and on the form of parametric distribution that will be used to model the expert's opinion. In this chapter, we review elicitation methods that have been reported in the literature.

We can categorise elicitation techniques in terms of the ranges of possible values of the variables whose distributions we are eliciting. For a single variable X, three principal cases arise.

1. Unrestricted. If X can take essentially any value, positive or negative, the family of distributions we choose to represent uncertainty must have the same property. Examples are the normal or t distributions. The normal distribution is the conjugate prior distribution when X is the mean of a normally distributed population.

2. Positive. Similarly, if X must be positive, or non-negative, then suitable distributions having this property are the gamma, inverse-gamma, log-normal and F families. This situation arises, for instance, when X is a population variance, and the inverse-gamma family is conjugate when X is the variance of a normally distributed population.

3. Bounded. The most common example of a bounded variable is when X is a probability or a population proportion, in which case it lies in the range [0, 1]. By far the most widely used distributions in this case are the beta family, which is also the conjugate family when X is the probability of success in independent trials.

For more than one variable, there are many more possible cases, but the following cover some of the most important in practice.

1. Mean and variance. If X comprises the mean and variance of a population, we have one unrestricted and one positive variable. If the data are samples from a normal distribution, then the conjugate family is the normal inverse-gamma family, which combines the normal conjugate for the mean with the inverse-gamma conjugate for the variance, but with a particular kind of association between the two variables.

2. Unrestricted vector. If the variables are all unrestricted, the most widely used distributions are the multivariate normal family, which is the conjugate family when the variables are the mean vector in multivariate normal sampling.

3. Unrestricted vector and variance. This situation arises, for example, when the variables are the parameters of a regression model, comprising an unrestricted vector of regression parameters and a positive error variance. The usual choice of family in this case is the multivariate normal inverse-gamma, which is conjugate for the regression model with normally distributed errors.

4. Variance matrix. When eliciting beliefs about a variance–covariance matrix, the usual distributional families are the Wishart and inverse Wishart. The latter is the conjugate family when X is the variance–covariance matrix in multivariate normal sampling.

5. Mean vector and variance matrix. Another extension of the normal inverse-gamma distribution, which is conjugate when the sampling is from a multivariate normal distribution, is the multivariate normal inverse-Wishart family.

6. Proportions in a classification. Another important case is when the variables comprise a set of probabilities or proportions which, in addition to individually lying in the range [0, 1], must sum to 1. The usual choice of distributions in this case are the members of the Dirichlet family, which are conjugate when the data are trials with outcomes classified into a number of distinct categories.

For all these situations, statisticians have at their disposal many more families of distributions than those listed above, but only a relatively small number of specific cases have been studied in the elicitation literature. The remainder of this chapter is organised according to the principal elicitation situations for which there is a body of research.

First, in Section 6.3, we describe methods of eliciting opinion about a proportion. This is an assessment task that has attracted a lot of attention, and it illustrates a variety of elicitation techniques. In Section 6.4, we consider more general approaches to the task of assessing a subjective distribution for a single quantity. We then review methods for eliciting opinion about multivariate distributions, considering the elicitation of opinion about a set of proportions in Section 6.5 and the parameters of a multivariate normal distribution in Section 6.6. As would be expected from their importance in statistics, regression problems have also been the focus of much attention. We review methods for eliciting opinion about the parameters of a normal linear regression model in Section 6.7 and generalised linear models in Section 6.8. Section 6.9 briefly reviews elicitation methods for an assorted collection of other problems, and some concluding comments are given in Sections 6.10 and 6.11.

This chapter makes greater use of statistical terms than previous chapters. Many of these terms are defined and/or described in the Glossary.

6.3 Eliciting opinion about a proportion

We consider situations where there is uncertainty as to whether an event will occur and we let P denote the probability that it will occur, where P is unknown. For instance, P might be the probability that a patient will have internal bleeding, as in the primary example in Section 5.1, or it might be the probability that a heart transplant will be successful or that a patient will have blood type A, and so on. In repeatable situations where a set of events can be considered, P is also a proportion, such as the proportion of heart transplants that are successful, or the proportion of patients with blood type A. An event that may or may not occur is referred to as a Bernoulli trial (see the Glossary). A set of independent Bernoulli trials (each with the same probability P) together form a binomial sampling model – the number of events that occur in n independent Bernoulli trials follows a binomial distribution with parameters n and p. For example, if ten patients are given heart transplants, then the number for whom it is successful follows a binomial distribution, Bin(10, p), assuming that the result of one transplant has no bearing on the results of other transplants. A large number of elicitation methods have been developed for quantifying opinion about a probability or proportion (P) when the underlying model is a Bernoulli or binomial distribution.

The conjugate prior distribution for Bernoulli and binomial sampling models is the beta distribution. It has two parameters, say α and β, and it can take a variety of shapes, as Figure 6.1 shows. Hence, it can portray a good variety of opinions reasonably accurately. As it is the conjugate distribution, the beta distribution is a particularly convenient form of prior distribution when expert opinion is to be combined with sample data from a Bernoulli trial or binomial data. If the prior distribution is a beta distribution, then combining it with sample data requires only simple arithmetic, and the posterior distribution is also a beta distribution. The usefulness of a beta distribution is not restricted to this case though, and because of its flexibility, it is also used to represent expert knowledge about a probability or proportion when no sample data is to be gathered. (It is also sometimes used for other types of elicitation problems – see Section 6.4.)

When opinion is to be modelled by a beta distribution, the elicitation problem is to determine values for α and β that reflect an expert's knowledge. Determining two parameters requires the expert to assess a minimum of two summaries and, although two summaries are enough in principle, overfitting brings benefits so it is better to elicit more assessments. Any apparent inconsistencies between the assessments may be reconciled mathematically (e.g., by taking the mean of the estimates that the various assessments imply) or by asking the expert to revise her assessments until they are internally consistent. Overfitting may also reveal that the beta distribution is a poor way of representing the expert's opinion, and then a different form of distribution (perhaps a non-parametric distribution) should be considered.

Winkler (1967) presented four different methods of assessing a prior distribution for a binomial sampling model and they split naturally into two pairs. In illustrating assessment questions, we shall suppose that p is the proportion of patients who have intestinal bleeding given certain symptoms. The first pair of methods were the *equivalent prior sample* (*EPS*) method and the *hypothetical future sample* (*HFS*) method. They both start by requiring a straight estimate of the proportion from the expert which is interpreted as the mean of the expert's distribution. To determine α and β, the expert must also assess at least one further quantity. The EPS method asks the expert to give an estimate of the sample size that she is basing her assessment upon. The larger the sample size, the more information the expert believes she has, obviously. If n is the assessed sample size and $\widehat{p}$ is the expert's estimate of p, then $\alpha = n\widehat{p}$ and $\beta = n(1 - \widehat{p})$.

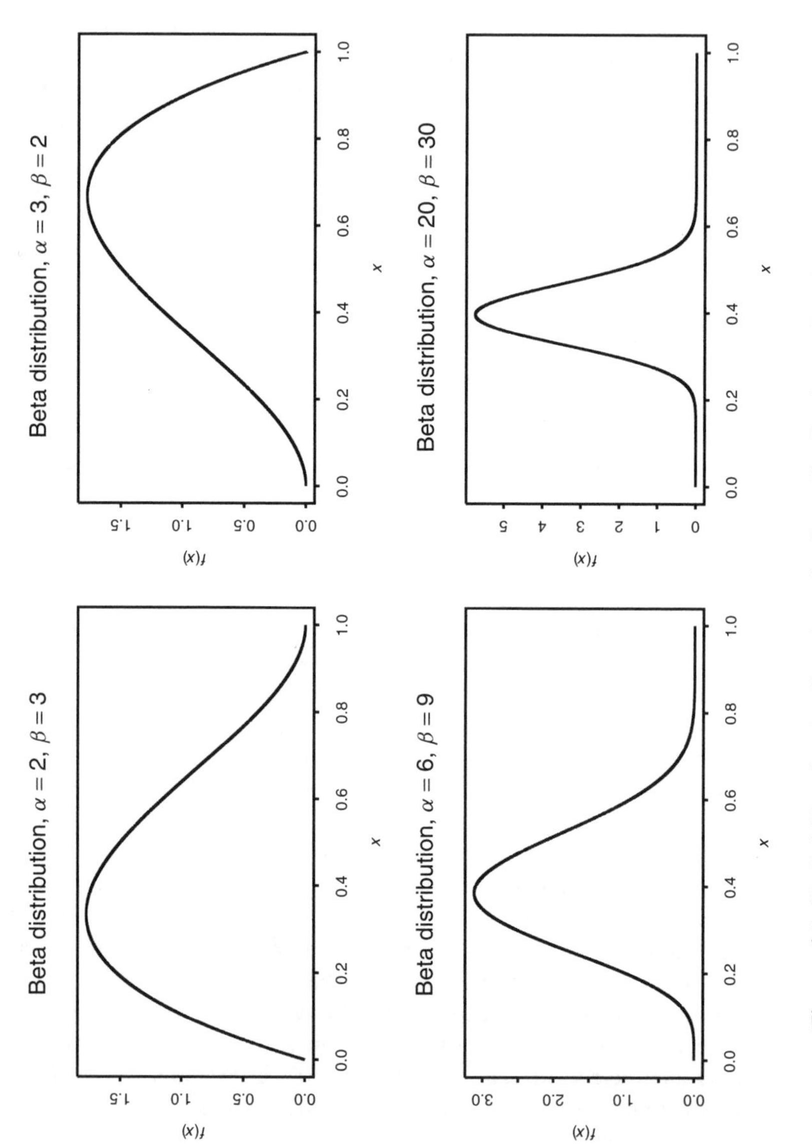

Figure 6.1: Probability density functions for Beta distributions of various shapes.

The HFS method asks the expert to update her estimate of p in the light of a hypothetical sample taken from the population of interest. In our example, for instance, this might be observing a sample of 30 patients with the given symptoms of whom 4 had internal bleeding. The expert is presumed to update her estimate of p by mentally using Bayes' theorem. If $\widehat{p_2}$ is the expert's estimate of p after observing r patients with the complaint from a sample of size n, then α and β are obtained by solving $\widehat{p} = \alpha/(\alpha + \beta)$ and $\widehat{p_2} = (\alpha + r)/(\alpha + \beta + n)$. Winkler referred to the EPS and HFS methods as indirect techniques because the subjective distribution that is implied by the expert's assessments is not clear to her at the time of elicitation.

The second pair of methods were the cumulative distribution function (CDF) method and the probability density function (PDF) method. In the CDF method, the expert is asked to give her median estimate of p and to give one or more quantiles (usually at least two) of her subjective distribution for p. Winkler then plotted the quantiles as a CDF and joined them with a smooth curve, giving a non-parametric representation of the expert's opinion. An alternative is to fit a beta distribution to the quantiles. The values of other quantiles can then be read off the CDF curve or calculated from the fitted distribution and fed back to the expert for confirmation. The CDF method is the elicitation method that was used in the example in Section 5.1.1. The PDF method elicits the density function rather than the distribution function. The expert's mode for the proportion is elicited and then values for the proportion that are half as likely as the modal value are also elicited. The density function is sketched by joining these points, or a beta function may be fitted. Other values and/or quantiles can be elicited to improve the shape and to feed back to the expert. Examples of a PDF curve for a variety of beta distributions have been given in Figure 6.1 and the CDF curves corresponding to these are given in Figure 6.2. The methods in this second pair were described by Winkler as direct techniques because the distribution becomes clear to the expert as the elicitation proceeds.

To illustrate the use of one of the indirect techniques and one of the direct techniques, suppose a doctor had to specify her prior distribution for the proportion of patients, p, who will experience an adverse side effect on a particular drug. The doctor was first asked to estimate the value of p and gave 0.4 as her mean. For the EPS method (the indirect technique), she was then asked 'Can you give a number n such that your knowledge would be roughly equivalent to having observed a random sample of n patients who took the drug (of whom 40% had an adverse side effect?)' The doctor thought that the extent of her knowledge was equivalent to a sample of size 25. If her opinion is to be modelled by a beta distribution, then, based on her assessments, the EPS method gives the parameter values $\alpha = 25 \times 0.4 = 10$ and $\beta = 25 \times 0.6 = 15$. The probability density function for this beta distribution is the solid line in Figure 6.3.

Suppose that the doctor also used the CDF method to quantify her opinion and gave 0.40 as her median estimate of p and 0.33 and 0.5 as her lower and upper quartiles, respectively. As a check, the doctor was asked whether p was equally

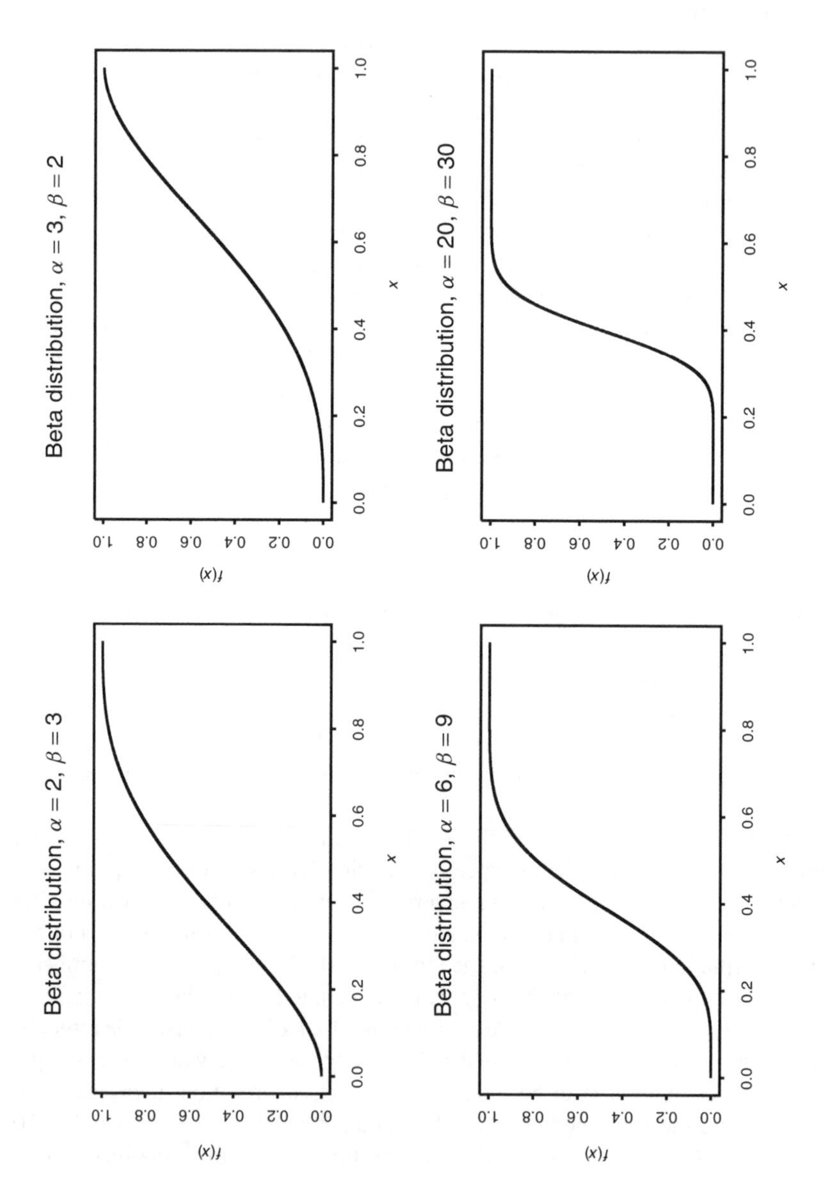

Figure 6.2: Cumulative distribution functions for Beta distributions of various shapes.

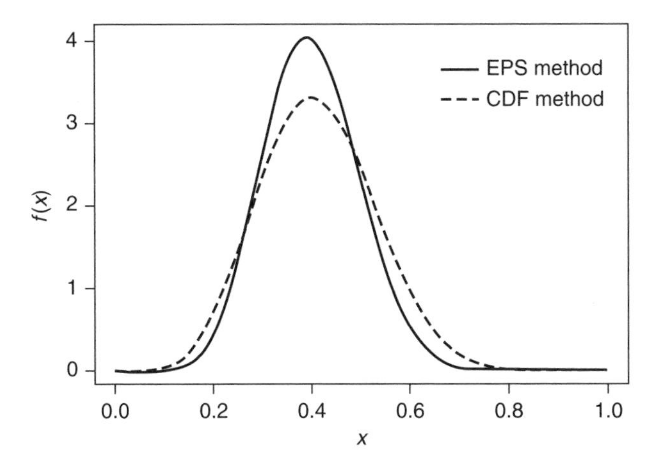

Figure 6.3: Beta probability densities elicited by the EPS method ($\alpha = 10$, $\beta = 15$) and the CDF method ($\alpha = 7$, $\beta = 10$).

likely to be in any of the four intervals (a) under 0.33, (b) between 0.33 and 0.4, (c) between 0.4 and 0.5 or (d) above 0.5. She thought it more likely that p would be in the interval from 0.4 to 0.5 and in further questioning she revised her median to 0.41 and her upper quartile to 0.49. On the basis of these assessments and a lower quartile of 0.33, the beta distribution with $\alpha = 7$ and $\beta = 10$ is a reasonable representation of her opinions. The probability density function for this distribution is the dotted line in Figure 6.3. With these assessments, the EPS method gives a distribution with a smaller variance than the CDF method, indicating that the extent of the doctor's knowledge is not equivalent to a sample as large as 25 or her lower and upper quartile assessments should have been closer together.

All four of the above methods given by Winkler tend to produce prior distributions that are unrealistically 'tight' (i.e., they have variances that are too small). For example, Schaefer and Borcherding (1973) conducted an experiment in which 22 subjects used the EPS and CDF methods to quantify their opinions about various proportions. Before any training, each subject assessed 50% central credible intervals for 18 different proportions. The percentage of these intervals that contained true values was 15.7% for the EPS method and 22.5% for the CDF method. The task in the HFS method, revising opinion in the light of additional data, is similar to the task in the 'bookbag-and-pokerchips' experiment (see Section 3.3.2), where conservatism has a marked effect. Insufficient revision of opinion when given the hypothetical data would result in a distribution that is too tight. Winkler (1967) felt that conservatism also influences the EPS method, hypothesizing that subjects equate their knowledge to too large a sample size, through not realizing the value of sample information. The CDF method tends to yield distributions that are again too tight, but slightly less tight than the PDF method and much less tight than the HFS and EPS methods (Winkler, 1967). In the example in Figure 6.3, the CDF method shows a slightly less tight distribution than the EPS method. On this basis the quantile method seems preferable, and it also seems the method of choice when judged by scoring rules. (Scoring rules are discussed in Section 8.2.) Some experiments have examined the effect of training and these found that the bias of tightness was reduced for all three methods, with particularly marked improvement for the EPS method (Staël von Holstein, 1971; Schaefer and Borcherding, 1973).

A further ten methods of eliciting the beta distribution's parameters were given in a review by Hughes and Madden (2002), who considered the task of eliciting expert opinion within the context of plant disease epidemics. Four of the methods have the same basic theme: obtain a location statistic and then identify a quantile, either by asking the expert to specify it directly or by specifying an interval within the range [0, 1] and asking the expert to assess her probability that the interval contains p. For the location statistic, a method by Fox (1966) requires the expert to give her estimate of the modal (maximum) value of p, while methods by Gross (1971), Weiler (1965) and Duran and Booker (1988) require her to give the mean of her distribution for p. They obtained quantile estimates as follows.

1. Let m^* denote the expert's assessed mode. Fox (1966) chose a value for K, calculated $m^* - Km^*$ and $m^* + Km^*$ and asked the expert to assess her subjective probability that the interval $(m^* - Km^*, m^* + Km^*)$ contains p.

2. Gross (1971) asked the expert to assess her subjective probability that p was in the interval $(0, K\overline{p})$, where $\overline{p}$ was the mean of the expert's distribution for p and $0 < K < 1$ was given to the expert by the facilitator.

3. Weiler (1965) also asked the expert to assess her subjective probability that p was in a specified interval, but specified the interval $(2\overline{p}, 1)$. (In the context of interest to Weiler, p was, the proportion of defective items in a batch and its value was small, so that $2\overline{p} < 1$.)

4. Duran and Booker (1988) gave the expert a probability, q, and then asked the expert to assess the qth quantile of her subjective distribution of p.

Each of the methods yields two equations involving α and β. These are solved using numerical methods, and the resulting values for α and β determine the expert's subjective distribution.

Weiler (1965) and Duran and Booker (1988) each gave a further method of eliciting α and β. Weiler gave the expert a probability, q, and asked the expert for two values K_1 and K_2, such that the probability that p was in the interval $(0, K_1)$ was q and the probability that it was in the interval $(K_2, 1)$ was also q. In a very similar vein, Duran and Booker (1988) also elicited two values, K_1 and K_2, but this time they represented the q_1th and q_2th quantiles of the subjective distribution of p, with q_1 and q_2 specified by the facilitator.

Two methods that use a different type of assessment task were given by Pham-Gia et al. (1992). The first method again elicits the mean of the subjective distribution of p, but the second quantity it elicits is the mean absolute deviation about the mean. In the second method discussed by Pham-Gia et al., the expert is asked to give her median estimate of p and an estimate of the mean absolute deviation about that median is then elicited. Eliciting a mean absolute deviation is somewhat unusual and empirical work does not seem to have been conducted that examines people's ability at performing such tasks or biases that affect the elicited assessments.

The above eight elicitation methods reported by Hughes and Madden can all be criticised for using only two assessments to determine two parameters. Eliciting additional assessments would enable some form of averaging to be used to estimate parameters, which should improve robustness, and allow odd assessments to be flagged. The final two methods given in the review by Hughes and Madden are better in this respect. They also differ from the other methods in that they avoid direct questions about p, which is not an observable quantity. Instead, they ask questions about sampling distributions, such as 'the number of patients who would have intestinal bleeding in a random sample of 20 patients with the specified symptoms'. Chaloner and Duncan (1983) used this form of question in a variant of the PDF method. The expert states the most likely number of patients in the sample, say x, and then assesses the relative likelihood of x patients rather than $x - 1$ patients, and of x patients rather than $x + 1$ patients. Assessments are used to estimate α and β and the implied endpoints of the shortest 50% predictive interval are then calculated. These endpoints are given as feedback to the expert, who comments on the length of the interval. If she finds it too short or too long, the parameter estimates are revised repeatedly until she is satisfied with its length. The process is repeated for a variety of sample sizes and the resulting parameter estimates are amalgamated in some way. Gavasakar (1988) also used questions about the sampling distribution in a variant of the HFS method. The expert first specifies the most likely number of males in a random sample of some specified size (as in the method of Chaloner and Duncan) and then revises her assessment after being given hypothetical sample data, as in the HFS method. Assessments are again used to determine the parameters of a beta distribution. This is done for a variety of sample sizes. Interestingly, Gavasakar conducted a computer simulation to assess the sensitivity of parameter estimates to errors in an expert's assessments. He compared his method with that of Chaloner and Duncan (1983) and found that his method was much less sensitive. However, in a simulation study of this nature it is difficult to choose appropriate error distributions for the different methods and he may have underestimated the magnitude of errors that are induced by hypothetical samples.

A further two methods of eliciting the parameters of a beta distribution are described in León et al. (2003). The 'least informative method' elicits the mean and the mode and solves the two equations this yields. A potential difficulty with this strategy is that the mean and mode may differ by very little, making the parameter estimates sensitive to small assessment errors. The 'most informative method' elicits the mean and the three quartiles. Then, beginning with a Beta(1,1) distribution, it iteratively varies the parameters until the distribution's median and upper quartile match the elicited values. The distribution's lower quartile and mean are then checked against the values elicited, with the mean and the quartiles being re-elicited if the values are not close. Finally, Gilless and Fried (2000) elicited the mode and the 90th percentile and solved for the parameters using numerical integration.

There has been lamentably little empirical work that compares the many different elicitation methods that have been proposed for quantifying opinion about proportions. Indeed, only the methods proposed by Winkler (1967) seem to have been studied in more than one experiment and much more work is needed before firm recommendations could be made as to which is the best method to use in practice.

6.4 Eliciting opinion about a general scalar quantity

A common elicitation problem is to quantify opinion about a scalar quantity, such as the number of days a patient will be in hospital, or the amount by which a particular drug will reduce a patient's blood pressure, and so on. The simplest approach to this problem is to ask the expert to specify a range $[a, b]$ in which the scalar quantity is believed to lie. If this is all that is elicited from the expert, then it is common to assume a uniform probability distribution over $[a, b]$. This can be criticised as too simplistic in at least two respects. First, the expert almost certainly would not believe that the unknown quantity in question is as likely to be very close to the limits a and b as to be at a more central point in the interval. Second, unless the range $[a, b]$ represents absolute physical limits to the possible values of the quantity (in which case the first criticism applies even more strongly), it is unreasonable to give zero probability to the event that the quantity lies outside the range.

As a simple response to the first criticism, another common practice is to use a triangular distribution. For this purpose the expert is asked also to specify a mode, say c. Then, the assumed distribution has the density

$$f(x) = \begin{cases} 2\frac{x-a}{(b-a)(c-a)} & \text{if} \quad a \le x \le c \\ 2\frac{b-x}{(b-a)(b-c)} & \text{if} \quad c \le x \le b \end{cases} .$$

The acceptability of uniform and triangular distributions as representations of uncertainty about model inputs in engineering applications is indicated by their featuring strongly in Oberkampf et al. (2004), but O'Hagan and Oakley (2004) criticise this practice as a failure to elicit adequately.

As well as being useful for quantifying opinion about proportions, the CDF and PDF methods may also be used to elicit the subjective distribution of any unknown scalar quantity. The quantiles that the CDF method requires are often elicited by asking the expert to give the ends of the intervals that split the range into sections of equal probability, as in the bisection method (see Section 5.2.1). The first bisection yields the 50% quantile, bisections on each side of this yield the

25% and 75% quantiles, and the next bisections yield the 12.5%, 37.5%, 62.5% and 87.5% quantiles. In principle, the process can be repeated to get assessments of small probabilities at the top and bottom of the range, but it has been found that people are poor at assessing extreme quantiles (Alpert and Raiffa, 1969). A technical discussion of the bisection method can be found in Garthwaite and Dickey (1985). Other quantiles that have been elicited are the tertiles ($33\frac{1}{3}$% and $66\frac{2}{3}$% percentiles) and the 17% and 83% percentiles. Tertiles are elicited in a manner similar to the bisection method, simply by asking the expert to divide the range of values for the variable in question into three so that she believes that the true value is equally likely to be in any one of the three intervals. In an experimental study in the UK water industry, Garthwaite and O'Hagan (2000) asked experts to make assessments of several quantiles and then tested the estimates for calibration. Their results suggested that in the design of an elicitation method, the use of tertiles, rather than quartiles, should be given more consideration. They also found that letting experts assess the quantile they preferred did not appear to improve the objective accuracy of their assessments, as judged by a scoring rule.

The CDF method is also known as the *variable interval* method, since the expert is asked to vary the interval within which she wishes to place a specified amount of her probability mass. The natural alternative to the variable interval method is to give the expert intervals of the random variable representing the event and ask her to determine how much of her probability mass should be allocated to each interval. This is known as the fixed interval method. The method was used in work for Her Majesty's Inspectorate of Pollution (O'Hagan, 1998) and drew on earlier work by Phillips and Wisbey (1993). The upper and lower bounds of the quantity of interest, X, were first elicited, and then the mode (labelled U, L and M respectively). The next stage was to elicit five probabilities from the expert, $p_1 = P(L < X < M)$, $p_2 = P(L < X < (L + M)/2)$, $p_3 = P((M + U)/2 < X < U)$, $p_4 = P(L < X < (L + 3M)/4)$, and $p_5 = P((3M + U)/4 < X < U)$. The probabilities were asked for in this order so as to avoid tendencies to anchor assessments and the choice of intervals did not require the expert to assess small probabilities. Quantiles were determined from these probability assessments.

O'Hagan (1998) reports an example where the method was used in the training of assessors for the work with Her Majesty's Inspectorate of Pollution. The experts were asked to quantify their opinions about the shortest distance by road from Birmingham to Newcastle-upon-Tyne. Let X denote this distance in miles. One expert gave her limits for X as 165 (L) and 250 (U) and 190 as her mode (M). Her probability assessments were as follows:

$$p_1 = P(165 < X < 190) = 0.5$$

$$p_2 = P(165 < X < 177.5) = 0.175$$

$$p_3 = P(220 < X < 250) = 0.05$$

$$p_4 = P(165 < X < 183.75) = 0.33$$

$$p_5 = P(205 < X < 250) = 0.25$$

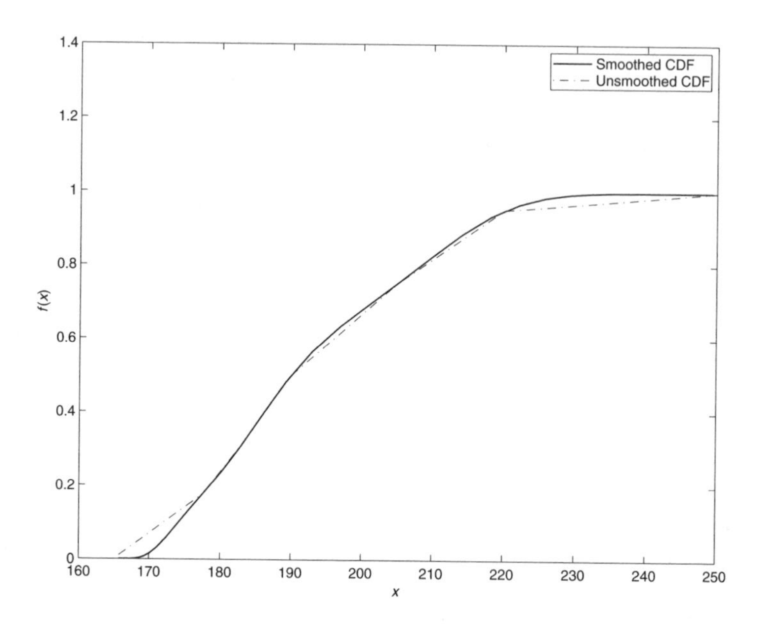

Figure 6.4: Expert's CDF for the shortest distance by road from Birmingham to Newcastle-upon-Tyne.

From these assessments, it follows that 165, 177.5, 183.75, 190, 205, 220 and 250 are the expert's 0.0, 0.175, 0.33, 0.5, 0.75, 0.95 and 1.0 quantiles, respectively. If these quantiles are plotted on a graph and joined with straight lines, it gives the dotted line in Figure 6.4.

Suppose, again, that the expert is certain that the scalar takes a value in the range $[a, b]$ and suppose her assessments give probabilities $p_1, p_2, \ldots, p_k$ that the uncertain quantity lies in the intervals $[a, c_1]$, $(c_1, c_2], \ldots, (c_{k-1}, b]$. (These may be obtained by the fixed or variable interval methods.) Another alternative is for the facilitator to simply assign the histogram distribution

$$f(x) = \frac{p_i}{c_i - c_{i-1}} \quad \text{if} \quad c_{i-1} \leq x \leq c_i, \ i = 1, 2, \ldots, k,$$

where $c_0 = a$ and $c_k = b$. An example of this type of histogram (with $k = 6$) is given in Figure 6.5. It may be an adequate representation of the expert's opinion, particularly if k is not small. The method is quite widely used and is advocated, for example, in Cooke (1991).

The histogram technique is equivalent to plotting the quantiles on a graph and forming a CDF curve by joining the quantiles with straight lines. However, the expert's beliefs would generally be better represented by a smooth CDF distribution, so it is preferable to draw a smooth CDF curve through the plotted quantiles. One method of determining a smooth CDF curve is the Bayesian approach of

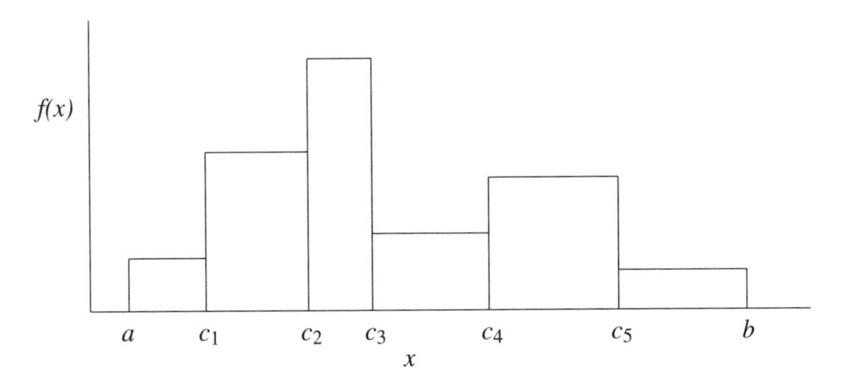

Figure 6.5: Subjective opinion represented by a histogram distribution.

Oakley and O'Hagan (2005). In Figure 6.4, the solid line is the posterior mean CDF using their method based on the expert's quantiles for the distance from Birmingham to Newcastle-upon-Tyne. It is probably a better representation of the expert's opinion than the unsmoothed CDF, as it seems unlikely that the expert's CDF should change abruptly at the arbitrary points chosen by the elicitation method.

Another way of representing an expert's opinion is to fit a standard distribution to the assessed quantiles. This usually yields a more useful representation of subjective opinion. Normal distributions have been fitted to quantile assessments (e.g., O'Hagan, 1998) and log-normal distributions have been fitted when the quantity of interest can only take positive values and the subjective assessments show positive skewness (O'Hagan, 1998). Garthwaite and O'Hagan (2000) fitted both normal and log-normal distributions to sets of subjective assessments that displayed skewness but found little clear benefit from using the log-normal distribution in terms of objective accuracy.

When there is a lower and an upper bound on the values the quantity can take, or the expert can specify such bounds, then a generalised beta distribution may also be fitted to quantile assessments. A generalised beta distribution is an ordinary beta distribution that is scaled to have support (L, U), where L and U are the bounds on the quantity of interest. (The *support* of a probability distribution $f(x)$ is the set of x-values for which $f(x)$ is non-zero.) O'Hagan (1998) used least squares to fit a generalised beta distribution to probabilities elicited by the fixed interval method, while Gilless and Fried (2000) fitted a generalised beta distribution to elicited values of L, U, the mode and the 90th percentile. In Figure 6.6, we give the histogram distribution corresponding to the unsmoothed CDF curve shown in Figure 6.4, together with a generalised beta distribution. The latter has a broadly similar shape to the histogram but it is, of course, both smoother and much easier to handle mathematically.

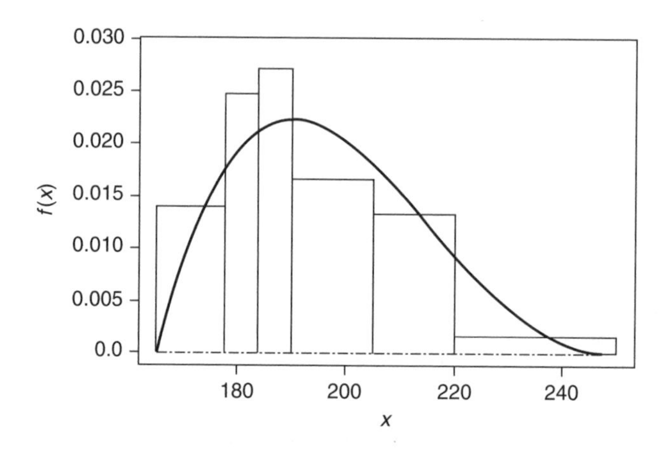

Figure 6.6: Histogram and generalised beta distribution representing the expert's opinion about the shortest distance by road from Birmingham to Newcastle-upon-Tyne.

The operational research technique of PERT (program evaluation and review technique) often makes reference to beta distributions. PERT is used for the management and scheduling of large projects and it allows for uncertainty in the time that will be taken to complete each individual component of the project. The method of allowing for uncertainty that is commonly described in textbooks (e.g., Dunn and Ramsing, 1981; Render and Stair, 2000) is to elicit the most likely time a job will take, together with the optimistic and pessimistic times. The optimistic time (L) and pessimistic time (U) are equated to the lower and upper limits of the support for a generalised beta distribution and the most likely value (M) is equated to its mode. Simple formulae are then given for the mean and standard deviation of the time the job will take, even though three assessments are insufficient to identify the four parameters of a generalised beta distribution. These simple formulae are as follows: mean $= (L + 4M + U)/6$ and standard deviation $= (U - L)/6$; presumably they are a useful rule of thumb. Lau and Somarajan (1995) discuss the use of subjective probability estimates, obtained through the variable interval technique, and consider them to be preferable to the above simplistic formulae for estimating the mean and standard deviation. This approach is further developed in Lau et al. (1996), who advocate eliciting seven quantiles (1%, 10%, 25%, 50%, 75%, 90%, 99%) or five quantiles (leaving out the 1% and 99% quantiles), and by Lau et al. (1997), who also discuss whether restricting attention to the beta family is necessary.

Approaches to eliciting the parameters of a distribution from a parametric family for an unrestricted or positive variable have received much less attention than eliciting the parameters of a beta distribution. In most cases, there are two parameters, as for instance with the normal, gamma or inverse-gamma families. The usual approach is to elicit the median or mode as a location measure and then one or more other quantiles to elicit a measure of dispersion. The parameters of the assumed family are then fitted to these elicited summaries. For instance, to elicit the parameters of a normal distribution, we might ask the expert to give her median, lower and upper quartiles, equating the assessed median to the normal mean and setting the normal variance to 0.550 times the square of the inter-quartile range. This method was used, for example, by Garthwaite (1989, 1994).

6.5 Eliciting opinion about a set of proportions

In this section, we consider situations where an event has more than two possible outcomes. For example, rather than considering whether a patient's blood type is 'A', suppose we want the probabilities that his/her blood type is 'A', 'B', 'AB' or 'O'. More generally, suppose an event takes exactly one of k possible outcomes and let p_i denote the probability that it takes the ith of these outcomes. Then, the sum of the p_i must equal 1, and this constraint means that each p_i cannot be considered in isolation. Hence, the task of quantifying opinion about the p_i is a multivariate problem. For repeatable events (or trials), p_i is also the proportion of outcomes in the ith category.

Suppose we have a set of independent trials, each of which results in one of k possible outcomes ($k \geq 2$). Then their overall outcome follows a multinomial distribution, provided the probabilities of each of the possible outcomes are the same in each trial. As an example, if we have a random sample of ten patients, then the multinomial distribution gives the probability that, for instance, three have blood type 'A', one has type 'B', two have type 'AB' and the remaining four have type 'O'.

When $k = 2$, the multinomial distribution reduces to a binomial distribution. Hence, in order to develop an elicitation method for a multinomial distribution, it is natural to try to extend methods that have been developed for binomial distributions. This strategy was followed by Chaloner and Duncan (1987) who extended their method of quantifying opinion about a proportion (Chaloner and Duncan, 1983). Their method elicits modal values from the expert and uses interactive graphics to help the expert quantify her opinion. It is assumed that her subjective distribution comes from the family of Dirichlet distributions, which is the conjugate family for multinomial distributions. The elicited assessments are used to determine parameter values. Use of the Dirichlet distribution makes it straightforward to combine sample data with the prior distribution, but the assumption that opinion corresponds to a Dirichlet distribution is quite restrictive. However, methods of eliciting a more flexible distribution do not seem to have been proposed. Perhaps for this reason, the Dirichlet distribution is also used when sample data will not become available.

An alternative elicitation method that uses hypothetical future samples was given by Dickey et al. (1983). Suppose X has a multinomial distribution with k possible outcomes and let $\mathbf{p} = (p_1, \ldots, p_k)$ denote the k-vector of probabilities that X falls into each category. They assumed that the subjective distribution for $\mathbf{p}$ is a Dirichlet distribution with overall weight parameter n and mean vector $\mathbf{b} = (b_1, \ldots, b_k)$. The first step in the elicitation process is to ask the expert for her probabilities that $X = i$ for $i = 1, \ldots, k$. Denoting these assessments by $\hat{P}(X = i)$, in principle $b_i = \hat{P}(X = i)$ for $i = 1, \ldots, k$. However, these probabilities should sum to 1 so, to make this hold, Dickey et al. set each b_i equal to $\hat{P}(X = i)/\sum \hat{P}(X = i)$. Determining n is trickier, and for this their method gives the expert a hypothetical future sample. For example, the expert might be told that eight observations were taken, of which three fell in the first category, two fell in the fourth category and one fell in each of the second, sixth and seventh categories. The expert's revised probabilities for several categories, conditional on this sample, are then elicited. For any given value of n, Bayes' theorem indicates the values that the revised probabilities should take. Comparison of these theoretical probabilities with the expert's assessments is used to estimate n. It would be possible to perform these calculations having elicited just one of the conditional probabilities after the hypothetical future sample has been given, but it is preferable to elicit the conditional probability for several categories and to average the values of n that each gives. If the estimates of n are very diverse, it would suggest that a more flexible form of distribution than the Dirichlet is required. Dickey et al. did not offer advice on how to choose the hypothetical samples that are given to the expert. Further theoretical detail on multinomial priors and the Dirichlet distribution were given by Regazzini and Sazonova (1999).

Dickey et al. (1983) give an example of using their method to quantify an expert's opinion about potential jurors' attitudes, assuming the availability of a death penalty. The expert was a social psychologist with an interest in legal matters and different attitudes of potential jurors were divided into the following four categories.

1. Would not decide guilt versus innocence in a fair and impartial manner.
2. Fair and impartial on guilt versus innocence and, on sentencing, would always vote for the death penalty, regardless of circumstances.
3. Fair and impartial and would never vote for the death penalty.
4. Fair and impartial and sometimes would and sometimes would not vote for the death penalty.

The psychologist's assessment of the proportion of jurors in each category, $\mathbf{p} = (p_1, p_2, p_3, p_4)$, was $(0.02, 0.08, 0.15, 0.75)$. As a hypothetical sample, she was told that 200 potential jurors had been questioned, of whom 16 had attitudes that fell in the first category, 20 in the second category, 32 in the third and 132 in the fourth. Given this information, the psychologist's revised assessment of $\mathbf{p}$ was $(0.05, 0.09, 0.16, 0.70)$. Figure 6.7 shows the psychologist's original probabilities for each category, the relative frequencies of each category and the psychologist's revised probabilities. If the psychologist, in revising her opinion, had given her

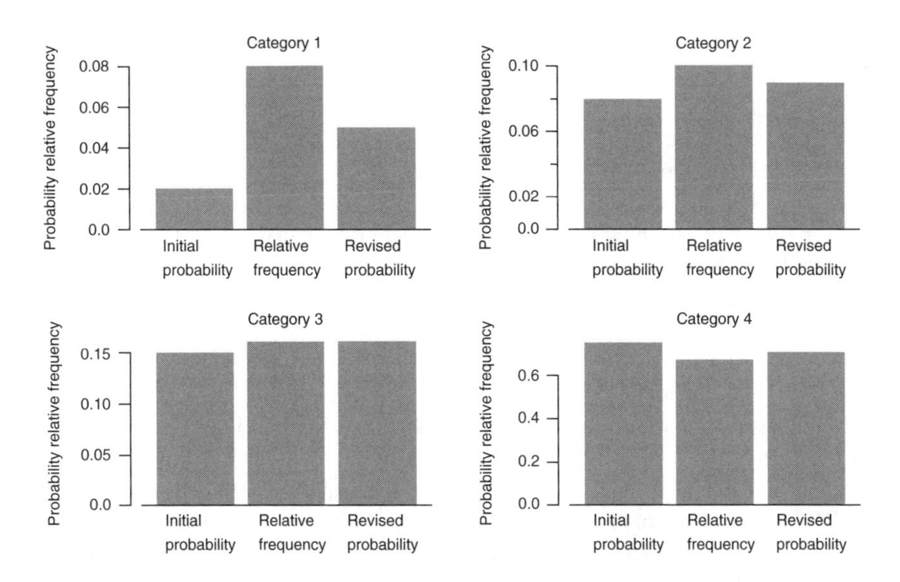

Figure 6.7: A psychologist's initial probabilities for four categories, the relative frequencies of the categories in a sample of 200 people and the expert's revised probabilities.

prior opinion and the data equal weight, then her weight parameter n would equal the size of the hypothetical data set, 200. In fact, as Figure 6.7 shows, the psychologist's revised assessments are closer to the hypothetical data than her original probabilities, so that she gave more weight to the data and hence n should be less than 200. From her assessments, Dickey et al. estimated n as 140.

Various other distributions that address some of the limitations of the Dirichlet family are discussed by O'Hagan and Forster (2004, Sections 12.11 to 12.19), but there is no literature addressing how to elicit such distributions.

6.6 Eliciting opinion about the parameters of a multivariate normal distribution

In this section we suppose that sample data come from a multivariate normal (MVN) distribution. For example, a datum could be one man's height, weight and blood pressure and the joint sampling distribution of these quantities might be an MVN distribution (approximately). The parameters of an MVN distribution are a mean vector, say μ, and a variance–covariance matrix, say Σ. Thus, in the above example, the mean vector would have three components that give a man's average height, average weight and average blood pressure, while Σ would give the variance of each quantity (e.g., the variance of a man's height) and their covariances (e.g., the covariance between a man's height and his weight). Some of the elicitation methods ask questions whose answers depend on both aleatory uncertainty (sampling variation) and epistemic uncertainty (imperfect knowledge on the part of the expert). Then, assumptions made about the sampling distribution are obviously important. Other elicitation methods, however, ask questions that do not involve aleatory uncertainty. These methods can be used to elicit beliefs about a vector of continuous-valued quantities without making assumptions about the sampling distribution.

Despite the importance of the MVN distribution, relatively few methods have been proposed for ways of quantifying opinion about its parameters. Gokhale and Press (1982) considered only a *bivariate* normal distribution and gave a method for quantifying opinion about its correlation coefficient. Dickey et al. (1985) considered an MVN distribution and addressed the problem of quantifying opinion about Σ, but they made the restrictive assumption that the expert's mean estimates of each diagonal element of Σ are all equal and that the expert's mean estimates of each off-diagonal element of Σ are also all equal. Dickey et al. (1986) assumed that the expert's opinion about an MVN distribution corresponds to a natural conjugate distribution. The greater flexibility of this form of prior distribution should, in theory, enable an expert's opinion to be represented better. However, the paper of Dickey et al. (1986) was wide ranging and the MVN distribution was considered only briefly.

A conjugate distribution for an MVN distribution has the form

$$f(\mu, \Sigma) \propto | \Sigma |^{-(d+p+2)/2} \exp\left[-\frac{c}{2}(\mu - \mathbf{m})'\Sigma^{-1}(\mu - \mathbf{m}) - \frac{1}{2}\mathrm{tr}(\Sigma^{-1}\mathbf{A})\right].$$

Here, $f(\mu, \Sigma)$ is the expert's subjective distribution for μ and Σ, m is the expert's mean estimate of μ, $\mathbf{A}/(d-2)$ is the expert's mean estimate of Σ, d and c are degrees of freedom parameters and $tr(\mathbf{Z})$ represents the trace of the matrix $\mathbf{Z}$, the sum of the lead diagonal components. Thus, four hyperparameters $(\mathbf{A}, d, \mathbf{m}, c)$ need to be elicited.

A method of eliciting these hyperparameters was given by Al-Awadhi and Garthwaite (1998). Let $\mathbf{X}$ denote the random variable that has the MVN distribution of interest. Eliciting $\mathbf{m}$ is straightforward, as $\mathbf{m}$ is the mean, median and mode of the expert's subjective distribution of $\mathbf{X}$; the expert may reasonably be asked to assess any of these summaries. In order to determine the degrees of freedom parameters, c and d, Al-Awadhi and Garthwaite elicited assessments from the expert after giving her hypothetical data. They used several different hypothetical data sets to obtain more than one estimate of each degrees of freedom parameter. Empirical work suggests that differing estimates of a degrees of freedom parameter should be reconciled by taking their geometric mean, rather than their arithmetic average (Al-Awadhi and Garthwaite, 2001).

Eliciting a variance–covariance matrix, such as $\mathbf{A}$, is a difficult problem, especially as the assessed matrix should be positive-definite in order to satisfy the laws of probability. However, the subjective distribution of **X** is a multivariate t distribution whose variance matrix equals $[1/(d-1-p)][1+1/c]\mathbf{A}$. (**A** is often referred to as the *spread* of the t distribution and is defined even if $0 < d < 1 + p$, when the variance would not be defined.) Kadane et al. (1980) gave an elegant sequential method for eliciting the variance (spread) of a t distribution. The method requires an expert to update her opinion on the basis of hypothetical data. Conditional on these data, the expert assesses conditional medians and other quantiles. The set of hypothetical data is increased by one datum at a time and the expert makes assessments after each increase. The carefully designed structure of the assessment method ensures that **A** is positive-definite. The method of Kadane et al. (1980) is complex and will not be described in detail here. In their method for quantifying opinion about an MVN distribution, Al-Awadhi and Garthwaite (1998) used the method of Kadane et al. but adapted it so that the diagonal elements of **A** are determined by unconditional assessments; the original method of Kadane et al. used conditional assessments to estimate these quantities, and this introduces marked bias (Al-Awadhi, 1997).

Al-Awadhi and Garthwaite (1998) showed that their elicitation method is a practical tool by using it to quantify a clinical psychologist's opinion about the effects of head injuries on various measures of intelligence and memory. These measures were *verbal comprehension* (VC) and *perceptual organisation* (PO) on the Wechsler Adult Intelligence Scale-revised test and *attention/concentration* (AC), *general memory* (GM) and *delayed memory* (DM) on the Wechsler Memory Scale-revised test. After eliciting his prior distribution, the values of its parameters were discussed with the psychologist and he was comfortable with their implications. In particular, the correlations between the five measures were as follows:

	VC	PO	AC	GM	DM
VC	1	0.36	0.17	0.29	0.05
PO	0.36	1	0.45	0.36	0.21
AC	0.17	0.45	1	0.67	0.58
GM	0.29	0.36	0.67	1	0.64
DM	0.05	0.21	0.58	0.64	1

These show that the five measures were all positively correlated in the elicited probability distribution, but in varying degree: those measures associated with memory (AC, GM and DM) had a higher correlation with each other than with the other two measures (VC and PO). The expert confirmed that this reflected his opinions.

A problem with the use of the conjugate prior to model expert opinion is that it forces a specific relationship between the mean vector and the covariance matrix. In a frequentist analysis for an MVN distribution, the sample variance–covariance matrix is both an estimate of the population variance and, after division by the sample size, it is also the estimated variance of the sample mean. In the standard conjugate distribution, a single variance matrix again fulfils both these roles. An experiment by Al-Awadhi and Garthwaite (2001) demonstrated that this is inappropriate. They examined different forms of assessment task and compared alternative ways of estimating hyperparameters. To quantify opinion about the vector of means, it proved preferable to ask directly about the means rather than individual observations while, to quantify opinion about the variance matrix, it was better to ask about deviations from the mean. One alternative to the conjugate distribution is to assume that the population mean and variance are independent in the prior distribution, an approach followed in Garthwaite and Al-Awadhi (2001). They gave μ an MVN distribution and let Σ have either an inverse-Wishart distribution or a generalised inverse-Wishart distribution. As μ and Σ are independent, the assessment method was structured so that elicitations used to determine the expert's distribution for one parameter were not used to determine her distribution for the other.

6.7 Eliciting opinion about the parameters of a linear regression model

The task of quantifying opinion about a linear regression model has attracted substantial attention. Suppose a response Y is related to a vector of explanatory variables $\mathbf{x}$ through the model $Y = \mathbf{x}'\beta + \epsilon$, where β is a vector of regression coefficients, and ϵ is random error, with $\epsilon \sim N(0, \sigma^2)$. For example, Y might be a patient's response to a drug while the explanatory variables are the dosage of the drug that the patient was given and the patient's sex and weight. The usual conjugate prior distribution specifies that (a) σ^2 equals $\omega\delta/\chi^2$ where ω and δ are constants and χ^2 is a chi-squared random variable with δ degrees of freedom and (b) given σ, β has a normal distribution with some mean $\mathbf{b}$ and some variance–covariance matrix $\sigma^2\mathbf{R}$. For this prior distribution, ω, δ, $\mathbf{b}$ and $\mathbf{R}$ are the hyperparameters that need to be determined via an expert's assessments.

To elicit opinion about these quantities, Zellner (1972) suggested questioning an expert about the regression coefficients. Some experts may be able to think about these coefficients directly. However, it is usually better to question people about observable quantities, such as Y, rather than ask direct questions about unobservable quantities, such as regression coefficients (Kadane and Wolfson, 1998). In general, elicitation methods for a linear regression model that have been proposed somewhat more recently ask the expert about observations $Y_1, \ldots, Y_m$ at values $\mathbf{x}_1, \ldots, \mathbf{x}_m$ of $\mathbf{x}$ (Kadane et al., 1980; Oman, 1985; Ibrahim and Laud, 1994). A potential disadvantage in questioning an expert about Y is that the variance of Y results both from the expert's uncertainty about the values of the regression coefficients *and* from random variation. To separate these sources of uncertainty, Garthwaite and Dickey (1988) asked questions about $\overline{Y_i}$, the average value of Y if a large number of observations were taken at a single value, $\mathbf{x}_i$. They argued that averages are quantities to which people can relate and that an expert can give assessments about $\overline{Y_i}$ without the need to consider random error. As $\overline{Y_i}$ is the average of a *large* number of observations, it can be equated to $\mathbf{x}_i'\beta$.

Let $\mathbf{X} = (\mathbf{x}_1, \ldots, \mathbf{x}_m)'$. We refer to $\mathbf{x}_i$ as a design point and $\mathbf{X}$ as a design matrix. We first consider the method of Kadane et al. To estimate the mean vector, $\mathbf{b}$, at each design point they elicited the median of Y which could be equated to the mean of Y as, by assumption, the distribution of Y is symmetrical. The $\mathbf{x}$-values that are used in the elicitation affect the quality with which expert opinion is captured and Kadane and Wolfson (1998) suggested a procedure for their selection. Letting $\mathbf{y}$ denote the vector of assessments, $\mathbf{b}$ is equated to $(\mathbf{X}'\mathbf{X})^{-1}\mathbf{X}'\mathbf{y}$, which would be the least squares estimate if $\mathbf{y}$ were a vector of observations at $\mathbf{X}$. To obtain δ, quantile assessments are elicited. Let y_q represent the qth quantile of Y. The expert assesses the 0.75 quantile, the 0.875 quantile and the 0.9375 quantile (by the bisection method) at each of two or more design points. The ratio $a(\mathbf{x}) = (y_{.9375} - y_{.50})/(y_{.75} - y_{.50})$ depends only on δ and its value is determined for each design point where the quantiles were assessed. The average value of the $a(\mathbf{x})$ is found (after adjustment for impossible assessments) and comparison with standard t distributions yields the estimate of δ.

Denote the vector of observations at the m design points by $\mathbf{Y} = (Y_1, \ldots, Y_m)$. Given the assumed conjugate distribution for β and σ^2, the expert's implied distribution of $\mathbf{Y}$ is a multivariate t distribution, and Kadane et al. elicited its parameters using the method discussed in Section 6.6. The expert is also questioned about the joint distribution of pairs of observations at a single design point, which yields an estimate of the hyperparameter ω. After obtaining ω, the parameters of the multivariate t distribution yield an estimate of the last hyperparameter that is required, $\mathbf{R}$. The method of Kadane et al. involves sophisticated distribution theory and matrix algebra, and the reader is referred to their paper for details.

Ibrahim and Laud (1994) used an approach similar to that of Kendal et al., including the same method of determining $\mathbf{b}$. They used assessments of the mean and precision to determine ω and δ and, in related work (Laud and Ibrahim, 1995), they used assessments of the median and the 95th percentile of the distribution of the precision. (Precision is defined to be 1/variance.) The context in which Ibrahim and Laud (1994) quantified opinion is the analysis of designed experiments. They let $\mathbf{X}$ be the design matrix for which data would be gathered, set $\mathbf{R}$ equal to $\tau(\mathbf{X}'\mathbf{X})^{-1}$ and chose the hyperparameter τ to reflect the weight that should be attached to the expert's opinion, relative to the weight that should be attached to the data from the experiment. Thus, $\mathbf{R}$ was chosen to reflect the expert's knowledge, but only in a limited way. They did not check whether this choice of $\mathbf{R}$ was a fair representation of the expert's opinions.

Oman (1985) used three different ways of determining $\mathbf{b}$. The first is to simply ask the expert to specify its value and the second is similar to the method of Kadane et al. (1980) and Ibrahim and Laud (1994); that is, the expert gives a central estimate of Y at several design points (Oman asked the expert to specify the mean of Y) and $\mathbf{b}$ is estimated by least squares. Oman's third way is to ask the expert to assess the covariance between Y and each independent variable in the regression model. Oman did not quantify opinion about ω and δ, and restricted his posterior analysis to inferences that depend only on a point estimate of σ^2, which he obtained using empirical Bayes methods.

Garthwaite and Dickey (1988) again used the same method as Kadane et al. (1980) to elicit **b**. To obtain ω and δ, they used assessments that depend only upon experimental error. The expert is first asked to suppose that two observations are taken at the same design point. It is pointed out to the expert that the observations will not be identical because of random variation, and the median of their absolute difference is elicited. The expert is then given a hypothetical datum and she then gives her updated median of their absolute difference. Garthwaite and Dickey (1988) gave equations for deriving estimates of ω and δ from these two assessments. In order to elicit **R**, they used a novel assessment task that requires the expert first to select the design point at which she can predict Y most accurately and then to repeat this task several times with an increasing set of restrictions on the x-values she can choose. The method was developed for the use of industrial chemists and exploits their experience of choosing design points to conduct experiments. It is not as flexible as the method of Kadane et al. (1980), in that it cannot be used with polynomial regression or with x-variables that are factors. However, it can be extended to elicit prior distributions that are suitable for variable-selection problems (Garthwaite and Dickey, 1992, 1996). A feature common to the methods of Kadane et al. and Garthwaite and Dickey (1988) is that a structured sequence of questions is used to ensure that **R** is a positive-definite matrix.

Illustrative examples where the above methods are used by a genuine expert to quantify her opinions about a regression model are given in Kadane et al. (1980), Winkler et al. (1978) and Garthwaite and Dickey (1991, 1992, 1996). Garthwaite (1994) conducted an experiment where participants used the method of Kadane et al. (1980) to quantify their opinions about four linear models each of which involved just one independent variable. The experiment found that the accuracy of the elicited distribution is improved if the design points are spaced far apart, but a greater number of design points did not give much additional benefit. It also found that the elicited variances of Y at a set of design points are affected by the order of the design points at which assessments are made.

Experimental research involving cue-weighting (see Section 5.3.4) is of direct relevance to multiple regression and should be taken into account when eliciting opinion about a regression model. For instance, considerable evidence has been amassed that people give predictive cues less weight when they are presented with non-predictive cues. This is termed the 'dilution effect' (Nisbett et al., 1981; Zukier, 1982) and warns that parsimony is important in defining a regression model to represent an expert's opinion. For example, Rakow et al. (2005) asked doctors to assess the likelihood of an important clinical outcome and, as in much of medical practice, there was a wealth of patient information available. Rakow et al. conjecture that the non-predictive information reduced doctors' ability to use predictive information to any great effect. The implication is that the expert should only be given information about variables that are likely to be relevant in the regression model. The temptation to include all possible variables in the model at the assessment stage (lest an important variable be omitted) can have adverse consequences.

6.8 Eliciting opinion about the parameters of a generalised linear model

The class of generalised linear models (GLMs) includes many useful statistical models: logistic regression and probit regression for the analysis of data in the form of proportions (counts and ratios), Poisson regression and log-linear models for data in the form of counts, some important models for the analysis of survival data and ordinary linear regression, among others. Work on the elicitation of prior distributions for a generalised linear model appears to start with Bedrick et al. (1996). They suggested an elicitation method in which predictive distributions at different design points are elicited and then combined to form a prior distribution. A method for eliciting the predictive distributions was not specified, but methods used for linear regression models might be applied with little change, provided the modelling assumptions enable each predictive distribution to be fully specified by eliciting its mean and variance. Bedrick et al. assumed that the predictive distributions are independent of each other and showed how this induces a prior distribution on the regression coefficients. They discussed properties of their method and showed that it has similarities to data augmentation. The method yields a prior distribution that cannot be written down in closed form, but combining the prior with sample data is straightforward using MCMC methods, as Bedrick et al. illustrated with examples.

There is a considerable amount of work on using past studies to construct prior distributions for GLMs. The task is tricky because typically the past studies are not completely comparable with the next study for which the prior distribution is required. For instance, the next study may examine additional variables that were not monitored in previous studies. One general approach is to use a rich prior model that contains more hyperparameters than a conjugate prior would contain; some hyperparameters are calculated from the past studies, others relate to variables for which past data are not available and still others influence the weight that is attached to the past studies relative to the weight attached to the new study. The latter hyperparameters could be chosen using expert opinion, but instead their values have typically been varied in sensitivity analyses. Chen et al. (1999) discussed variable selection for GLMs, Chen et al. (2003) extended this to generalised linear mixed models and Chen and Shao (1999) discussed prior distributions for multivariate categorical response data models. One way of controlling the influence that the prior distribution has over the posterior distribution is through power prior distributions (Chen et al., 2000), which raise the prior density to a power, $0 \leq a_0 \leq 1$.

Some specific forms of GLM have also received attention. In particular, forming prior distributions for log-linear modelling for contingency tables has been considered (King and Brooks, 2001). This work focused on the reconciliation of two forms of information: opinion about the probability that an observation will fall in a particular cell and opinion about the regression parameters of the model. King and Brooks (2001) also considered the task of using partial information, where the expert has knowledge about some coefficients but virtually no knowledge about others. Also, practical methods of quantifying expert opinion about a logistic regression have been developed and used by subject-matter experts. Garthwaite and Al-Awadhi (2006) give a method that models the sampling distribution by a continuous piecewise-linear logistic regression model, which is more flexible than the standard logistic regression model. The greater flexibility requires more parameters, but by exploiting expert opinion the model is still practicable even when there is limited sampling data. An example of a continuous piecewise-linear regression model is given in Figure 6.8. The method developed by Garthwaite and Al-Awadhi uses interactive graphics to elicit the prior distribution. The expert assesses conditional and unconditional medians and quartiles by clicking a mouse on graphs and bar charts displayed on a computer screen. Ecologists used the method to quantify their opinions about the habitat distribution of various rare and endangered animals in Queensland, Australia. Figure 6.9 shows a graph after the expert has assessed two sets of conditional quartiles for the left-hand points on the graph. (She will do the same for the right-hand points next.) The solid line shows her assessed medians.

Kynn (2005) modified the method of Garthwaite and Al-Awadhi (2006) so as to allow the expert a choice of assessment tasks and used a simpler form of the prior model that does not require covariances to be assessed. Her method gives the expert feedback in the form of credible intervals and PDFs. It has been implemented in freely available software that can be downloaded from www.maths.qut.edu.au/~whateley. Ecologists have also used her method to quantify their opinions about rare and endangered flora and fauna in Queensland.

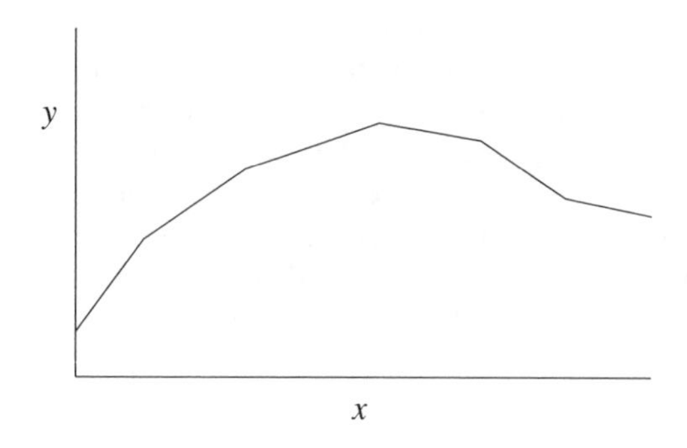

Figure 6.8: A continuous piecewise-linear model.

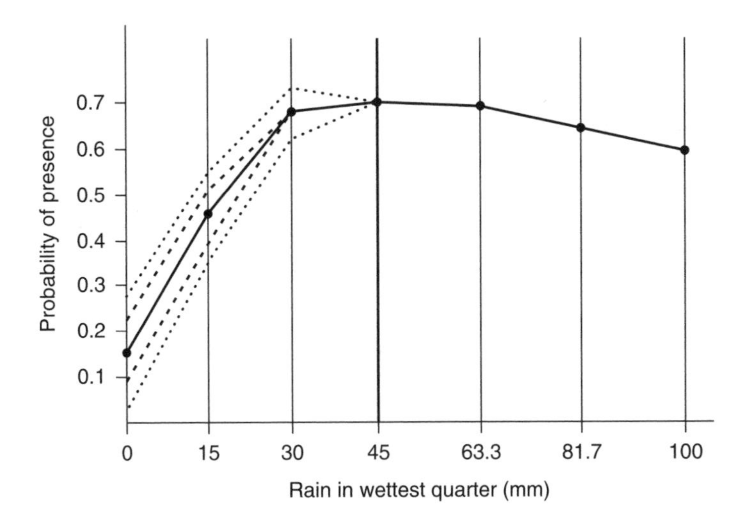

Figure 6.9: A screen showing median and conditional quartile assessments using the interactive graphical method of Garthwaite and Al-Awadhi (2006).

6.9 Elicitation methods for other problems

Some other elicitation problems have attracted a little attention, but less than one might expect. For example, we found no published work on eliciting a conjugate prior distribution for a Poisson distribution. However, there has been some work on quantifying opinion about Poisson processes. Smith and Faria (2000) addressed situations in which fault-free machines arrive as a Poisson process at an unknown rate, but they translated the problem into a binomial model by considering the number of machines that would be fault free when a specified number arrived. Parent and Bernier (2003) considered an extension of a Poisson point process to threshold models. They developed a prior-posterior model within the Bayesian framework that includes eliciting opinion for an informative semi-conjugate prior. Their work was motivated by an application in hydrology and they examined alternative ways of quantifying opinion. In particular, they found it preferable to elicit quantiles rather than ask direct questions about the parameters of the prior distribution. Further work on eliciting prior distributions for Poisson processes is found in Campodónico and Singpurwalla (1994). They elicit measures of location and spread for the expected number of failures in two time periods. In constructing a prior distribution from these assessments, the facilitator treats the assessments as correlated data that may be subject to bias.

A significant area of statistics is aimed at modelling the occurrences of extreme values. Naturally, in cases of extreme values there is less data available than for non-extreme values, so that the potential benefit from exploiting expert opinion is substantial. Lenox and Haimes (1996) considered fitting separate distributions to the tails of the main distribution of observations as a method of understanding extreme values. They used the Von Mises-Jenkinson distribution to model the tails of the 'parent distribution' and obtained subjective estimates of its parameters by eliciting probabilities with the fixed interval method. There has also been work on eliciting values for the generalised extreme value distribution, in order to fit a gamma distribution to expert opinion for use as a prior distribution (Coles and Powell, 1996). The general idea that quantiles (extreme quantiles in this case) should be elicited rather than the parameters of the prior distribution was supported.

The task of quantifying opinion about the regression coefficients in a proportional hazards model was carefully addressed by Chaloner et al. (1993). In their motivating example, two treatments for AIDS sufferers were to be compared with a placebo, and survival time was defined as the time before the development of toxoplasmosis encephalitis. Opinion about the regression parameters was modelled by an extreme value distribution rather than an MVN distribution, as the use of MVN distributions in this context can have undesirable implications that are unlikely to represent an expert's opinions (see Meinhold and Singpurwalla, 1987). So as to ask questions about observable quantities, Chaloner et al. questioned the expert about the two-year survival probability under each treatment regime. Let p_0 denote this probability for the placebo (so that p_0 determines the baseline hazard rate) and let p_1 and p_2 denote the probabilities for the two treatments. The first quantity to be elicited was a 'best guess' of p_0, denoted by $\hat{p}_0$. Then, the joint distribution of p_1 and p_2 was elicited conditional on $p_0 = \hat{p}_0$. This was done by eliciting (a) quantiles of the marginal distributions of $p_1 \mid (p_0 = \hat{p}_0)$ and $p_2 \mid (p_0 = \hat{p}_0)$ and (b) the expert's probability that both p_1 and p_2 are larger than their median values. The assessments were used to determine the parameters of a type B bivariate extreme value distribution, as given by Johnson and Kotz (1972). An interactive computer program was used to elicit opinion and graphical feedback was given to the expert in the form of density plots of marginal distributions and a contour plot of the joint prior distribution of p_1 and p_2. The expert could also use a graphical dialog box to adjust the means and standard deviations of the marginal distributions and a quantity that determined the correlation between p_1 and p_2. Figure 6.10 shows the computer screen at one stage during the elicitation session. Shown on the screen are plots of two marginal distributions, a contour plot of a joint bivariate probability distribution and a graphical display for adjusting the joint distribution.

Regression structures can also feature in time series modelling. For example, the first-order autoregressive, AR(1), time series model has the form

$$y_t = \rho y_{t-1} + \beta \mathbf{x}_t + \epsilon_t,$$

where β is a vector of regression coefficients, $\mathbf{x}_t$ is a vector of values taken by a variable at different time points and $\varepsilon_t \sim N(0, \sigma^2)$ is random error. Kadane et al. (1996) consider the *unit root* problem, where the task is to distinguish between the case where $\rho = 1$ and the case where ρ is close to one. They develop a method of eliciting a prior distribution for this problem that is very similar to the elicitation

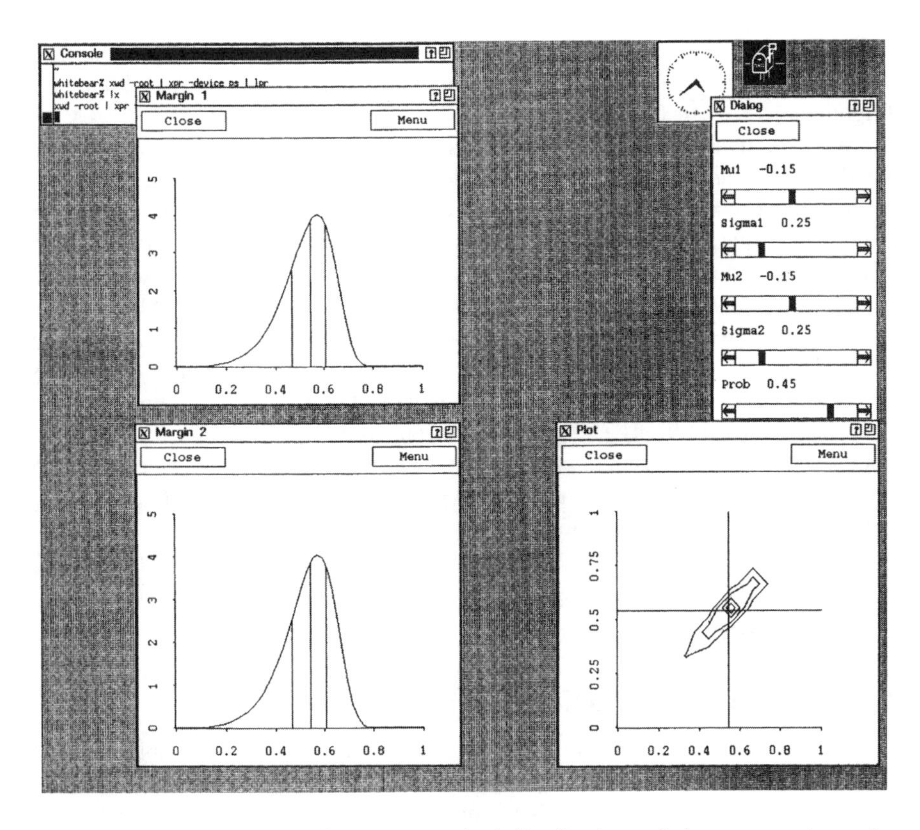

Figure 6.10: A screen showing two marginal distributions, their contour plot and a graphical display device for adjusting their parameters.

method for normal linear models that was given by Kadane et al. (1980). However, they elicit a more flexible prior distribution that models the expert's opinion with a piecewise version of the conjugate prior for a normal linear model. The prior contains three components to represent $\rho < 1$, $\rho = 1$ and $\rho > 1$. They illustrate the use of the method by eliciting the opinion of a macroeconomist on the projected real domestic gross national product for the next five quarters.

6.10 Deficiencies in existing research

This chapter has reviewed many methods for eliciting a parametric distribution. Despite the large quantity of research that has been reviewed here, there are striking gaps. Ideally, methods would have been examined empirically by using them to quantify people's opinions both in experimental tests and in real applications. Regrettably, some of the methods have never been used for either purpose, and the great majority have only been used in a single set of trials. This makes it difficult for the area of elicitation to evolve – without a reasonable amount of empirical work it is hard to identify which elicitation methods work well and which are best forgotten.

With multivariate sampling models, such as multiple regression, one reason for the limited applied work is the complexity of the elicitation methods, making an interactive computer program almost essential as the medium through which an expert is questioned. As long back as 1980, software was developed to elicit opinion about normal linear models and was made freely available (see Kadane et al., 1980), but it was probably before its time, as the Markov Chain Monte Carlo methods that have brought Bayesian statistics to prominence were not available in 1980. There is now a much greater need for user-friendly software that elicits prior distributions. Ideally, such software would offer a choice in the assessment tasks that are used for any particular elicitation, so that users could vary tasks and compare the elicited distributions. Then the software would help advance research into assessment methods as well as providing a practical means of quantifying expert opinion in real applications.

6.11 Conclusions

6.11.1 Elicitation practice

- In complex situations, structure must be imposed on the probability distribution used to represent the expert's opinion for the elicitation problem to be manageable. Usually it is assumed that her opinion can be modelled by some specified parametric distribution, when the elicitation problem reduces to estimating the parameters of the distribution.

- For multivariate problems, it is typically assumed that opinion can be modelled by a conjugate prior distribution. This assumption is also commonly made with univariate problems, especially if it is envisaged that sample data will become available.

- Assumptions about the prior distribution should be checked with the expert as the purpose of the distribution is to represent her opinions. Feedback and overfitting are recommended and can help identify an inappropriate distribution.

- Feedback and overfitting can also highlight assessments that are out of line. Inconsistencies in an expert's assessments may be reconciled through reassessment by the expert or through some form of averaging.

- The most widely used tasks in elicitation methods are the assessment of a central measure (a mean, median or mode), the assessment of quantiles and the revision of opinion when given hypothetical sample data. To elicit quantiles, the variable interval method is most generally used. The most commonly elicited quantiles are the lower and upper quartiles, which can be assessed using the method of bisection, but there is some evidence that assessment of 0.33 and 0.67 quantiles leads to distributions that are better calibrated.

- Many elicitation methods reported in the literature have only been devised theoretically and have never been used to elicit anyone's opinions. In contrast, other methods have been developed with careful consideration given to practical details and they have been tested and used in real problems by genuine experts. When there is a choice of elicitation method, in practice it is clearly better to use one of the latter methods.

- Many distributions are available to statisticians that are more flexible than the commonly studied conjugate families, but in most cases no elicitation methods for these have been proposed in the literature.

- Interactive computing is almost essential if an elicitation method requires a sequence of questions in which some questions are determined by an expert's earlier answers. Benefits from interactive computing include the facility to provide feedback to the expert and to identify apparent inconsistencies in her assessments as they are made. Interactive graphics have also been used in various problems and seem to be particularly useful in regression problems.

- The development of a decent elicitation method requires knowledge of certain areas of the psychology literature and, especially in multivariate problems, it also requires a degree of statistical expertise. In addition, implementing an elicitation method can require computing skills. Thus, inter-disciplinary teams are best suited to their development.

6.11.2 Research questions

- Empirical research is needed to examine and compare some of the many elicitation methods that have been proposed but never tested.

- Effective elicitation methods need to be developed for several standard problems. For instance, methods are needed for quantifying opinion about a positive scalar, such as the parameter of a Poisson distribution, the parameters of a univariate normal distribution and those of a gamma distribution.

- Flexible user-friendly software for eliciting distributions needs to be developed by inter-disciplinary collaboration between experts in statistics, psychology and programming.

Chapter 7

Eliciting Distributions – Uncertainty and Imprecision

7.1 Introduction

Eliciting a probability density function is an inherently imprecise process. This imprecision is due to two factors. Firstly, it is difficult for any expert to give precise numerical values for their probabilities (or other summaries), though this is a requirement in most elicitation schemes. Secondly, an expert can only provide a finite number of probability judgements, yet in most cases this will not be sufficient to specify a unique probability distribution. Consequently, in most elicitations the facilitator cannot claim that the resulting distribution *perfectly* describes the expert's beliefs. We consider these two issues in turn.

7.2 Imprecise probabilities

Consider an example of a political expert stating her probability that candidate A will win an election. The expert believes that it is 'very likely' that candidate A will win, but 'not a certainty'. In providing a probability $P(A \text{ wins})$, the expert must now attach a numerical value to her feeling that the event $\{A \text{ wins}\}$ is 'very likely'. Clearly, her stated value will be subject to rounding error; the expert may only consider the first two decimal places in her probability, for example, and will certainly not state $P(A \text{ wins}) = 0.9243461$. However, the issue of imprecision goes

Uncertain Judgements – Eliciting Experts' Probabilities A. O'Hagan, C. E. Buck, A. Daneshkhah, J. R. Eiser, P. H. Garthwaite, D. J. Jenkinson, J. E. Oakley and T. Rakow

beyond this. The expert is unlikely to be in the habit of stating numbers to represent feelings, and may be unsure if her feeling that an event is 'very likely' is most appropriately described with a probability of 0.85, 0.9, 0.99, and so on. Olson and Budescu (1997) report that when representing aleatory uncertainty, individuals may be content to use precise numerical methods (i.e., probabilities), but when representing epistemic uncertainty, there is a preference to use imprecise methods, such as verbal descriptions of uncertainty.

Theoretically, this imprecision can be removed through the consideration of betting preferences. The expert can be asked to consider her preference for the following two gambles:

G_1: Gain £100 if A wins the election, gain nothing otherwise.

G_2: Gain £100 if a red ball is (randomly) selected from a bag containing r red balls out of n in total, gain nothing otherwise.

(Note that it is not necessary to consider a financial reward such as gaining £100 in this argument; any outcome that the expert considers to be preferable to the status quo can be used.) It is assumed that the expert is content to assign a precise probability of r/n to the event that a red ball is selected on the assumption that each ball has the same chance of being selected. If the expert prefers gamble G_1 over G_2, then it can be deduced that her probability of the event $\{A \text{ wins}\}$ is greater than r/n. If the expert is indifferent between these two gambles, then it can be deduced that her probability for A winning is precisely equal to r/n. In principle, the values of r and n can be refined to obtain a precise numerical probability, but in practice this approach will have limitations. Consider for example $r = 55$ and $n = 100$. The expert may be quite content to select G_1 if she feels that "A is very likely to win; I would be surprised if A lost" rather than "I am expecting A to win but would not be overly surprised if A lost"; she feels that the number 0.55 is too small in describing her degree of belief in the event $\{A \text{ wins}\}$. However, with $r = 90$ and $n = 100$, the expert may legitimately respond, "I cannot give an answer; I think the event $\{A \text{ wins}\}$ is very likely and would choose G_1 if I thought $P(A \text{ wins}) > 0.9$, but I am unsure if my degree of belief in the event $\{A \text{ wins}\}$ is appropriately represented by the number 0.9". The fact that the expert is unable to express a preference cannot be taken to mean that her probability really is 0.9, as she may have the same difficulty in expressing a preference when $r = 95$ and $n = 100$. Simply understanding what it means to *be* 90% certain is not sufficient for identifying when one *feels* 90% certain.

One view is that the reason we cannot, in general, provide a precise probability is that such an entity does not exist; our uncertainties are inherently vague. Accepting this position affects the nature of elicitation; rather than attempting to elicit the expert's *true* prior distribution, the objective is to elicit *a* prior distribution that is an acceptable (approximate) representation of the expert's views. Winkler (1967) describes the notion of a 'satisficing' prior distribution that an expert is "content to live with at a particular moment of time." An alternative view is given in Lindley et al. (1979) who consider the possibility that "the subject has, in some sense, a set of coherent probabilities that are distorted in the elicitation process." O'Hagan (1988) defines a 'true probability' (i.e., a true description of an individual's beliefs) as the probability that the individual would obtain if they were "capable of arbitrarily fine comparisons and possessed perfect reasoning powers," though he also states that "true probabilities are unattainable in practice." See the discussion in Section 1.5.3. Whether we take the view that there is a unique 'true' distribution or believe that there can be many 'good' or 'satisficing' distributions to represent an expert's beliefs, it is certainly the case that the expert's elicited probabilities must be treated as imprecise. The stated value is only one of a range of values that the expert would regard as more or less equally good representations of her beliefs.

There has been little research in formally addressing this difficulty within conventional probability elicitation. Walley (1991) proposed bounding a probability P with upper and lower probabilities, $\overline{P}$ and $\underline{P}$ respectively. This still leaves the issue of how to specify $\overline{P}$ and $\underline{P}$ with absolute precision unresolved. Walley (1991) also reports that the theory resulting from the use of upper and lower probabilities leads to some counter-intuitive results. One such result is the phenomenon of *dilation*, where observing one event reduces one's level of knowledge about another. Walley (1991) gives the example where H_1 and H_2 are the events of heads on two tosses of a coin. An expert judges the coin to be fair, and so asserts $P(H_1) = P(H_2) = 0.5$ as precise probabilities. However, she believes that the tosses may not be independent, and so is unwilling to specify the probability that *both* H_1 and H_2 will occur, except to say that this probability is less than 0.5 (as it must logically be). If we now observe that H_1 occurs (the first toss is a head), the expert becomes unwilling to give a precise probability for H_2, since $P(H_2 \mid H_1) = \frac{P(H_1, H_2)}{P(H_1)}$ may have any value between 0 and 1. Thus, learning the outcome of the first toss changes her belief about the probability of H_2 from a precise 0.5 to complete imprecision. Despite such unhelpful implications, Walley (1991; 1996) argues strongly that probabilities should be represented as intervals rather than precise values.

Lindley et al. (1979) model the relationship between the expert's 'true' probability P and her stated probability P^*. They assume a normally distributed error δ on the log-odds scale, that is,

$$\log\left(\frac{P^*}{1-P^*}\right) = \log\left(\frac{P}{1-P}\right) + \delta.$$

Lindley et al. were concerned with reconciling incoherent probabilities rather than explicitly considering imprecision in probability assessments, but, clearly, incoherence could be a direct consequence of imprecision. In a similar vein, Oakley and O'Hagan (2005) also allow for 'noise' in the expert's probabilities to account for the imprecision. For reasons of mathematical convenience, in their formulation $P^* = P + \varepsilon$, where ε is again a normal random variable with zero mean that represents the imprecision in the probability assessment. Oakley and O'Hagan assume a normal distribution for ε, though this is a somewhat arbitrary choice. The variance of this normal distribution can be chosen in consultation with the expert to capture the level of imprecision in her beliefs.

An elicitation scheme that does not specifically ask for precise probabilities is the 'trial roulette' approach introduced in Gore (1987). Here, the expert is given a certain number of gaming chips and asked to distribute them amongst the bins of a histogram. The proportion of chips allocated to a particular bin gives the expert's probability of the parameter lying in that bin, though clearly this probability is then subject to rounding error. A parametric density function can then be fitted to the resulting histogram. Increasing the width of the bins and decreasing the total number of chips will simplify the elicitation task for the expert, in terms of the level of precision required, at the expense of accuracy in representing the expert's 'true' beliefs.

One might also consider increasing the number of chips and the number of bins, to obtain more detailed assessments from the expert. In principle, with an infinite number of chips, we could obtain a precise distribution from her. Of course, this is unrealistic. Whereas the expert might be able to confidently assign 20 chips so that the probability in each bin is assessed to the nearest 0.05, she surely could not assign 200 chips so as to be confident of exactly how many to place in each bin – assessment of probabilities to the nearest 0.005 will rarely be feasible. There has been no empirical research on how precisely an expert can assess probabilities by such means. Ultimately, imprecision cannot be removed from elicitation. Asking for more summaries will have limited value because the expert will tire and become increasingly anchored to previous assessments. We have to acknowledge that an expert's probability distribution can never be elicited precisely in practice.

Some authors regard this difficulty in expressing probabilistic beliefs as grounds for abandoning the use of probability in describing epistemic uncertainty altogether and instead using non-probabilistic measures such as fuzzy membership functions or the Dempster–Shafer belief functions (see Helton and Oberkampf, 2004). O'Hagan and Oakley (2004) argue that probability is *uniquely* suitable for describing uncertainty (epistemic or aleatory), and so the issue of how to elicit it as reliably and precisely as possible needs further research. This is our view and the fundamental rationale for this book.

7.3 Incomplete information

Suppose a facilitator is eliciting an expert's probability distribution for an uncertain parameter θ. The expert states that she believes $P(-1.96 < \theta < 1.96) = 0.95$. The facilitator may then suppose that the expert believes that $\theta \sim N(0, 1)$, as this is consistent with her stated probability. This choice of distribution then implies that $P(-1.645 < \theta < 1.645) = 0.9$, but the expert has not stated this belief and may have a different value for this probability. It is clear that any distribution chosen to represent the expert's beliefs will express uncertainty about the parameter in more detail than the expert has provided.

Note that it is standard procedure in elicitation to use feedback; in the above example, the facilitator might feed back the statement $P(-1.645 < \theta < 1.645) = 0.9$ to the expert to see if she agrees with the stated probability of 0.9. If her probability is substantially different from 0.9, then she will reject this value and provide her own probability. This will then inform the facilitator that a normal distribution is not an appropriate distribution for representing the expert's beliefs (assuming the expert does not revise her earlier assessments). However, given the earlier discussion of imprecision in probability assessments, it is likely that if the expert did accept the statement $P(-1.645 < \theta < 1.645) = 0.9$, she would have also accepted alternative values in the region of 0.9, corresponding to alternative distributional assumptions.

The implication of this is that the facilitator will continue to have uncertainty about the expert's density function after the elicitation has been conducted (with the exception of some trivial cases, for example, where the expert chooses to provide a discrete distribution). In theory, the more probability judgements an expert makes, the less the uncertainty about her distribution. Unfortunately, increasing the number of judgements typically increases the difficulty of the elicitation task for the expert. Hora et al. (1992) describe a study involving the bisection method. In this approach, the expert first divides the sample space for the uncertain parameter into two equally probable intervals. The expert then subdivides each interval into two further equally probable intervals (which has the effect of stating the inter-quartile range). Hora et al. reported that the experts found this subdivision process particularly difficult.

Two schools of thought have emerged in the literature for dealing with this issue. In the Bayes Linear approach, the attempt to find a complete probability distribution is abandoned in favour of a partial specification of the expert's beliefs. Bayes linear methods are specifically concerned with how to update beliefs after observing new data. The expert is asked to provide prior means and covariances for all the relevant variables of interest (uncertain parameters and observational data). Once the data have been observed, adjusted means and covariances for the uncertain parameters are then derived. The adjusted mean can then be interpreted as a posterior estimate of the uncertain parameter, with the adjusted variance giving the expected mean square error of this estimate. This limited specification of uncertainty can be sufficient for some practical problems, and examples of applications are given in Craig et al. (1996) and O'Hagan et al. (1992). However, the elicitation itself is more demanding on the expert, as the expert has to provide moments from her distribution. Peterson and Miller (1964) reported inaccuracies in subjects' assessments of the means of highly skewed distributions, and Hofstatter (1939), Lathrop (1967) and Beach and Scopp (1967) give evidence of difficulties in assessing variances.

A second approach is to quantify formally the uncertainty about the expert's distribution following the elicitation. This in effect treats elicitation as any other exercise in Bayesian inference. The facilitator first considers his prior beliefs about the expert's density function. During the elicitation, the expert provides the facilitator with 'data' by providing various summaries from her distribution. The facilitator then updates his beliefs about the expert's distribution given this data and is then able to consider his remaining (posterior) uncertainty about the expert's distribution. Note that this is distinct from the problem of a decision maker updating his *own* beliefs about an unknown parameter given probability judgements from an expert (see Section 5.4).

Berger and O'Hagan (1988) consider the case when they believe that f is unimodal. Given some quantiles from the expert's distributions, they are then able to derive minimum and maximum posterior probabilities given data, by considering all unimodal prior distributions that have the specified quantiles. This approach can be somewhat conservative in that 'unrealistic' prior distributions are permitted; there will be some unimodal distributions with the specified quantiles which the expert would not recognise as representing her beliefs appropriately.

Oakley and O'Hagan (2005) make the assumption that $f(\theta)$ is a smooth function. In particular, they suppose that a priori $f(\theta)$ may be similar to some parametric density function $g(\theta)$ (e.g., a normal density), but that $f(\theta)$ may deviate from $g(\theta)$ in some arbitrary manner. In particular, the ratio $f(\theta)/g(\theta)$ is described by a stationary Gaussian process. The belief of smoothness is implemented in a correlation function which states that if θ is close to θ' then $f(\theta)/g(\theta)$ should be highly correlated with $f(\theta')/g(\theta')$. Following the elicitation, the facilitator can then derive his posterior beliefs about the expert's density function $f(\theta)$. For example, Figure 7.1 shows the facilitator's posterior distribution for the expert's $f(\theta)$ after eliciting the four probabilities shown in Figure 5.1. The solid line is the expert's posterior mean of $f(\theta)$, which could be used as a 'best' fitted distribution, while the dotted lines give 99% pointwise credible intervals for $f(\theta)$. In this case, the good fit of the

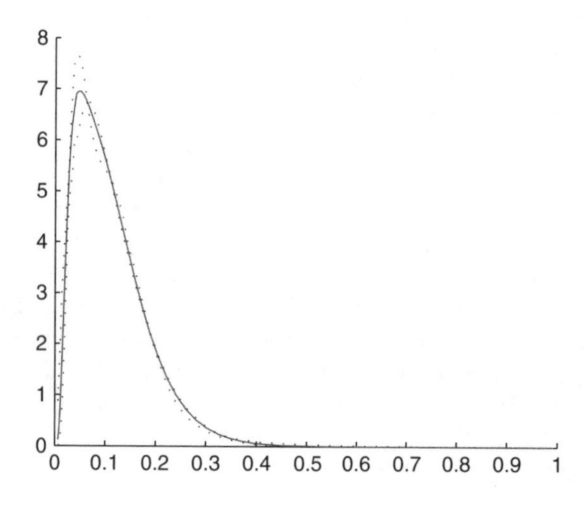

Figure 7.1: The facilitator's posterior mean (solid line) and pointwise 99% intervals for the expert's density function (dotted lines) given the probabilities shown in Figure 5.1.

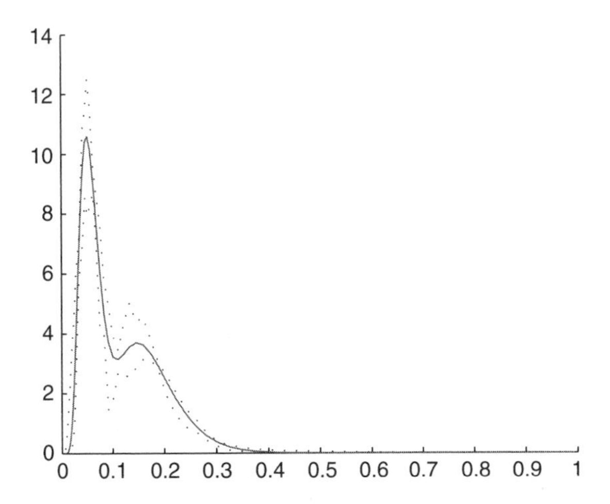

Figure 7.2: The facilitator's posterior mean (solid line) and pointwise 99% intervals for the expert's density function (dotted lines) when the expert's stated probabilities are less consistent with a simple beta distribution than those used in Figure 7.1.

expert's probabilities to a simple beta distribution results in negligible uncertainty about her distribution.

Figure 7.2 shows the posterior distribution in a case where the expert's stated probabilities were less consistent with a beta distribution.

In both Berger and O'Hagan (1988) and Oakley and O'Hagan (2005), the methods given are non-parametric in that the expert's beliefs are not forced to fit any particular family of parametric density functions. However, the main advantage in both approaches is the ability to consider the *uncertainty* about the expert's distribution. In particular, a sensitivity analysis can be conducted to investigate the consequences of this uncertainty, as described in Section 5.4.2. Additionally, the probabilistic formulation of Oakley and O'Hagan (2005) allows a probabilistic sensitivity analysis to be undertaken; thus rather than simply obtaining bounds on posterior inferences as in Berger and O'Hagan (1988), a probability distribution of posterior inferences is obtained, representing the facilitator's posterior uncertainty.

In most cases, it will be possible to find a parametric family with sufficient flexibility to give an acceptable fit to the elicited probability judgements. A non-parametric approach may improve the fit, but possibly only at a level of accuracy too precise for the expert to contemplate. This is illustrated in an example in O'Hagan and Oakley (2004). An initial set of elicited probabilities suggests a bi-modal density function, so that no standard parametric distribution can fit the elicited probabilities exactly, and the non-parametric method of Oakley and O'Hagan (2005) is used to provide a posterior distribution for the facilitator, treating the expert's assessments as exact.

7.4 Summary

In this chapter, we have discussed why one should exercise caution in interpreting any elicited probability distribution as a perfect representation of an expert's uncertainty. There cannot be a fully probabilistic solution to the problem of imprecision in probability assessments, as the notion of an imprecise probability itself is in violation of an axiom of subjective probability. Consequently, any suggestion for dealing with this problem is likely to attract some controversy. The technique of feedback for addressing the uncertainty in the expert's complete density function is more established, though not ideal because any feedback probability judgements are likely to influence the expert. We believe that more research is needed in both these areas.

In the context of a decision problem, uncertainty about the expert's 'true' distribution will not always be important; the optimal decision may be robust to small changes in the adopted probability distribution. Nevertheless, we argue that it is important to investigate the consequences of varying elicited distributions on any subsequent decisions whenever possible.

7.5 Conclusions

7.5.1 Elicitation practice

- The accuracy of any fitted distribution as a representation of the expert's opinion is compromised by (a) imprecision in the expert's stated summaries and (b) the fact that only a limited number of summaries can be elicited in practice.

- While there is as yet no consensus on how to express the implied uncertainty in the elicited distribution (e.g., probabilistically or using upper and lower bounds), it is important to acknowledge it explicitly when asserting a fitted distribution.

7.5.2 Research questions

- Much more research is needed into formal expressions of imprecision in elicited probability distributions.

- To what extent is it possible to determine a limit on the achievable accuracy of any elicited distribution? Is it lower when eliciting joint distributions than marginal distributions?

Chapter 8

Evaluating Elicitation

8.1 Introduction

8.1.1 Good elicitation

This chapter addresses the fundamental question of how we evaluate the quality of a set of probabilities or probability distributions elicited from an expert. How successful has the elicitation been? The aim of the elicitation is to formulate the expert's knowledge and opinions faithfully in probabilistic form, so any evaluation of the elicitation should measure the extent to which this aim has been met.

Unfortunately, it is basically impossible to measure this because we do not know what the expert's 'true' underlying beliefs are, except as revealed by the elicitation.

8.1.2 Inaccurate knowledge

Calibration has been mentioned several times in previous chapters, and it has clearly been widely used in psychology experiments to assess the accuracy of elicitation methods and the presence of biases. However, comparing the elicited probabilities or distributions with the true values of the events or variables in question does not evaluate the success of elicitation; it does not compare the elicited probabilities with the expert's underlying opinions. Calibration measures the agreement between the expert's stated probabilities and reality.

Failure of an expert to be calibrated could have two quite different causes. She might hold beliefs that are actually well calibrated to reality, but fails to express those beliefs accurately, in which case mis-calibration signifies *poor elicitation*. However, she might have expressed her beliefs accurately in probabilistic form but

Uncertain Judgements – Eliciting Experts' Probabilities A. O'Hagan, C. E. Buck, A. Daneshkhah, J. R. Eiser, P. H. Garthwaite, D. J. Jenkinson, J. E. Oakley and T. Rakow

those stated probabilities fail to calibrate because her beliefs correspond poorly to reality. In this case, the failure is due to *inaccurate knowledge*.

If an expert makes perfect use of her knowledge, including making due allowance for any biases that may affect that knowledge (such as selective memory), then she should be well calibrated. But where an expert's probabilities are not well calibrated, then poor elicitation, inaccurate knowledge and poor appreciation of her knowledge may all be present, and it is not possible to say which is more responsible for mis-calibration. Hogarth (1975) notes that people are imperfect processors of information and proposes that "... assessment techniques should be designed both to be compatible with man's abilities and to counteract his deficiencies" (Hogarth, 1975, p. 284). This suggests that in designing an elicitation method, one aim should be to achieve good calibration.

8.1.3 Automatic calibration

In a trivial sense, we can show that experts are always well calibrated. Consider the set of events to which an expert gives probability $\frac{1}{2}$. Either the target event to which a probability (of 0.5) was formally assigned or its converse will occur (and not both). Therefore, one can create an enlarged class of events consisting of these target events and their converse events to which probability 0.5 has been given (explicitly in the case of the target events or implicitly in the case of the converse events). In this enlarged set, exactly half of the events will occur. The expert is therefore automatically calibrated at the point $p = \frac{1}{2}$. Now consider the events to which probability $\frac{1}{3}$ is given. To each such event, we can add two more events, that the converse occurs and a toss of a fair coin comes down heads and that the converse occurs and the coin comes down tails. These events also implicitly have probabilities $\frac{1}{3}$, and in the larger set of events exactly one-third will occur, so calibration is automatic also at $p = \frac{1}{3}$. The argument can be extended to show calibration arising automatically at every possible value of p.

The important implication of this highly artificial demonstration is that it only makes sense to check for calibration over subsets of events. If we find that experts' 50% credible intervals calibrate badly, with perhaps the true value lying inside these intervals 70% of the time in a particular experiment, this is meaningful. It is true that by combining these with the events of the true value falling outside the stated 50% intervals, we would get automatic calibration, but the two subsets of events are different. The inference that, in the experiment, the experts' 50% intervals were too wide is a useful conclusion (and the complementary finding that the true value fell outside the intervals too rarely is so obvious as to be barely worth mentioning).

The choice of appropriate sets of events over which to check for calibration can help with separating poor elicitation from inaccurate knowledge. If the events are characterised by being based on a common area of knowledge, then poor calibration may be due to the expert's knowledge being deficient, or perhaps misleading. If, however, the events are characterised by a common task or form of reasoning, then we may be able to attribute poor calibration to poor elicitation, in the sense that the expert is conceptualising the task badly or processing knowledge badly.

8.1.4 Lessons of the psychological literature

If we now refer back to the extensive psychological literature in which the success of elicitation is judged by calibration, to what extent can we infer poor elicitation from poor calibration? Ideally, each individual research finding should be critically assessed in this light, but some general observations may be in order.

- Where experts show consistent biases in performing some task, across a range of knowledge areas, this indicates poor elicitation. Psychologists have described the heuristics that they believe to be responsible for these biases – such as anchoring and adjustment, representativeness or availability – in terms of tendencies to process information quickly, often without using all the available information. These psychological theories therefore support the interpretation of such experiments as indicating poor elicitation, as opposed to inaccurate knowledge.
- Where one formulation of a task gives better calibration performance than another, again across a range of knowledge areas, this suggests that more accurate elicitation is achieved with the better-performing method.
- Where a consistent bias is found across a range of experts, poor elicitation ought to be considered as a possible source of bias. Sometimes experts may hold beliefs that mismatch reality in the same way – however, it is unwise to assume that this is always the source of a bias.

It is through the testing of a range of experts across a range of knowledge areas that the empirical literature provides useful lessons about how to achieve elicited probabilities that accurately reflect the expert's beliefs.

8.1.5 Outline of this chapter

To what extent can we use this kind of reasoning to judge the success of a particular elicitation or series of elicitations from a single expert? In Section 8.2, we develop the use of proper scoring rules to compare the expert's performance against reality, bearing in mind that poor scores will inevitably confound poor elicitation with inaccurate knowledge. Section 8.3 considers how lack of coherence may indicate poor elicitation and examines the roles of feedback and overfitting.

8.2 Scoring rules

In empirical work, probability distributions may be elicited for uncertain quantities whose actual values are known to the experimenters. In other circumstances, such as weather forecasting, predictive distributions may be assessed for quantities whose values become known subsequently. In both cases, it can be useful to compare the assessed probability distributions with the observed data to provide an objective measure of their accuracy. This is the purpose of a scoring rule.

Formally, a scoring rule is a formula for awarding a score to the expert, which can be thought of as a reward. It is a function which measures, in some sense, the 'degree of association' between the expert's elicited probability distribution for some unknown event and the true value of that quantity. Various scoring rules have been devised and the different rules measure the 'degree of association' in different senses, so that they do not always agree as to which of two distributions has the greater degree of association with the relevant quantity's true value. However, the objective accuracy of different assessors, or various assessment methods, should not be based on the outcome of a single event since, for example, an event which an assessor considers unlikely should occasionally occur. Over a set of trials, the scoring rules that are commonly used are reasonably consistent in the order in which they rank assessors, or assessment methods (Winkler and Murphy, 1968; Staël von Holstein, 1970a,b).

As well as being useful for comparing assessors or assessment methods, the other purposes of a scoring rule are to provide an incentive for experts to record their opinions well and to help train experts to quantify their opinions accurately. To this end, it is important that a scoring rule should encourage experts to record their true beliefs. More precisely, "The scoring rule is constructed according to the basic idea that the resulting device should oblige each participant to express his true feelings, because any departure from his own personal probability results in a diminution of his own average score as he sees it." (de Finetti, 1962, p. 359). A scoring rule with this property is termed *proper*.

There are scoring rules which intuitively appear reasonable but which are not proper. As an example, suppose an event has two possible outcomes, E_1 and E_2, exactly one of which will happen. Also suppose that an expert assesses the probability that E_1 will happen as p_1 and the probability that E_2 will happen as p_2 (where $p_2 = 1 - p_1$). Then an apparently sensible scoring rule gives the expert a score of p_1 if E_1 occurs and p_2 if E_2 occurs; that is, the score is the probability assigned to the actual outcome. This scoring rule is called the *linear scoring rule*. To see that this is not a proper scoring rule, suppose 0.7 is the expert's true probability that E_1 occurs and 0.3 is her probability for E_2. If she states these probabilities, then her expectation is that 70% of the time she will score 0.7 and 30% of the time she will score 0.3. Hence, her expectation of her score is $0.7(0.7) + 0.3(0.3) = 0.58$. If, rather than reporting her true probabilities, she instead says that her probability that E_1 occurs is 1 and her probability that E_2 occurs is 0, then her expectation is that 70% of the time she will score 1 and 30% of the time she will score 0, so her expectation of the score is $0.7(1) + 0.3(0) = 0.7$. This is higher than when she reports her true beliefs, illustrating that the scoring rule is not proper.

Scoring rules have been extensively used in experiments in a broad variety of areas, such as weather forecasting (Murphy and Winkler, 1984), medicine (Rakow et al., 2005), academic testing (Echternacht, 1972), the engineering of intelligent computer systems (Druzdzel and van der Gaag, 1995), economics experiments (Braun and Yaniv, 1992), risk analysis (De Wispelare et al., 1995) and stock market forecasting (Staël von Holstein, 1972). They have also been used in numerous experiments in which students quantified their opinions about general knowledge questions (e.g., Lichtenstein and Fischhoff, 1980). In most of this work, scoring rules have been used to compare assessment methods or different groups of assessors. However, scores can also be given as feedback to experts in order to motivate them to perform well and to help them improve their assessments. Results concerning their use as feedback have been mixed. For example, in one of the classic 'book-bags and poker-chips' studies, Phillips and Edwards (1966) found that subjects who received scoring-rule feedback gave better forecasts than subjects who received no feedback. Fischer (1982) found that the promise of incentive pay based on scores had a positive effect on judgement accuracy, but trial-by-trial scoring-rule feedback did not. In an experiment involving stock market forecasting (Staël von Holstein, 1972), subjects received scoring-rule feedback to try to help them improve the accuracy of their predictions of stock price movements, but there was no discernible change, perhaps because the prediction task is hard. Similarly though, Echternacht (1972) reviewed the use of scoring rules in academic testing and concluded that there is little evidence of their effectiveness. In interpreting these findings, it is worth remembering that poor scores may be due, in part, to inaccurate knowledge.

Scores have also been used as feed-forward in order to provide another aid for assessors. An elicitation method, ELI (van Lenthe, 1993a,b) asks assessors to draw probability density functions using graphical interactive computer software. The corresponding scores that would result from possible actual values of the quantity of interest are given as feed-forward to assessors, who can repeatedly change their probability density functions if they are unhappy with the implied distribution of scores. In experiments, this appeared to improve both the calibration and the objective accuracy of probability judgements (van Lenthe, 1994).

8.2.1 Scoring rules for discrete probability distributions

The development of scoring rules has focused on assigning scores to assessed probability distributions that are discrete, rather than continuous, and we first consider this case. In many situations, of course, the quantities of interest are discrete, or discrete approximations can be used. To introduce some notation, suppose an event E takes exactly one of the n outcomes, $O_1, \ldots, O_n$. Let p_i be the probability that the expert states she attaches to $O_i (i = 1, \ldots, n)$. It is required that the p_i satisfies the usual laws of probability, so each p_i must be non-negative and $\sum p_i = 1$. Let $d_i = 1$ if O_i occurs and 0 if O_i does not occur, so that exactly one of $d_1, \ldots, d_n$ is non-zero. As scoring rules are concerned with the degree of association between assessments and the outcome that happens, scores are functions of $p_1, \ldots, p_n$ and $d_1, \ldots, d_n$.

The most common scoring rule is the *probability scoring rule*, which was introduced by Brier (1950) to measure the objective accuracy of weather forecasters. It is also referred to as the *quadratic scoring rule* and, in the context of meteorology, as the Brier Score. The probability score (*PS*) is defined by

$$PS = \sum_{j=1}^{n} (p_j - d_j)^2.$$

Notice that, like other scoring rules that we shall see later, PS measures the discrepancy between assessments and reality, and so should be viewed as a penalty rather than a reward to the expert. High scores are bad and low scores are good.

Useful comparisons between sets of probability judgements can be made by calculating the *mean probability score* ($\overline{PS}$), simply the arithmetic mean of the individual probability scores. A score of zero represents perfect accuracy: faultless prediction of the outcome of every event. A score of 1 represents counter-perfect accuracy. Comparison is often made with the performance of a 'uniform judge', who always states the same probability over a set of probability judgements. For instance, a judge who assessed each outcome of a set of dichotomous events to be 0.5 would obtain a mean *PS* of 0.25.

Various schemes have been proposed which decompose measures of overall accuracy into meaningful sub-components of accuracy. One purpose of such decomposition is to try to explain how and why judges achieve the levels of accuracy that their probability scores indicate. Also, scores can be given as feedback to help train assessors, but the information given by probability scores may be too coarse to inform an expert precisely how to improve her judgements (Yates, 1994). This may underlie the mixed results (reported earlier) when scores have been given as feedback.

The most widely used decomposition of the mean probability score is the Murphy decomposition (Murphy, 1973). It is assumed that there is a set of events $E_1, \ldots, E_m$, each of which has the same set of possible outcomes $O_1, \ldots, O_n$. For example, E_i could be the ith patient who goes to his doctor because of persistent stomach ache, and $O_1, \ldots, O_n$ could be the illnesses that the patient might have (assuming he or she has only one of these illnesses). Each patient has the same set of possible outcomes. Murphy's decomposition partitions the mean probability score into three components:

$$\overline{PS} = \sum_{j=1}^{n} \text{var}(d_j) + \sum_{j=1}^{n} \text{calibration}(j) - \sum_{j=1}^{n} \text{resolution}(j), \tag{8.1}$$

where $\text{var}(d_j)$, calibration (j) and resolution(j) are defined for the jth outcome as follows.

$\text{Var}(d_j)$: Let $\overline{d}_j$ be the proportion of events for which outcome j occurred (e.g., the proportion of patients with illness j). Then $\text{var}(d_j)$ is set equal to $\overline{d}_j(1 - \overline{d}_j)$

(If d_j is a zero-one indicator that denotes whether outcome j occurred, then $\text{var}(d_j)$ is the maximum likelihood estimate of the variance of d_j).

Calibration(j): This reflects the calibration of the probability assessments for outcome j. The probability assessments for this outcome are partitioned into categories according to their value. For example, the first category might contain probability assessments for the outcome that are in the range 0 to 0.1; the next category might contain those in the range 0.1 to 0.2; ...; the last category, those in the range 0.9 to 1. We suppose that there are T categories. Restricting our attention to outcome j, let $\overline{f}_j^k$ denote the average of the probability assessments in the kth category and let n_j^k denote the number of times the assessments for this outcome fell in the kth category. Also let $\overline{d}_j^k$ be the proportion of times that outcome j occurred when the probability assessment for this outcome was in the kth category. For good calibration, $\overline{f}_j^k$ should be similar to $\overline{d}_j^k$. To reflect this, calibration(j) is defined by

$$\text{calibration}(j) = \frac{1}{n}\sum_{k=1}^{T} n_j^k(\overline{f}_j^k - \overline{d}_j^k)^2.$$

Resolution(j): Resolution is defined by

$$\text{resolution}(j) = \frac{1}{n}\sum_{k=1}^{T} n_j^k(\overline{d}_j^k - \overline{d}_j)^2$$

where n, n_j^k, $\overline{d}_j^k$ and $\overline{d_j}$ have been defined above.

The first term in (8.1), $\text{var}(d_j)$, is determined purely by the outcomes and not by the probability assessments. The second term, calibration(j), indicates how well the probability assessments related to reality for outcome j. The third term, resolution(j), differs from the first two terms in that a larger value improves $\overline{PS}$. It indicates whether the categories discriminate between the occurrence/non-occurrence of outcome j, that is, is the occurrence of outcome j associated with the probability assessment of its occurrence? Note that while good probability assessments imply high resolution, the converse does not hold. For instance, suppose it always rains when a particular meteorologist gives a high probability that it will be dry, and it is always dry when the meteorologist gives a high probability that it will rain. Then the meteorologist has poor calibration but good resolution. (Indeed, the meteorologist's forecasts would be useful, as long as you are aware that the likely occurrence is the converse of what is predicted.) Further helpful discussions of Murphy's decomposition may be found in Yates (1990), Benson and Whitcomb (1993), and Wright and Ayton (1992). DeGroot and Fienberg (1983) discuss the similar idea of the *refinement* of forecasters.

The extent to which Murphy's decomposition is useful is unclear. Benson and Whitcomb (1993) used the decomposition to analyse subjects' assessments in finer detail, in the hope of identifying differences between three alternative elicitation methods. However, the decomposition did not reveal any large differences. Similar negative results were obtained by Wright and Ayton (1992) when they compared different elicitation methods. In contrast, Browne et al. (1999) found that both resolution and calibration varied with task difficulty. Also, calibration, which is a component of the Murphy decomposition, has often been used as feedback to help train assessors, one aim being to temper bias caused by overconfidence. (This work has been reviewed in Section 4.2.)

Murphy's is not the only decomposition that has been proposed for the mean probability score. Stewart and Lusk (1994) gave a decomposition that is based on the lens model paradigm. It gives seven components and was used to provide a framework for research on the forecasting process. Yates (1994) proposed another decomposition of $\overline{PS}$ into three components: bias, slope and scatter. Bias is a measure of the tendency of probability judgements to be higher or lower than the proportion of times the target event actually occurs (i.e., the sample base rate). The slope reflects a combination of calibration and discrimination, while scatter refers to the variability (i.e., random error) in a person's judgements. Yates' decomposition has been used quite frequently (Yates et al., 1991; Benson and Whitcomb, 1993; Whitecotton et al., 1998; Browne et al., 1999; Stone and Opel, 2000).

The logarithmic and spherical scoring rules are other proper scoring rules that have been proposed. As before, let p_i denote the assessed probability that the outcome will be O_i and let $d_i = 1$ if O_i occurs and $d_i = 0$ if O_i does not occur ($i = 1, \ldots, n$). Then the logarithmic score is

$$LS = \ln\left(\sum_{i=1}^{n} p_i d_i\right)$$

and the spherical score is

$$SS = \sum_{i=1}^{n} p_i d_i \Big/ \left(\sum_{i=1}^{n} p_i^2\right)^{1/2}.$$

The spherical scoring rule has been little used but empirical work has compared the probability scoring rule with the logarithmic scoring rule. In experiments involving probability revision, such as book-bags and poker-chips experiments, the logarithmic rule seems more effective (Phillips and Edwards, 1966; Schum et al., 1967). In other contexts, though, results have typically been inconclusive (Murphy and Winkler, 1970; Jensen and Peterson, 1973). Meteorologists, who are the group that makes the most use of scoring rules, favour the probability scoring rule, suggesting that it may be the preferable scoring rule for judgements related to prediction.

A criticism of all the above scoring rules is that they do not always value assessments that are nearly right more than assessments that are far from right. For example, suppose the range of possible values that an unknown quantity will take has been partitioned into four intervals and that two assessments assign the following probabilities to these intervals.

First assessment: (0.5, 0.3, 0.1, 0.1)
Second assessment: (0.1, 0.3, 0.5, 0.1).

If the true value of the quantity actually falls in the fourth interval, then the second assessment might well be regarded as better than the first one, but both assessments would obtain identical scores when evaluated under any of the scoring rules described so far. One rule that will assign the two assessments different scores is known as the *ranked probability score* (Epstein, 1969; Murphy, 1970, 1971; Staël von Holstein, 1970a). Let $p_1, \ldots, p_n$ be the assessed probabilities for n ordered categories. If the correct category is category j, then the ranked probability score is

$$\frac{3}{2} - \frac{1}{2(n-1)} \sum_{i=1}^{n-1} \left\{ \left(\sum_{k=1}^{i} p_k \right)^2 + \left(\sum_{k=i+1}^{n} p_k \right)^2 \right\} - \frac{1}{n-1} \sum_{i=1}^{n} |i-j| p_i .$$

The scoring rule is proper and has merit but it has been little used.

8.2.2 Scoring rules for continuous probability distributions

In this section, we consider scoring rules for evaluating the accuracy of a subjective *continuous* probability distribution. For example, an expert might assess the time a specific operation will take, the average change in blood pressure that will be observed in a particular drug trial, or the future change in the weight of a person who has recently developed diabetes. We let θ denote the quantity of interest and $f(\theta)$ denote the expert's subjective probability density function for the value that θ will take. For a scoring rule to be used, the value of θ must be known to the facilitator, or become known, and we let θ^* denote its actual value. Scoring rules have been used far less frequently with continuous distributions than with discrete distributions, so our discussion is consequently briefer.

Scoring rules for discrete distributions can be extended to the continuous case by limiting arguments, where the number of categories is increased indefinitely and the size of each category is shrunk towards 0. For continuous distributions, the analogs of the linear, quadratic (probability), logarithmic, spherical and ranked probability scoring rules are, respectively,

$$\text{linear score} = f(\theta^*),$$

$$\text{quadratic score} = 2f(\theta^*) - \int_{-\infty}^{\infty} \{f(\theta)\}^2 \, d\theta, \tag{8.2}$$

$$\text{logarithmic score} = \ln[f(\theta^*)], \tag{8.3}$$

$$\text{spherical score} = f(\theta^*) \Big/ \left(\int_{-\infty}^{\infty} \{f(\theta)\}^2 \, d\theta \right)^{1/2}, \tag{8.4}$$

and

$$\text{ranked probability score} = \int_{-\infty}^{\theta^*} \{F(\theta)\}^2 \, d\theta + \int_{\theta^*}^{\infty} \{1 - F(\theta)\}^2 \, d\theta$$

(see Buehler, 1971; Staël von Holstein, 1977; van Lenthe, 1994), where $F(\theta)$ denotes the expert's cumulative distribution function. All these scoring rules are proper, apart from the linear scoring rule. Matheson and Winkler (1976) develop a rich class of scoring rules for continuous distributions that includes the quadratic, logarithmic and spherical scoring rules given in equations (8.2), (8.3) and (8.4).

The logarithmic scoring rule has the attractive property that it is simple to use, in that the score attached to a probability density function depends only on the value of the function at θ^*. Bernardo (1979) shows that, among proper scoring rules, only the logarithmic score, or a linear transformation of this score, has this property. The property is useful when $f(\theta)$ is not a simple mathematical expression, while an estimate of $f(\theta)$ is available at each iteration of a Markov chain Monte Carlo sampler. Then $f(\theta^*)$ can be evaluated at each iteration and the logarithm of their average is the logarithmic score. As an example, suppose μ and σ^2 have an intractable joint distribution and $\theta \sim N(\mu, \sigma^2)$, so that $f(\theta)$ has no closed form. If each iteration of a Markov chain gives an estimate of μ and σ^2, then $f(\theta^*)$ can be evaluated at each iteration and the score is obtained by determining the logarithm of their average.

As mentioned earlier, van Lenthe (1993a,b) developed an elicitation method that uses feed-forward to help assessors. An expert assesses a probability distribution for θ and the elicitation method calculates the logarithmic scores that would result from each of a large range of potential values of θ. The scores are given as feed-forward to the expert who is then given the option of revising (or further revising) her assessed probability distribution until she feels that she has optimised her distribution of potential scores. The logarithmic scoring rule gives very low scores if $f(\theta^*)$ is small, but van Lenthe (1994) found that this feature did not have any special effect on experts' assessed distributions, in that there were no clear differences from using the logarithmic scoring rule, rather than other proper scoring rules whose scores have a bounded range. Specifically, he found little to choose between logarithmic, quadratic and spherical scores as a feed-forward device. However, he also used the (improper) linear scoring rule for this purpose and it led to subjective distributions that were more poorly calibrated. Garthwaite (1994) and Garthwaite and O'Hagan (2000) used logarithmic scores to compare different ways of implementing an elicitation method.

One aim of a proper scoring rule is to encourage honesty. Under such a rule, an expert's optimal strategy is to give an assessed distribution that corresponds exactly to her true beliefs. However, one question that arises is, *What if the expert is not allowed to report her true beliefs?* For example, suppose that the expert's opinions are to be modelled by a normal distribution but her opinions do not correspond to any normal distribution. In such circumstances (which arise in practice), we might like a scoring rule that encourages the expert to report a distribution that is as close as possible to her real subjective distribution. Friedman (1983) argues that a scoring rule should not only be proper but also be *effective*. A scoring rule has this property if, for some chosen metric, a person's expectation of her score is a strictly decreasing function (or increasing function if a low score is good) of the distance between her reported and actual distributions. Then, it is in an expert's interests to assess the permissible distribution that is as close as possible to her true subjective distribution. (The measure of 'closeness' depends upon the metric.) Friedman shows that the quadratic rule is effective with respect to the L_2 metric (root mean squared difference) and the spherical scoring rule is effective with respect to a 'normalised' L_2 metric. Any effective scoring rule is also proper, but the converse does not hold. The logarithmic scoring rule is an example of a rule that is proper but is not effective for any metric (Nau, 1985), which might be considered a severe criticism of that scoring rule.

Suppose that the reason for quantifying an expert's opinion is to form a prior distribution that will be combined with sample data to form a posterior distribution. Also, suppose that the posterior distribution will be used to estimate some quantity of interest. A relevant question is whether the assessed prior distribution is likely to be misleading or helpful; that is, should we expect the posterior distribution to yield a better estimate than the sample data on its own? Garthwaite (1992) supposes that inaccuracy in estimating the quantity is penalise by a quadratic loss function and that a sampling scheme has been fixed but the sample data have yet to be gathered. He shows that the preposterior expected loss gives a proper scoring rule and that, in a variety of situations, the posterior distribution is expected to have a smaller loss for any sampling scheme if $(\overline{\theta} - \theta)^2 < 2\text{var}(\theta)$, where θ is the quantity to be estimated, and $\overline{\theta}$ and $\text{var}(\theta)$ are the prior mean and prior variance of θ. A similar result can be found for linear models (Garthwaite, 1992), and in an experiment it was found that 75% of the assessed distributions led to posterior distributions that would be more useful, for any sample, than those obtained from a non-informative prior distribution (Garthwaite, 1994).

8.3 Coherence, feedback and overfitting

8.3.1 Coherence and calibration

A set of probability statements is *coherent* if they are collectively consistent with the laws of probability. (There are only three axioms of probability, but many laws follow from these axioms.) If a set of statements violate any of these laws, they are said to be incoherent although, if only minor adjustments to the assessed probabilities will make them coherent, they might be considered to be 'basically' coherent but imprecise (see Chapter 7). For example, suppose exactly one of the outcomes A, B or C will occur. If an expert's assessed probabilities are $P(A) = 0.5$, $P(B) = 0.3$ and $P(C) = 0.4$, then the set of probabilities are incoherent as they should sum to 1, while $P(A) + P(B) + P(C) = 1.2$. On the other hand, if $P(A) = 0.45$, $P(B) = 0.275$ and $P(C) = 0.3$, they are imprecise rather than completely incoherent, as $P(A) + P(B) + P(C) = 1.025$.

Should a set of probability assessments be coherent? On the one hand, if the incoherence is blatant, then the expert's opinion seems to be poorly represented. For example, if an expert assesses the probability of outcome A as 1.4, or any other value greater than 1, then the assessment does not represent her opinion, assuming she appreciates that probabilities must lie between 0 and 1. Similarly, if an expert assesses the probability that an operation will be successful as 0.8 and the probability that it will not be successful as 0.3, then again the inconsistency in these assessments should be apparent to the expert, so that they cannot represent her opinions. On the other hand, it is often far less obvious whether the laws of probability have been violated. For instance, if the correlation between A and B is denoted by ρ_1 and those between A and C and between B and C are denoted by ρ_2 and ρ_3, then $\{\rho_1 = 0.8, \rho_2 = 0.8, \rho_3 = 0.29\}$ is a coherent set of assessments, while $\{\rho_1 = 0.8, \rho_2 = 0.8, \rho_3 = 0.27\}$ is not. (In the latter case the correlation matrix is not positive-definite.) With a complex set of assessments, an expert may well have no 'gut feeling' for how they should interrelate and then it seems plausible that her assessments might be incoherent but still a good reflection of her opinions. An expert might argue, "You tell me that my assessments are not consistent, but each of them seems OK to me, and I do not know how I should change them to make them consistent." In a similar spirit, Todd and Gigerenzer (2000, p. 737) write, "We do not compare human judgement with the laws of logic or probability, but rather examine how it fares in real-world environments. The function of heuristics is not to be coherent. Rather, their function is to make reasonable, adaptive inferences about the real social and physical world given limited time and knowledge."

Alternative points of view are tenable. As noted in Chapter 7, Lindley et al. (1979) suppose that "the subject has, in some sense, a set of coherent probabilities that are distorted in the elicitation process." The work of Savage (1954), DeGroot (1970) and others discussed in Section 1.5 demonstrate that a person wishing to make rational decisions under uncertainty must act as if they have a coherent set of probabilities, and these normative theories underlie the objectives of most elicitation. Thus, coherence may be considered a requirement for a set of probability assessments to accurately portray a person's opinions, and some measure of the degree of any incoherence could indicate how well an expert's opinion has been captured. At a more pragmatic level, if an expert's assessments are incoherent, then in important applications it will usually be necessary to adjust them so that they satisfy the laws of probability – often assessments will be transformed into a distribution that satisfies these laws. If the amount of adjustment is substantial, then either her original assessments were a poor reflection of her opinions, or the adjusted assessments are a poor reflection of her opinions, or both. Any of these possibilities is unwelcome, making coherence a desirable quality in a set of assessments.

Some empirical work has been conducted that examines whether coherence relates to objective accuracy, measuring the latter by calibration. One hypothesis is that the same heuristics may underlie findings of both incoherences and mis-calibration, in which case a clear correlation between coherence and calibration measures would be predicted (Bolger and Wright, 1993). Results have been disappointing. Wright et al. (1988) asked untutored student subjects to assess the individual probabilities of pairs of events that were similar and related, such as 'the pound sterling will be below one US dollar in February 1985' and 'the pound sterling will be below one US dollar in March 1985'. Subjects also assessed the intersection, disjunction and union of each pair of events. No evidence was found of a relationship between a coherence measure that Wright et al. constructed and subsequent calibration. Wright and Ayton (1987) also conducted an experiment examining the empirical relationship between coherence and calibration, but they too found no evidence of the anticipated relationship. Wright et al. (1994) found slightly more positive results in an experiment where subjects predicted the outcomes of snooker matches in the 1989 World Championship; there was some indication of a relationship between coherence and calibration, but with no clear pattern. They concluded that their results still suggested "the absence of a systematic general relationship between degree of (in)coherence and degree of subsequent (mis)calibration in the present, more ecologically valid, forecasting situation" (Wright et al., 1994, p. 21).

8.3.2 Feedback and overfitting

Feedback comes in various forms. To help train assessors to be better calibrated, a common approach is to elicit the probability of an event and then give the assessor feedback by telling her whether the event occurred (see Section 4.2) and perhaps give a score for her assessment. This is the most common form of feedback in the psychological literature. A different use of the term occurs when statisticians describe an elicitation method, where a common approach for assessing a probability distribution is as follows:

(i) Elicit sufficient assessments to determine the probability distribution.

(ii) Use this distribution to estimate other quantities that should correspond to the expert's opinion if the assessed probability distribution adequately represents her opinion.

(iii) Inform the expert about these estimated quantities (perhaps using interactive graphics) and ask if they represent her opinion.

(iv) If they do not represent her opinion, ask her either to give estimates of the quantities that do or to revise some of her earlier assessments.

Steps (ii) to (iv) are repeated until a probability distribution is determined that seems to represent the expert's opinion adequately. In this context, the estimated quantities told to the expert in step (iii) are referred to as feedback.

If the accuracy of an assessed probability distribution is to be judged by the closeness of the distribution to the expert's actual opinion, then feedback to the expert is the most natural way of evaluating the distribution – the expert is in the best position to judge whether something corresponds to her opinion. Also, she can say which of her opinions are least clear cut and hence can be adjusted without contradicting her opinions. If the expert cannot refine her assessments in such a way that she is comfortable with the feedback values, then a more flexible probability model is needed to represent her opinions (assuming her opinions are sufficiently coherent that they can be represented by a probability model).

A problem with feedback is that an expert may be too willing to accept that a proposed value is representative of her opinion, since this is the easiest option and avoids having to revise the value. Moreover, if she revises it, then the proposed value may act as an anchor and lead to her adjustment being insufficient. As noted in Section 5.4.3, overfitting is an alternative that avoids these issues. The expert is asked to make more assessments than are necessary for her subjective probability distribution to be estimated. Then a distribution is fitted to these assessments through some form of reconciliation. The elicitation procedure may stop there but, ideally, the differences (residuals) between her assessments and those given by the fitted distribution would be calculated and fed back to the expert, with large residuals highlighted. The expert is then asked to modify assessments and a probability distribution is refitted. This has the important benefit that large assessment errors are noticed and corrected. (Experts sometimes make comments like, "When I made that assessment I wasn't thinking properly.") The process is repeated until a distribution is found that is close to all her revised assessments or it is decided that a broader class of probability distributions is needed. The latter rarely happens in practice, no doubt because typically a range of probability distributions will adequately represent an expert's opinions, but also because widening the scope of an elicitation method is usually a time-consuming task, especially if the method is implemented in an interactive computer program, which is almost essential if the cycle 'revise assessments/refit probability distribution/give feedback' is to be viable.

Various forms of feedback have been used in practice. Kadane et al. (1980) develop an elicitation method for a normal linear model, $y = \beta'\mathbf{x} + \epsilon$, and elicit medians, conditional medians and quantiles of the predictive distribution of y at more $\mathbf{x}$-values than are necessary to estimate the parameters of the prior distribution (cf. Section 6.7). The method is implemented in an interactive computer program. After fitting a prior distribution to the assessments, the computer calculates the median of y at each $\mathbf{x}$-value where assessments were elicited and then feeds back this value, together with the difference (residual) between this value and the assessed median. Large residuals are flagged and the expert is asked to reconcile differences. In a similar vein, the prior distribution involves a degrees of freedom parameter, δ, and this can be estimated from quantile assessments at a single design point, but Kadane et al. recommend assessing it at more than one design point. The computer calculates separate estimates of δ, signals to the expert the existence of widely discrepant values and allows re-assessment if the expert wishes. A different form of feedback is used by Garthwaite and Dickey (1988, 1992), who also quantify opinion about the parameters of a normal linear model through an interactive computer program. The computer questions the expert about the differences in y between pairs of $\mathbf{x}$-values, eliciting medians and quantiles of the differences. From these assessments, the median and quantiles of y at individual $\mathbf{x}$-values are calculated and fed back to the expert, who may revise them, although in practice they almost always proved to be acceptable (Garthwaite and Dickey, 1996). In a software system developed by O'Hagan (1998), experts assess 67% central credible intervals for scalar quantities and the computer infers 50% and 99% central credible intervals under the assumption that opinion corresponds to a normal or log-normal distribution. These intervals are given as feedback to the expert, who can alter them and/or re-assess the 67% interval. As well as credible intervals, credible *regions* have also been used as feedback, such as in the context of quantifying opinion as a Dirichlet distribution (Chaloner and Duncan, 1987).

Graphical feedback has also been used quite widely. To elicit a beta distribution for a probability, Chaloner and Duncan (1983) use an interactive computer program to question an expert about the number of successes that would be observed in a binomial experiment of, say, 20 trials. An initial estimate of the beta distribution is elicited and then a shortest 50% prediction interval is displayed, which the expert asks the computer to lengthen or shorten until it is an acceptable representation of her views. Chaloner et al. (1993) gave feedback in the form of marginal distributions for individual parameters and contour plots for the joint distribution of a pair of parameters. The probability distributions could be adjusted by the expert using sliders displayed on a computer screen. Craig et al. (1998) report an elicitation tool developed for one important application (matching hydrocarbon reservoir history). This provides a variety of visual feedback, such as diagrams to indicate the joint implication of beliefs that have been individually assessed, and application-specific maps.

Before finishing this section, it should be mentioned that the view has occasionally been expressed that feedback is unhelpful and that mechanical calculation should be used to obtain coherence from a set of elicited assessments, rather than involving the expert in the reconciliation of any inconsistencies. For example, Osherson et al. (1997) develop a method for reconciling inconsistent assessments and, in explaining the rationale behind their method, they write (1997, p. 2),

> '... within the present perspective there is no attempt to change the informant's mind about any aspect of uncertainty. Rather, human judgement is elicited in its natural state, and then massaged into coherency in a second step. The goal of such a two-step procedure is to preserve the insight that informs human judgement of chance, despite its incoherence. In contrast, influencing judgement before it is formulated

risks denaturing it, diminishing the knowledge content before it can be exploited'.

While few would agree completely with this statement, milder forms of some of its sentiments influence the thinking behind several elicitation methods. Thus, over-fitting coupled with feedback is generally preferred to simple feedback, so as to avoid influencing judgement before it has been formulated, and the reconciliation of assessments through some form of averaging is widely used when the task is too complex for an expert to handle. However, an assessed distribution is meant to represent the expert's opinion so, where possible, inconsistencies should be reconciled by her rather than through an algorithm. In addition, feedback encourages an expert to think carefully about her assessments and, among other benefits, it can highlight apparent errors, thus providing an opportunity for the expert to correct them, which seems the most sensible way of dealing with assessments that are out of line.

8.4 Conclusions

8.4.1 Elicitation practice

- One form of accuracy for an elicited probability distribution is that it should accurately reflect the assessor's opinion. Another is that it should accurately reflect reality. The latter is easier to measure but the former should be paramount, as it is the defining quality of a subjective distribution.
- Poor calibration of an expert's judgements, or poor performance on scoring rules, may reflect inaccurate knowledge rather than poor elicitation. Although most of the psychological literature is based on such measures, to the extent that consistent biases are found across a variety of experts and knowledge areas, they point to sources of poor elicitation.
- Proper scoring rules are useful for comparing assessors and assessment methods. They also provide an incentive for experts to record their opinions well and can play a part in training experts to quantify their opinions accurately.
- On the basis of the frequency of use, the probability scoring rule (also called the *quadratic score* or the *Brier score*) is the preferred scoring rule for judging subjectively assessed *discrete* distributions and the logarithmic scoring rule is the preferred rule for continuous distributions.
- In principle, various decompositions of the probability scoring rule can be used to focus on different features of a set of assessments (such as calibration), but, in practice, their efficacy is unclear.
- There has been only limited empirical success at relating coherence to better calibration.

- The most natural way of improving the accuracy with which a subjective distribution represents an expert's opinion is through feedback.
- Where possible, overfitting should be coupled with feedback to avoid 'leading questions'.
- Interactive software is almost essential for the effective use of feedback.

8.4.2 Research questions

- The score given to an assessed distribution is largely determined by the accuracy of the mean of the distribution, with only a minor role being played by the variances, covariances and degrees of freedom parameters. Methods of decomposing scores so as to focus on these parameters are needed so that ways of eliciting them can be evaluated effectively.
- Work on whether agreement between an expert's probabilities and reality related to coherence has been limited and should be extended. In the same vein, work is needed which examines whether coherence relates to the accuracy with which an expert's opinions are represented.
- Using feedback in elicitation methods appears to work well in practice. However, its benefits should be demonstrated rigorously by experiment and the usefulness of different forms of feedback compared. Also, the presumed superiority of overfitting over simple feedback should be tested empirically.

Chapter 9

Multiple Experts

9.1 Introduction

Decision makers often have access to more than one expert when decisions are made in the presence of uncertainty. It may then be necessary or desirable to obtain a single distribution that encapsulates the beliefs of several experts. There are two options available for obtaining this single distribution. The first is to elicit one distribution from each expert individually and independently of the others and then mathematically combine the resulting distributions into a single distribution. The second is to create an interaction between the group of experts, through which a single distribution is elicited from the group as a whole. These two approaches are referred to as *mathematical* and *behavioural* aggregation respectively. Within each broad approach, there are a number of specific methods and variations.

There is substantial research reported in the psychology literature relating to group decision-making which, obviously, is relevant to behavioural aggregation. Often, this research also compares group decisions with mathematical aggregation methods, with a view to establishing whether the interaction in group decision-making leads to better judgements than those obtained by aggregating the individuals' judgements mathematically. As with the cognitive psychology literature on individual probability judgements, the literature on group judgements (which is more in the tradition of social psychology) has relied heavily upon experiments using students and other groups with little or no substantive expertise in the quantities for which their judgements are elicited. As before, the limited experience with substantive experts generally indicates better performance.

We consider methods of mathematical aggregation in Section 9.2 and behavioural aggregation approaches in Section 9.3. Many more details and references for some of these topics may be found in the review of Clemen and Winkler (1999).

Uncertain Judgements – Eliciting Experts' Probabilities A. O'Hagan, C. E. Buck, A. Daneshkhah, J. R. Eiser, P. H. Garthwaite, D. J. Jenkinson, J. E. Oakley and T. Rakow

9.2 Mathematical aggregation

We first consider the situation where separate probability distributions have been elicited from a number of experts. A decision maker now wishes to combine these to produce a single aggregated distribution. It is worth noting that, whereas the individual distributions are clearly supposed to represent the actual views and knowledge of the corresponding experts, the aggregation may not have such an interpretation. We first consider methods in which the aggregate distribution *does* have the interpretation that it represents the decision maker's beliefs. In keeping with the convention used in the preceding chapters, we will suppose that the experts are female and that the decision maker is male.

9.2.1 Bayesian methods

Suppose that each of a group of n experts is asked individually for her beliefs about some unknown quantity θ, so that a distribution $f_i(\theta)$ is elicited from expert i. The decision maker now wishes to combine these to obtain a single distribution. From the Bayesian perspective, the correct procedure for doing so is for the decision maker to treat each expert's distribution as data and to update his own distribution in the light of these data, since in Bayesian decision theory it is the decision maker's beliefs that are ultimately required, rather than the experts'. The decision maker is sometimes described as a *supra-Bayesian* in this context. This approach was proposed by Morris (1974) and developments can be found in Lindley et al. (1979), French (1980), Winkler (1981), Lindley (1983), Genest and Schervish (1985), West and Crosse (1992) and Gelfand et al. (1995). A full review of this work is provided by Clemen and Winkler (1999). An application in pesticide use in apple production is described in Roosen and Hennessy (2001).

Formally, the decision maker begins with his own prior distribution $f(\theta)$ for θ and the result of incorporating the expert opinions is his posterior distribution $f(\theta \mid D)$, where $D = \{f_1(\theta), \ldots, f_n(\theta)\}$ is the set of the experts' elicited distributions. This is derived using Bayes' Theorem, according to which $f(\theta \mid D)$ is proportional to $f(\theta)$ multiplied by the likelihood term $f(D \mid \theta)$.

In practice, this approach is complex and difficult to implement, as in addition to eliciting his own prior distribution, $f(\theta)$, the decision maker must construct $f(D \mid \theta)$, which formulates his prior beliefs about what the experts are going to tell him, conditional on the true value of θ. Consider, for example, the simplest possible situation in which the unknown θ is a simple event, which we will denote by E, and we have only two experts. The decision maker's prior distribution is now just his probability, $P(E)$, that E will occur, and the experts' distributions become just the two probabilities, P_1 and P_2, each representing that expert's probability that E will occur. To specify $f(D \mid \theta)$, the decision maker must specify his knowledge and beliefs about the two experts' knowledge in great detail, through the following components:

- If E really will occur, what values of P_1 and P_2 does he expect the experts to give? He needs to specify a joint probability distribution $f(P_1, P_2 \mid E)$ for these. If he believes the experts to be good, then each marginal probability distribution, $f(P_i \mid E)$, should be concentrated on high values of P_i, but the decision maker must specify complete probability distributions. More than this, he must give a joint probability distribution that reflects also how similar he thinks the two experts' probabilities will be.

- If E really will *not* occur, what values of P_1 and P_2 does he expect the experts to give? He needs to specify a joint probability distribution $f(P_1, P_2 \mid E^c)$ for these. If he believes the experts to be good, then each marginal probability distribution $f(P_i \mid E^c)$ should now be concentrated on *low* values of P_i, but again the decision maker must specify a full joint probability distribution that represents in detail his beliefs about what the experts will say and how similar he expects their judgements to be.

It is clear, even in this simplest of all cases, that the decision maker needs to conduct a very sophisticated elicitation exercise on his own beliefs, in order to apply the supra-Bayesian method of aggregation. To deal with a situation where θ is a continuous random variable (or the even more complex case of several random variables) and/or where there are more than two experts, the task becomes substantially more complex still. The literature described at the beginning of this subsection has tried to formulate simple but realistic models for the decision maker's beliefs, in order to simplify the task. For instance, in the simple situation described above, Lindley (1985) supposes that the decision maker's distributions for $\log \frac{P_1}{1-P_1}$ and $\log \frac{P_2}{1-P_2}$ given E and E^c are bivariate normal, so that he only needs to specify means, variances and covariances (although this already entails him specifying 10 values on an unfamiliar 'log-odds' scale). See Clemen and Winkler (1987) for an application in weather forecasting.

9.2.2 Opinion pooling

Given the practical difficulties with the supra-Bayesian approach, a simpler and widely used technique is opinion pooling. Here, a *consensus distribution* $f(\theta)$ is obtained as some function of the individual distributions $\{f_1(\theta), \ldots, f_n(\theta)\}$, with the consensus distribution then used for decision-making purposes.

The simplest such function is the *linear opinion pool*

$$f(\theta) = \sum_{i=1}^{n} w_i f_i(\theta),$$

which is just a weighted average of the individual distributions with weights w_i summing to 1. For instance, the decision maker may choose to give each expert equal weight, so that $w_i = 1/n$ (for all i) and $f(\theta)$ is the simple average of the $f_i(\theta)$s. Alternatively, he may choose to give some experts more weight than others, depending on his beliefs about the expertise of the experts. Ways of assigning the weights are discussed by Genest and McConway (1990), and we discuss one method in some detail in subsection 9.2.3.

One drawback with this approach is that it is not *externally Bayesian* (Madansky, 1978). When new data are obtained, one could either update each expert's distribution individually and then combine the resulting posterior distributions using linear pooling, or first combine the expert's distributions as before and then update the consensus distribution. With the exception of some trivial cases, these two options do not yield the same result. Hence, a decision maker could be confronted with two different posterior distributions implying two different optimal decisions, but without any justifiable means of choosing between them. Similarly, when eliciting a joint distribution for two random variables X and Y, it may be that all the experts agree that X and Y are independent, yet a linear pooling of their joint distributions will generally give a distribution in which they are no longer independent.

An alternative is the *logarithmic opinion pool*, in which the consensus distribution is obtained by taking a weighted *geometric* mean of the n individual distributions,

$$f(\theta) = k \prod_{i=1}^{n} f_i(\theta)^{w_i},$$

where k is a normalising constant that ensures that $f(\theta)$ integrates to 1. The interpretation of the weights is as before, and if the decision maker is equally confident in the abilities of the experts, it may be natural to choose $w_i = 1/n$ for all i.

The logarithmic opinion pool is externally Bayesian when the weights w_i sum to 1 and is also consistent with regard to judgements of independence, but does not have another desirable property – that of (coherent) *marginalisation*. Suppose we have two mutually exclusive events, A and B, and define C to be the event 'A or B', so that (for any coherent individual) $P(C) = P(A) + P(B)$. Now suppose that each expert states her own probabilities for the events A and B. To obtain a pooled probability $P(C)$, one could either compute $P(C)$ for each expert and pool the results or first pool the probabilities for A and B separately and then sum the results. Under logarithmic pooling, these two options do not yield the same result (again, apart from some trivial cases). McConway (1981) shows that only the linear pool satisfies this marginalisation requirement. Hence, neither the linear nor the logarithmic pool satisfies both the marginalisation and externally Bayesian requirements. Indeed, no formula for mathematical aggregation can satisfy both requirements. Although the decision maker may not wish to marginalise, or to update with new information, so that one or the other of the criteria would be irrelevant, the fact that both linear and logarithmic rules have these fundamental inconsistencies is relevant. Note, however, that the criteria have been criticised by French (1985),

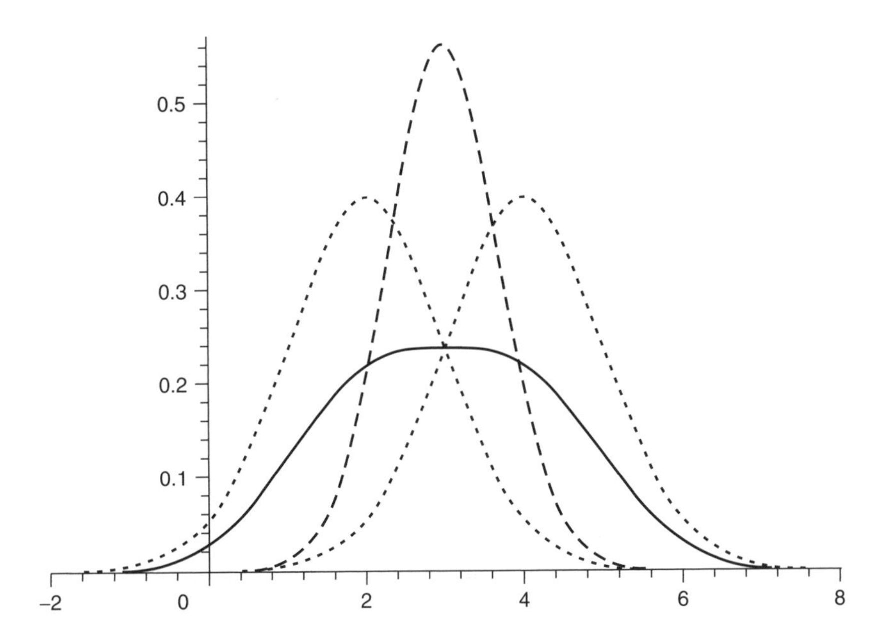

Figure 9.1: Example of opinion pooling. Dotted lines are expert distributions; dashed line is the equal-weighted logarithmic pool; solid line is the equal-weighted linear pool.

Lindley (1985) and Genest and Zidek (1986). Unlike the supra-Bayesian approach, the result of opinion pooling does not represent the actual beliefs of any individual and so may not behave as one would expect a probability distribution to behave.

The linear and logarithmic opinion pools lead to quite different aggregated distributions. Figure 9.1 shows an example with two experts, in which equal weighting is used for both methods.

The expert distributions are the two dotted lines in Figure 9.1, while the linear and logarithmic opinion pools are, respectively, the solid and dashed lines. This example is typical, in the sense that when pooling distributions we usually find that the logarithmic pool provides a much narrower distribution, implying much stronger aggregate knowledge than the linear pool. In particular, the logarithmic pool implies stronger information than that given by either expert separately, whereas the linear opinion pool represents less knowledge than either expert alone.

We might regard the logarithmic pool as appropriate if both experts were held to be good judges of θ on the basis of the information available to them, but they have different, independent sources of information. In that case, the decision maker's beliefs would concentrate on values of θ that are well supported by *both* experts, which is how the logarithmic pool behaves. With the logarithmic pool, the decision maker regards as implausible any values of θ that are considered implausible by any single expert. One expert could thereby effectively 'veto' another expert's distribution by giving zero probability to the range of values of θ supported by that expert. The linear opinion pool, on the other hand, is more appropriate when the decision maker's beliefs encompass the full range of values of θ that either expert considers reasonable. It concentrates more in the area where they overlap, but it does not rule out the low or high values of θ that are supported by only one expert. In doing so, it treats the experts more as if they are fallible than the logarithmic pool.

In general, while the linear opinion pool has been quite widely used in practice, the logarithmic opinion pool has been largely ignored, perhaps because it is perceived to lead to unrealistically strong aggregated beliefs.

9.2.3 Cooke's method

In choosing weights for an opinion pool, it is desirable to weight highly those experts who are perceived to be 'better'.

Perhaps the most sophisticated way of choosing the weights w_i is that of Cooke (1991), who advocates assigning them on the basis of the performance of each expert in a separate elicitation exercise utilising 'seeding variables'. The seeding variables are quantities that are chosen from the same subject area as the uncertain quantity of interest. The true values of the seeding variables are known to the facilitator, but not to the experts, and the experts are then asked to give various quantiles for these variables (typically the 5th, 50th and 95th percentiles). Experts are downweighted for poor performance relative to others in two aspects: if their probabilities are poorly calibrated or if their distributions are wide so that they give little information. The weight for expert j is proportional to the product of a calibration component C_j and an information component K_j.

Both components are based on the idea of the Kullback-Leibler (K-L) distance between two discrete probability distributions. Let $\mathbf{p} = \{p_1, p_2, \ldots, p_m\}$ and $\mathbf{q} = \{q_1, q_2, \ldots, q_m\}$ be two probability distributions for a discrete random variable taking m possible values. Then the K-L distance between them is

$$I(\mathbf{p}, \mathbf{q}) = \sum_{i=1}^{m} p_i \ln(p_i / q_i).$$

The distance is 0 if $\mathbf{p} = \mathbf{q}$, otherwise it is positive. The elicited quantiles define regions with fixed elicited probabilities $\mathbf{q}$. For instance, if expert j provides the 5th, 50th and 95th percentiles for each seeding variable, then $\mathbf{q}_j = \{0.05, 0.45, 0.45, 0.05\}$. The calibration component C_j in Cooke's weighting system is based on $I(\mathbf{p}_j, \mathbf{q}_j)$, where $\mathbf{p}_j$ is the proportion of true values of the seeding variables that fall in each of the four regions elicited from expert j. This is related to the logarithmic scoring rule (see Section 8.2.1). Cooke notes that if n seeding variables are used, then $2nI(\mathbf{p}_j, \mathbf{q}_j)$ has approximately a chi-square distribution if the expert is perfectly calibrated, and he defines C_j to be the probability that such a chi-square random variable would exceed the observed value of $2nI(\mathbf{p}_j, \mathbf{q}_j)$. Poor calibration will lead to a small value of this weight component.

The information component K_j is defined by comparing the expert's distribution for each seeding variable with a uniform distribution (or where appropriate, a log-uniform distribution). A more informative distribution will be far from uniform, placing concentrations of probability on relatively short ranges. So K_j is the average value, over the n seeding variables, of the K-L distance between the expert's distribution and a uniform distribution, thereby giving more weight to more informative experts.

Cooke and Goossens (2000) give empirical evidence to show that this can improve the overall performance of the elicitation. The use of seeding variables will also give some indication of the likely performance of the elicitation for the unknown variable of interest.

However well methods such as Cooke's assign weights to the experts, they neglect another relevant factor, which is the degree of correlation between experts. Two experts who have very similar information and views will be equally 'good', and will both be similarly weighted in the aggregate. However, their high correlation means that one expert is redundant given the other, and so their information is weighted twice as much as it should be. In general, groups of similar experts will receive too much weight and minority views will be under-represented.

For more detailed reviews and discussion of opinion pooling, see Genest and Zidek (1986) and Clemen and Winkler (1990, 1999).

9.2.4 Performance of mathematical aggregation

There is little literature studying the performance of mathematical aggregation methods for probability distributions. There is a little more concerning the aggregation of individual probabilities for events and a more substantial literature on the use of analogous methods for combining point estimates (usually referred to in the psychology literature as forecasts). The latter has tended to show that simple averaging of estimates (analogous to the equal-weighted linear pool) performs as well as more complex weighted averages (Clemen, 1989). Generally, the average performs almost as well as the best individual forecaster (which is good, in view of the fact that the best forecaster is of course not known to the decision maker when he needs to combine the estimates). Studies of the combination of event probabilities reviewed by Ferrell (1985) gave similar findings. Winkler and Poses (1993) considered various combinations of probabilities from four medical experts for the survival of patients in an intensive care unit, comparing the performance of each individual expert and simple averages of two, three or all four. They found that the best combination was an average of two experts, who were the two most experienced experts and also the least similar pairing. Clemen et al. (1996) compared two methods of Bayesian aggregation with simply asking decision makers to form a probability distribution for θ intuitively in the light of the experts' distributions. They found that both methods outperformed the intuitive decision makers, but that the more complex Bayesian method was no better than the simpler one.

There is scope for more research, particularly on the aggregation of probability distributions, but the general message of the literature is that simple aggregation methods work well in comparison with more complex methods. It is clear that more sophisticated methods can beat a simple equal-weighted linear pool, but the conditions under which this improved performance can be achieved are not clear. The study of Winkler and Poses (1993) may be a pointer in the sense that they found with genuine substantive experts the best method used the most experienced experts and also avoided overlap of their expertise.

Cooke and Goossens (2006) report very positive results with the Cooke (1991) method using the elicitation of quantiles for seeding variables to construct weights. The method has been used in 45 separate elicitation studies for risk analyses at Delft University of Technology (Netherlands). The studies used a total of more than 500 subject-matter experts in application areas ranging across nuclear safety, volcanoes, dams, infectious diseases, pollution and many others. The performance-weighted average generally calibrated as well as or better than either the best individual expert or the equal-weighted average on the seeding variables. It is not possible to say whether this success carried over to the substantive variables that were the subject of the elicitations, but the results are encouraging and support the idea that more complex methods might be justified in practical elicitation with real experts.

However, it is likely that the elicitation needs to be well constructed in other ways in order to achieve the benefits of more complex mathematical aggregation methods. In particular, experts should be carefully chosen, to obtain good coverage and minimal redundancy, and given training in the elicitation techniques to be used.

9.3 Behavioural aggregation

The alternative to combining distributions is to combine the experts. Methods of behavioural aggregation include simply bringing the experts together in a group and eliciting their beliefs as if they were a single expert, as considered in Section 9.3.1. Others involve various more restricted forms of interaction between the experts, as in Section 9.3.2. If these methods do not succeed in obtaining a single consensus distribution for all the experts, whether because the experts cannot agree or because the technique does not produce convergence of opinion, then a mathematical aggregation may be applied at the end. The purpose of the interaction in all of these methods is to share knowledge and interpretations between the experts.

9.3.1 Group elicitation

Readers may be able to recall many examples of when working in a group yielded outcomes superior to those that would have been achieved had they worked individually. Real and genuine benefits will often accrue through group interaction; however, the kinds of tasks that are typically the focus of elicitation exercises are not necessarily those for which group interaction yields the greatest benefits.

Group interaction typically helps most with tasks with a definitive and demonstrable correct answer or which rely upon a sequence of stages that can be 'solved' (Hill, 1982). Thus, group interaction is of greatest benefit when having a group of people (as opposed to an individual) means that there is an increased probability that at least one individual finds the correct solution to part, or all, of a problem (and can convince the group that this is so). Group elicitation of probability distributions is a different kind of task, and the rationale for how it might improve on individual elicitation is much less clear.

Nevertheless, the approach of bringing the experts together and eliciting a single distribution from the group as a whole is attractive. It should avoid the choice of a method of mathematical aggregation which, no matter how well argued each method may be, necessarily embodies an element of subjectivity and arbitrariness on the part of the decision maker. It should also have the merit of pooling knowledge and allowing the experts to bring their combined expertise to bear on the question. In principle, this synthesis should lead to more informed 'aggregated' distributions than the mechanistic approaches of mathematical aggregation. Group elicitation is advocated by a number of practitioners of elicitation (see e.g., O'Hagan, 2005, and particularly Phillips, 1999).

Although in group elicitations there is the objective (whether explicit or implicit) of reaching a single consensus distribution, it is not clear whether it is wise for the facilitator to push for consensus 'at any cost'. In cases where the experts are reluctant to agree, it may be better to allow disparate views to crystallise into a number of different distributions, which might then be aggregated mathematically.

The challenges with this approach are of a psychological rather than a mathematical nature. One risk is that any individual may have undue influence because of the force of personality rather than actual expertise. There is some evidence in the psychology literature to suggest that groups (unsupervised) do not always pool their knowledge effectively. Stasser and Titus (1985) find that individuals with unique information can be ineffective at integrating this information into the group. Wittenbaum and Stasser (1996) observe that groups can be poor at identifying and pooling specialist information held by individuals. Additionally, Gigone and Hastie (1993) find that the influence of a single piece of information will increase with the number of experts who know it. The role of the facilitator in managing the group discussion is crucial and demanding.

There is mixed evidence regarding whether and how the psychological biases associated with elicitation, discussed in Chapters 3 and 4, apply in group elicitation. For instance, Sniezek (1992) suggests that anchoring effects can be reduced as the group will not be starting from a single value; each member will have their own anchor. On the other hand, groups may be more subject to the representativeness heuristic and more likely to commit conjunction fallacies: Kerr et al. (1996) review the evidence for whether groups are more or less influenced by different biases and heuristics than individuals.

Another possibility is that group elicitation will lead to overconfidence, through a phenomenon known in the studies of group judgements generally as 'group polarisation' (Plous, 1993; Sniezek, 1992). This refers to the observation that the final judgement of a group is usually more extreme than the average of the group members' initial judgements. It applies particularly to situations where it is possible to define opinions/beliefs as lying to one side or another of some neutral position or above/below the mid-point of a scale. In probability elicitation, for instance, this neutral position or mid-point might be represented by a probability of 0.5. When individuals' initial judgements mostly lie above the neutral point, the final consensus judgement is typically even higher than the mean or median of these initial values. When most initial judgements are below the mid-point of a scale, the final consensus judgement is typically even further below the scale mid-point. Thus, in the context of probability judgements, we would expect group discussion to increase confidence (e.g., increasing the frequency with which probabilities close to 0 and 1 are quoted). Heath and Gonzalez (1995) found that interactive decision-making did indeed increase participants' confidence in their decisions, though, for their particular task, there was no corresponding improvement in accuracy. Sniezek and Henry (1990) reported that if individuals are asked for their judgements again after leaving the group they show less overconfidence, but still more overconfidence than they had shown before the group interaction.

Although group polarisation has been characterised in the psychology literature as representing overconfidence by the group, it is not necessarily unreasonable. For instance, if a supra-Bayesian decision maker receives probabilities from a number of experts for some event that are consistently above 0.5, this might very well lead him to assign a probability for that event that is even higher than any single expert has given. This might arise particularly if the decision maker believes that the experts are making independent judgements on the basis of different evidence. Then the weight of evidence coming from all the experts, in favour of this event occurring, is stronger than from any one expert. Similarly, in group elicitation where the experts are able to pool differing sources of evidence and experience, it is not unreasonable for their consensus distribution to be more 'polarised' than any of their individual distributions. Nevertheless, it seems that group polarisation can easily arise for psychological reasons quite different from any genuine pooling of evidence, and the result would then indeed represent overconfidence.

Although these various studies do not relate directly to the elicitation of probability distributions, they suggest that group elicitation may be subject to recognisable biases and may, in particular, produce narrower distributions than would be warranted by the evidence (aggregating more like a logarithmic than a linear opinion pool).

9.3.2 Other methods of behavioural aggregation

Since some of the findings on group judgement implicate the nature of the group interaction as a cause of poor performance, there has been interest in methods of behavioural aggregation in which the interaction between the experts is more controlled.

One procedure for formally managing a group elicitation is the Delphi method. The general Delphi approach is applicable for various kinds of group judgement and decision-making, but in the case of eliciting a group consensus probability distribution, each individual first provides her own distribution together with some explanation of her views. This is then supplied to all the other experts in the group. The experts are then given the option to revise their own beliefs in light of the views they have heard from the other experts. This process may be iterated several times, until either the experts converge to a single distribution or the facilitator decides to mathematically aggregate the different distributions. Further details can be found in Linstone and Turoff (1975), Pill (1971), Parenté and Anderson-Parenté (1987) and Rowe and Wright (1999). In the Delphi method, the experts interact only through the written exchange of judgements and reasons. Often, the interaction is anonymous, so the experts do not know which other experts have contributed which opinions. Such interaction relies on the better experts being able to explain their views more cogently (which is clearly not necessarily true, particularly if the protocol allows only limited scope for explanations), rather than their reputation being a factor in persuading the other experts.

The Nominal Group Technique, described in Delbecq et al. (1975), is a variant of Delphi in which each expert presents her views to the group for group discussion. Experts then individually revise their beliefs as before. Another modification is given in DeGroot (1974), in which each expert revises her beliefs in the light of the other views by applying a linear opinion pool to all the distributions. The expert may choose how much weight to give to each distribution on the basis of how she rates the importance of the others' views. This process can be iterated, with the revised distributions being given to all the experts for further opinion pooling.

Kaplan (1992) argues that the emphasis of group elicitation should be on combining the expert's evidence rather than their opinions. The experts are invited to present and discuss their evidence in a group meeting, but the facilitator then, proposes a distribution to the experts on the basis of the combined evidence. The justification for this is that the facilitator will have the statistical expertise required for formulating an appropriate probability distribution, whereas the experts may not. However, in our experience, the nature of the evidence typically presented by experts is such that a statistician would not be able to derive a suitable probability distribution without further input from the experts; the experts will always need to make their own probability judgements on the basis of their evidence.

Clemen and Winkler (1999) state that, with the exception of the methods discussed above, "behavioral combination approaches are ... for the most part, not a set of fixed procedures." They add,

> "There are, however, a number of important issues relating to behavioral combination: the type of interaction (e.g. face to face, via computer, anonymous); the nature of the interaction (e.g., sharing information on relevant issues, sharing probabilities, trying to assess "group probabilities"); the possibility of individual reassessment after interaction; and the role of the assessment team (e.g., as facilitators)."

9.3.3 Performance of behavioural methods

As we have noted above, much of the evidence relating to behavioural aggregation has dealt with decision-making or point estimation, rather than probability elicitation. However, the findings are again generally that the group often performs less well than a simple average of the individual judgements. Situations in which the potential for the group to perform better is realised depend very much on the nature of the interaction. There is evidence that successful groups recognise their most knowledgeable members and make decisions that give substantial weight to the best members (Littlepage et al., 1997; Sniezek, 1990). An interesting study, which seems to combine the messages arising from other research is that of Reagan-Cirincione (1994). She found that small group elicitations could outperform their best individual member if the elicitation combined three features:

- Impartial facilitation that is responsive to the potential for biases in the group interaction;
- A well-designed protocol that involved careful structuring and decomposition of the elicitation task;
- Continuous feedback via computer technology of the implications of the experts' judgements.

Reagan-Cirincione was working with students on tasks for which they did not have appreciable substantive expertise. Other findings indicate that good group elicitation performance may be found more consistently with genuine experts.

9.4 Discussion

The debate over mathematical versus behavioural aggregation has not been (and is unlikely to be) resolved, with no evidence to favour one method conclusively over the other. Phillips (1999) describes an experiment comparing individual, pooled and consensus distributions, and finds evidence to support behavioural over mathematical aggregation. Conversely, Mosleh et al. (1987) report empirical evidence that mathematical aggregation yields better results than behavioural methods. In general, it seems that a simple, equally weighted, linear opinion pool is hard to beat in practice. But the advocates of other methods, such as Bayesian methods, Cooke's method of weighting a linear pool or the methods of group elicitation, Delphi, and so on, are right to believe that they have the potential to extract more information through unequal weighting or group synergy. That this potential has not been consistently realised in experimental studies does not mean that it is not there, only that the conditions under which the benefits will be found have not yet been clearly delineated.

We have argued in previous chapters that the literature on how badly individuals perform in assessing probabilities and uncertainty does not mean that one should not try to do elicitation. On the contrary, it is necessary to learn from experimental and theoretical research, in order to build elicitation procedures that are robust and lead to the best possible representation of the expert's knowledge. By the same token, the fact that sophisticated methods of aggregating expert opinion sometimes perform badly does not mean that we should not use them. Difficulties with managing the interaction in group elicitation should not necessarily mean that we abandon face-to-face interaction and resort to Delphi methods. Again, we need to learn the best way to use the methods.

In the area of mathematical aggregation, Cooke's method seems to have the greatest potential, and there is evidence that embedding it in a very thorough and structured elicitation protocol is successful. Similarly, there is evidence that group elicitation, which arguably has even more potential, can also be successful if embedded in a well-structured process that manages the interaction through careful facilitation and strong feedback. The importance of the whole elicitation process was stressed in Section 2.2 and is reinforced by both these conclusions in the context of eliciting from multiple experts. Examples of the care that goes into constructing good elicitation protocols for multiple experts can be found in Winkler et al. (1995) and Cooke et al. (2001).

9.5 Elicitation practice

- The simple average (equal-weighted linear opinion pool) of distributions from a number of experts provides a simple, robust, general method for aggregating expert knowledge.
- More complex mathematical aggregation, such as the method of Cooke (1991), has the potential to perform better but its success depends on having a well-structured elicitation process. Experts should be chosen to have a broad range of substantive knowledge without undue overlap.
- Group elicitation probably has even greater potential, since it can bring better synthesis and analysis of knowledge through the group interaction. However, success depends even more on the abilities of the facilitator, who must encourage (a) the sharing of knowledge (as opposed to opinions), (b) the recognition of expertise, (c) the study of feedback, but must avoid (d) the group being dominated by shared knowledge or over-strong opinions, (e) the kinds of biases found in individual assessments and particularly (f) the tendency of groups towards overconfidence.

9.6 Research questions

- There is a need for more research into behavioural aggregation of probability judgements and probability distributions (as opposed to decisions or estimates). As with individual elicitation, the use of substantive experts in such experiments is much preferred.

- There is still much to be done to identify what characteristics of the experts, the facilitation and other aspects of the elicitation process are conducive to success with the more complex mathematical aggregation methods or group elicitation.

Chapter 10

Published Examples of the Formal Elicitation of Expert Opinion

10.1 Some applications

During our review of the literature on elicitation of subjective probabilities, we have encountered an enormous number of case studies in which researchers have sought to elicit subjective probabilistic information and represent it in a rigorous way as part of a formal investigation. As part of this review, we cannot hope to summarise (or even cite) all of the interesting and relevant case studies in the published literature. Nonetheless, since the application areas are so varied and the range of elicitation methods so impressive, we felt it was worth including an illustrative selection here. In what follows, we summarise a selection of case studies from a range of disciplines and then provide a list of other examples that readers might find useful.

10.2 An example of an elicitation interview – eliciting engine sales

To help focus ideas at the start of this chapter, we feel that it may be helpful to provide a description of an elicitation interview that is reported in full in the published literature. Shephard and Kirkwood (1994) base their paper upon a transcript of an

Uncertain Judgements – Eliciting Experts' Probabilities A. O'Hagan, C. E. Buck, A. Daneshkhah, J. R. Eiser, P. H. Garthwaite, D. J. Jenkinson, J. E. Oakley and T. Rakow

elicitation interview between a manager (the expert) and an analyst (Kirkwood). We recommend that anyone who is considering undertaking elicitation for the first time should read this paper in its entirety. The authors offer considerable insight about what happened at each point in the process and are highly informative about how to conduct a successful elicitation interview. Here we simply summarise the crucial stages of the approach they took. They use a five-stage protocol that was (according to Merkhofer, 1987, whom Shephard and Kirkwood cite as their source) developed at the Stanford Research Institute. Further descriptions of the protocol can be found in Morgan and Henrion (1990) and Spetzler and Staël von Holstein (1975).

1. Motivating

   This first step involved the analyst trying to bring out some of the issues that might bias assessments, so that the manager could be aware of them and try to avoid them. The quantity that they were trying to elicit was the number of engines that the company would sell in that year. The analyst draws out from the manager that the sales are already forecast and that he feels that there are incentives for employees to meet these forecasts.

2. Structuring

   The analyst established with the manager the specific event that they were trying to elicit – the manager's (prior) probability distribution for the number of sales to a single customer in the current calendar year. They also discussed the market conditions, the manager believing that the market was increasing at the time.

3. Conditioning

   Here the analyst tries to bring to the manager's mind all the relevant information and knowledge, which will help him to overcome psychological biases such as overconfidence and availability. The manager displayed little reaction to the explanation of overconfidence, but reacted strongly to that of availability because he recognised an instance of the phenomenon when he had been discussing engine sale forecasts with others in the company.

4. Encoding

   At the start of the process of actually obtaining the probabilities, Shephard and Kirkwood suggest that eliciting the extreme quantiles before the central ones reduces the effects of some biases, particularly overconfidence. The elicitation was undertaken using a probability wheel, a method with which the manager happened to be familiar. The analyst encouraged the manager to think about the conditions that they had agreed would result in a large number of engine sales. On getting a response of between 60 and 70 for the 99th percentile, the analyst moved quickly to the first percentile, rather than trying to refine the estimate, in order to try to avoid an anchoring effect. The manager gave a range of 10 to 20 for this percentile.

In the elicitation of the centre of the distribution, the analyst changed to asking the manager for a probability that the number of sales would exceed a particular value, 30 in the first case. They used the probability wheel to determine it, the manager indicating that 30 was his 25th percentile. This was repeated, from the other end of the scale, establishing that 50 was his 75th percentile. The analyst elicited the manager's median by suggesting values to him and asking him to respond by saying whether they were likely to sell more or less than that number. This process narrowed the median down to 44.

5. Verifying

   The final step of the protocol aimed at increasing the accuracy of the assessments. The analyst used the bisection method; the manager was asked to suppose that it was known that sales would be below his median value, 44, and to specify where to divide the range from 0 to 44 such that the two parts would now be equally probable. The answer, 28, was reasonably close to the value for the lower quartile, 30, elicited by the previous method. The same method for the upper quartile also produced a result similar to that obtained previously.

   The analyst plotted the values on a line and, indicating the inter-quartile range, enquired of the manager whether the number of sales was more likely to be inside or outside that range. The manager recognised that (if his assessments were correct) the answer should be that it is equally likely to be inside as to be outside. However, he felt that it was more likely to be inside, so they adjusted the quartiles inwards. To complete the checking, the analyst sketched the cumulative distribution function and the probability density function, which were confirmed by the manager.

Throughout the process, Shephard and Kirkwood are keen to emphasise the interpersonal skills needed by the analyst, such as when not to press an issue, when and when not to use technical terminology (the manager had some probability training and so was familiar with terms such as 'cumulative distribution function') and keeping to time.

10.3 Medicine

Before discussing examples of elicitation in the medical field, it is worth noting that good explanations of Bayesian methods, their application to medicine and comparisons with frequentist techniques can be found in Burton (1994), Burton et al. (1998) and Spiegelhalter et al. (2004).

10.3.1 Diagnosis and treatment decisions

On reviewing the medical literature, we found that there is a well-established sense of the value of decision-making based on probabilities in the sphere of diagnosing illness and disease. This is associated with an interest in looking at formal, mathematical methods and applying them to diagnosis. Of particular note for this review is the use of Bayesian networks, where signs and symptoms are mapped out and probabilities are assigned to them. Discussion of such a network being used for oesophageal cancer is found in van der Gaag et al. (2002) and this paper also includes a detailed account of the elicitation exercise that the authors undertook to obtain the probabilities for the network. The method they adopted involves asking the experts about conditional events: given that a certain symptom is observed, how likely is another symptom to be observed? An important point of their methodology is that the descriptions of the events were given to the experts in written form, with a probability scale and associated verbal anchors, as described in Renooij and Witteman (1999). An example of a verbal statement of a conditional event is

> "Consider a patient with a *polypoid* oesophageal tumour; the tumour has a length of *less than 5 cm*. How likely is it that this tumour invades into the *lamina propria* (T1) of the wall of the patient's oesophagus, but not beyond?"

The authors made a conscious choice to use the phrase 'how likely', which invites the expert to think in terms of a one-off event rather than in terms of frequency, "to prevent difficulties with the assessment of conditional probabilities for which the conditioning context is quite rare." The questions are laid out on the page in such a way that those based upon the same conditioning event are placed together. This encourages the experts to assess them at the same time and enables the experts to check the coherence of their assessments.

When the authors implemented the technique, they used two experts, who were required to give a joint assessment of the probabilities after discussion with each other. With this method, the experts were able to give their responses "at a rate of 150–175 probabilities per hour". In their feedback to the authors, the experts reported that the method was easier than the other methods of elicitation that they had been exposed to before. Regarding the verbal anchors, which they found helpful, the experts were more inclined to use the probabilistic words when they were most uncertain about the probability they were assessing. (The authors also noted that the experts found that it became easier for them to use the numbers the more assessments they had made.) van der Gaag et al. (2002) also report an effect of the differing understanding of probabilistic words. The experts said that they rarely used the word 'impossible', with 'improbable' being used (particularly with patients) for events with a probability of around or less than 5%, rather than for events with a probability of 15% as is indicated by the anchors on the scale that they used. This phenomenon is very important when using probabilistic words. Phillips (2004) gives a further example of this in relation to the assessment of whether a repair job would be finished on time.

> "The repairman's 'probably', meaning to him 30 percent, was heard by his first line manager as better than 50 percent, so he reported 'good chance' to his departmental manager, who passed on 'highly likely'.

> By the time it reached the top, the generator was 'almost certain' to be ready in time."

The use, by van der Gaag et al. (2002), of a scale which combines words and numbers may provide an answer to some of these problems, but see Section 4.4.1.2 for more discussion of these matters.

The use of subjective probabilities has also been explored within diagnostic radiology. Berbaum et al. (2002) compare asking for a probability that a feature is abnormal using a quasi-continuous scale (101 points from 0 to 100%) with the use of the standard five-point confidence rating (from 'suspicious, but likely normal' to 'definitely abnormal') for calculating receiver operating characteristic (ROC) curves. Their main conclusions are about the relative success of both the methods regarding the production of useful ROC curves; the use of the subjective probability ratings did not, for their data, give ROC points across more of the range of the probability of a false positive assessment than the discrete five-point method. However, the authors make some points that refer to the elicitation itself. The method used for eliciting the probabilities is just to ask for the (whole) number between 0 and 100 that represents the individual's probability, expressed as a percentage. The authors presented the assessments of two of their experts and note that they restricted themselves to using only a few of the 'categories' (i.e., the integers between 0 and 100, inclusive) on their subjective probability scale (one expert used 15 different values, the other used 9). This effect was quite common across the experts they assessed. The categories used by each of the six experts whose opinions are given in the paper are all multiples of 5, except for the use of low probabilities: 2, 3 and 8%. A quote that the authors give from one of their experts is quite informative about how people sometimes think about probability. The assessment the experts are being asked to make is of their probability that the abdominal examination they are looking at is abnormal. One expert says:

> "2%, I was pretty darn sure it was normal, 5% is my standard 'I think it is normal and am not going to think about it anymore' number, and 3% is between them, and the single case of 10% was probably one where I wanted to call it normal but there was some nagging suspicion that I couldn't put a handle on."

This highlights the fact that (at least when the elicitation method used is as direct as the one used by Berbaum et al.) people do not like to be too specific, rounding their assessment to the nearest five percentage points, and even attaching specific meanings to some specific values.

Further work on subjective probabilities being used in ROC studies within radiology is found in Rockette et al. (1992) . They use a similar method, comparing the discrete five-point scale (based on verbal definitions) with a continuous scale requiring a numerical response. Comparing their results from five radiologists, each assessing 95 cases of abdominal computed tomographic examinations for the presence of mass lesions by both the above methods, they found that the continuous scale could be split into five intervals which corresponded to the five discrete categories: 0 to 9% being 'definitely absent', 10 to 39% being 'probably absent', 40 to 75% being 'possibly present', 76 to 89% being 'probably present' and 90 to 100% being 'definitely present'. They also found that for the cases that were positive, the response interval used most often was 95 to 100%, whereas for negative cases it was 6 to 11%. Comparing the ROC curves obtained from each method for each radiologist, there was no significant difference between the curves obtained by either method. The authors conclude that there are some disadvantages to the discrete method which are not present with the continuous method, such as producing degenerate data-sets, the need for additional training and its loss of some accuracy (compared to the continuous scale). Thus, they recommended that the continuous method should be used.

In another example of probability elicitation in medical diagnosis, Chaloner and Rhame (2001) record an elicitation of the opinions of physicians regarding AIDS sufferers; specifically, whether the patients involved in a particular clinical trial will experience a particular condition within two years given that they take different treatments. The physicians were asked to assess the percentage of people that they expected to experience the condition within two years for two treatment regimes of the same drug, to note the difference between the two and then to give their 95% probability interval of that difference. A similar process was used to compare the two alternative drugs.

In a similar elicitation process, Must et al. (2002) elicited responses from 37 experts who were asked to look at drawings of seven girls and to assess their body mass index (BMI) – specifically, the percentile of the population that each girl's BMI fell in. The results showed that there was more variability within the experts when the assessments were around the 50% point than when they were at the top or bottom of the range.

Subjective probability also appears in literature about the diagnosis of pulmonary embolism (PE). Sostman et al. (1994) look into ventilation-perfusion scintigraphy and compare the standard Prospective Investigation of Pulmonary Embolism Diagnosis (PIOPED) categorical probabilities with subjective probabilities elicited from physicians, using 'a continuous (percentage) scale'. The PIOPED system has four categories which are each considered to represent a range of probabilities: high probability ($\geq$80%), intermediate probability (20–79%), low probability ($\leq$19%) and a normal scan. They encourage the use of probability assessments, noting the possibly different interpretations of a 15% chance and a 5% chance (both of which would be classified as 'low' probability under PIOPED) and that the range of probabilities, 20–79%, that makes up the intermediate range is large. Christiansen et al. (1997) also compared PIOPED categorical probabilities with subjective probabilities, asking three nuclear medical physicians to assess both (first PIOPED, then probability) for a set of lung scintigrams. The elicitation was performed using a probability scale, with 0% and 100% as the only values actually indicated, described as a *visual linear scale* (VLS). The physicians were asked to re-assess their subjective probabilities from the scintigrams six months later. ROC curves were calculated from both the VLS scale and the categorical assessments, with no significant difference found between the two. The authors also noted that they found the 'repeatability' (across the six-month period between assessments) of mid-range (20–80%) probability assessments to be poor. Overall, Christiansen et al. suggest that using subjective probabilities "does not significantly improve the overall accuracy of diagnosis compared to the categorical PIOPED probability assessment alone." Further work in this area can be found in Gottschalk et al. (1993) and Perrier et al. (1996). Bigaroni et al. (2000) consider the diagnosis of deep vein thrombosis, comparing probability assessments made by doctors with the results from Wells' score.

The management of severe head injuries is considered by Harmanec et al. (1999). Using a decision model with subjective probabilities, their model required 2808 probabilities to be elicited. Elicitation was undertaken by asking doctors directly about the probabilities of interest. It was noted that the doctors reviewed their assessments as they went along, which the authors suggest was useful for ensuring the consistency of the probabilities. In their discussion, Harmanec et al. (1999) suggest that doctors do not think about numerical probabilities when making treatment decisions, and so the authors would prefer to use elicitation methods that use qualitative descriptions which can be converted into probabilities. However, they also advocate providing the expert with (computer) graphics that will help them in "visualizing and manipulating the relevant numerical parameters".

In a rather different study, Timmermans et al. (1996) elicit, from surgeons, probabilities of operative mortality for patients with abdominal aneurysm, given different patient profiles, and compare them to probabilities produced by a model. They found that the surgeons' assessments were similar to those produced by the model, but the standard deviations for three out of the four surgeons were smaller than those from the model. Comparing how they used the profiles, the authors conclude that the surgeons "weigh the clinical information differently than the model when estimating operative mortality." Timmermans (1999) also gives a good review of studies involving probability judgements, both those made about patients and those given to them. Her conclusions are that doctors, like most people, have problems with giving good probability assessments and that probabilities should be given as numbers as well as by using probabilistic phrases.

10.3.2 Clinical trials

Statistical methods have long been a part of clinical trials. A discussion of the issues surrounding the use of Bayesian methods, and, in particular, genuine prior information from experts, is given in Stevens and O'Hagan (2002). They counter the argument that using expert opinion introduces subjectivity into a scientific process by arguing that all statistical modelling requires decisions to be made by the statistician, which are, ultimately, subjective. Indeed all statistical methods, classical/frequentist or Bayesian with either non-informative priors or expert opinion priors, have some amount of subjectivity within them. Stevens and O'Hagan (2002) also discourage the use of non-informative priors since, they argue, it is unlikely that those involved in the trial have no knowledge of the possible treatment effects. This knowledge, perhaps from previous stages of the trial, should be included in any model used for the trial. They also recognise the need for suitable guidelines and practices to be written and employed. The authors suggest that, "submissions of evidence on the cost-effectiveness of new interventions using the Bayesian approach must include supporting documentation that demonstrates clearly that a formal process of elicitation has been followed if the prior information is to be accepted as credible."

Both classical and Bayesian methods are used for deciding the trial size. A robust Bayesian approach to clinical trials is described by Kadane (1994), who uses elicitation of priors from experts. The example clinical trial he uses compares two alternative drug treatments, with the success measure being systolic blood pressure (SBP) after open-heart surgery. The method involves selecting a panel of experts, who then identify covariates that they deem to be important in determining which drug to use (unanimity across the experts was not required for inclusion). A normal linear model of the change in SBP with these covariates is used, with the conjugate normal-inverse-gamma prior being assumed for each expert. The experts then have their prior beliefs elicited using the method of Kadane et al. (1980); see Section 6.7. When results from patients are received, each expert's distribution is updated using Bayes' Theorem. Then for each new patient, her values for the covariates are measured and are fed into the current linear models of each expert. If all experts (through their models) 'agree' that one drug is better than the other (i.e., would reduce the patient's SBP by a greater amount), then the treatment of using the other drug is considered to be *inadmissible* (and it would be considered to be unethical to use such a treatment). Other methods of determining admissibility are mentioned. Otherwise, a treatment is selected using an agreed method. Kadane argues that this method of conducting a clinical trial avoids some of the problems of determining when to stop a clinical trial when using a 'classical randomised design'. Under this procedure for running a clinical trial, a patient cannot be given a treatment that all of the experts involved consider to be worse than an alternative (at the time the treatment decision is made). As a consequence, the need to stop a trial because evidence is emerging that one treatment is worse than another does not arise.

Within the context of deciding the sample size for clinical trials, Oremus et al. (2002) elicited success probabilities of two treatments for female stress urinary incontinence: the standard surgery and a new treatment of collagen. The elicitation involved using both a postal questionnaire and individual interviews. Some of the questions were presented in relative frequency terms, for example, "When success is defined by patient satisfaction at one year after completion of treatment, what percentage of patients do you expect will have their stress urinary incontinence successfully treated with..." followed by the two options: surgery and collagen. The other approach used was to ask for the maximum percentage reduction in the efficacy of collagen (compared to surgery) for which the medic would still use collagen as her first treatment (selecting an option from a set of percentage point ranges: 0–10%, 11–20%, ..., 90–100%). This was followed up by asking for the medic's likelihood (expressed as a percentage) that each treatment will, in the end, be a satisfactory treatment for the condition. The results of the postal questionnaire suggest that respondents would prefer to use collagen (owing to its benefits of being less invasive than surgery etc.) when the reduction in efficacy is no more than 23%.

Tan et al. (2003) describe the elicitation of prior distributions for randomised controlled trials, the context being whether a radioactive iodine (RAI) treatment should be given (as well as surgery) to patients with hepatocellular carcinoma. A previous trial had been stopped because results were showing a much lower success rate for those treated with RAI than for those who were not. The authors consider whether a Bayesian approach would have been preferable. The elicitation involved 14 experts. They were given a questionnaire in which they were asked to consider the difference in the two-year recurrence-free survival rate (and the percentage by which the difference was better or worse) when RAI was used. Given intervals of five percentage points, the experts were asked to give their 'weight of belief' for each interval, with the weights adding to 100. They were also asked for their estimate of the two-year survival rate for surgery alone. Following Spiegelhalter et al. (1994), the mean probability values for each interval (across the experts) were calculated, with an overall histogram and a skeptical histogram (using just the values of the five most skeptical experts as to the use of RAI) being drawn. Each histogram was then transformed onto a log-hazard ratio scale and normal distributions were fitted.

10.3.3 Survival analysis

The human survival probabilities that are used in demographic and economic models traditionally come from historical data, put into, so-called, life tables. The alternative use of subjective probabilities of survival is explored by Hurd and McGarry (2002) (in particular in the context of their predictive validity). The Health and Retirement Survey (conducted in the United States) asks people questions such as the following:

> "Using any number from 0 to 10 where 0 equals *absolutely no chance* and 10 equals *absolutely certain*, what do you think are the chances you will live to be 75 or more?"

The question was repeated but using 85 years instead of 75. Hurd and McGarry used this type of elicitation for people aged between 46 and 65 and examined how subjective probabilities of survival changed when new information became available. The survey was repeated two years later with essentially the same respondents. The authors found that the estimates that individuals made of their own survival probability reduced after a parent had died. They also concluded that the subjective survival probabilities did have some predictive power. They found that those who survived long enough to take part in a second phase of the experiment had originally given probabilities of their survival that were approximately 50% higher than those that had been given by people who died before the second phase.

A Bayesian approach to survival models is also discussed by Coolen (1996), although not in a medical context. In a discussion about eliciting survival times of equipment from engineers, Coolen notes that if a system is always replaced after a certain time the expert has no knowledge of the system beyond that time and would have to make 'imaginary' assessments of failure time. By applying the theory of censored data (using survival distributions from the exponential family) he elicits from the expert (a) the number of systems (out of a sample) that would fail before the cut-off time and (b) the total working time of the failed systems. He then uses these two properties to construct conjugate prior distributions for use in future analyses.

In a rather different study, the elicitation probabilities of survival for patients in intensive care units (ICUs) was performed and reported by Winkler and Poses (1993). This study was also briefly discussed in Section 9.2.4. Winkler and Poses classified physicians into four levels of expertise: intern (patient's house officer), critical care fellow, critical care attending physician (on duty in the ICU) and the patient's primary attending physician. Each physician was then asked to give their probability that a patient entering the ICU would survive until hospital discharge (i.e., using the direct method of probability assessment by just asking for a number). The study showed that the most reliable estimates came from the physicians with the most ICU experience. The patient's primary attending physician tended to offer higher probabilities for survival (than the others), which Winkler and Poses suggest might be due to them having more emotional investment in the patient. Critical care attending physicians gave probabilities more in the middle of the range (which Winkler and Poses suggest perhaps reflects the fact that they spend more of their time dealing with 'life-and-death situations' in the ICU). All four types of physician were well calibrated, but the primary attending and the critical care attending physicians were better able to discriminate between those who would live and those who died. The authors also explored combining the opinions. They concluded that obtaining a 'second opinion' increased the overall performance (especially if it was combining the primary attending and critical care attending physicians), but that further opinions did not add much more. Winkler and Poses also give a good description of the different types of uncertainty, within this context:

> "We speculate that physicians may be better able psychologically to consider uncertainty about prognosis (because, after all, no one can predict the future perfectly) than about diagnosis (because theoretically, at least, it is possible to know every patient's diagnosis)."

10.3.4 Clinical psychology

There has been some research in clinical psychology which has asked patients (and controls) to estimate probabilities. For example, MacLeod and Tarbuck (1994) investigate parasuicides; investigating whether they think that negative events are more likely to happen to them than to others. The elicitation method used was to ask for a number from 1 to 10, where 1 represented 'not at all' and 10 represented 'certain' in response to a question about how likely a specific negative event, for example, "your health will deteriorate", was to happen to them within the next two years. They found that if the individual was asked to give reasons as to why she thought the event would not occur, this produced lower probabilities. They thus concluded that parasuicides would consider a negative event to be likely if it was brought to their attention, but they would not necessarily expect it otherwise.

Similarly, Woods et al. (2002) considered obsessive compulsive (OC) symptoms, asking people to describe a "negative event that you worry might happen to you" and then to give a probability that it will occur (at some point in the future) using an 11-point scale: <10, 10, 20, ..., 80, 90, $>90\%$. They also asked questions designed to test their level of obsessive compulsiveness. The results were that, while the group of students (with no previous history of OC symptoms) showed an increase in the probability of the negative events occurring when their score for OC symptoms was higher, this pattern was not apparent in the group of patients with OC symptoms. A possible reason for this, according to the authors, is that OC patients may have developed better probability assessment skills through their therapy, if not better severity judgement or coping skills, two other factors that influence anxiety.

Probabilities were elicited from clinically depressed and anxious children and adolescents by Dalgleish et al. (1997). Using the three groups: depressed, anxious and a control group, each person was asked to estimate the likelihood of a given negative event happening either to the subject (self-referent) or to another child (other-referent). A probability scale was used with no numbers; it just had the ends marked 'definitely won't happen' and 'definitely will happen', the respondents marking the scale to indicate their probabilities. The result was that those in the anxious and control groups gave higher probabilities to the event happening to another person than to themselves, whereas those in the depressed group gave similar probabilities for themselves and the other person. The authors use a similar method of probability elicitation in a study of recovered depressed children in Dalgleish et al. (1998).

The much reported overconfidence effect is challenged by Gambara and León (1996), particularly in the context of clinical judgement making. Their experiment involves diagnosing depression or anxiety from a list of symptoms (which are similar for both conditions) and consisted of two components. In the first, the respondents were asked to rate 34 symptoms as to how characteristic of each condition the symptom is, on a three-point scale. In the second part, the respondents were given the symptoms of 20 patients and they were required to give their diagnosis (depression or anxiety) and their probability of the diagnosis being correct (between 50 and 100%). The authors created an 'evidence' score for each patient, based on the patient's symptoms and the assessment of how characteristic the symptom was for each condition. They conclude that there is a negative relationship between this evidence score and the confidence of the assessor (defined by the difference between her subjective probability and the proportion of correct diagnoses). Thus, there were signs of underconfidence when there was a high evidence score in favour of one condition. The authors do, however, query whether the relative simplicity of the assessment, compared to other studies, has an effect on this aspect of their findings.

10.4 The nuclear industry

There is some enthusiasm within the nuclear industry for using elicited probabilities, because of the sensitivity of the science involved and the need to think about time scales far beyond those of a practical scientific experiment. The following examples of elicitations in the nuclear industry are illuminating.

De Wispelare et al. (1995) discuss probability elicitation within the context of nuclear waste, eliciting distributions of the change in temperature during seven discrete periods of time, up to 10,000 years, in the vicinity of Yucca Mountain, Nevada. The elicitation process included a period (of a few weeks) where the selected experts were given the opportunity to do some work on data and read relevant literature before their views were sought. The elicitations were done using the variable interval method, with between five and nine quantiles being elicited from each expert for each time period. The cumulative distribution functions and probability density functions of the distributions were then produced. The paper also discusses the aggregation of the experts' opinions (there were five experts in the study), by both behavioural and mathematical methods. Four exercises were conducted in an attempt to gain a consensus view: (1) giving the experts their five individual distributions, (2) giving them mathematically aggregated distributions (using four different methods), (3) offering a mixture of the two and (4) allowing discussion of the substantive issues where there was disagreement between the experts. In the first exercise, consensus was reached quite easily but in the others, it was harder. The experts felt that they needed to have a number of distribution curves in order to represent the range of opinion for the second method. For the third, a consensus distribution could not be agreed upon, but a solution was proposed that involved taking points from the three different distribution curves. Similarly, in the fourth method, the experts could not agree about the size of the effects, with three separate curves being agreed upon in the end. Further analysis is given in Miklas et al. (1995) and further work on the use of expert opinion with regard to hazards in the nuclear power industry from volcanic activity in the same region can be found in Ho and Smith (1997).

In their paper discussing probability safety assessments (PSAs) for nuclear power plants Ang and Buttery (1997) adopt a map between numerical probability values and verbal descriptions of uncertainty, which they call a 'degree of belief scheme'. The values used, shown in Table 10.1, are inconsistent with the mapping used by Renooij and Witteman (1999), particularly in the conflicting numerical values ascribed to a 'probable' event. This highlights one of the dangers of using verbal descriptions of uncertainty in probability elicitations.

Table 10.1: The map between numerical probability and verbal descriptions of uncertainty used by Ang and Buttery (1997) and Renooij and Witteman (1999). Note that the word 'probable' represents a probability of 0.85 on one scale and 0.1 on the other.

Renooij and Witteman Verbal Descriptor	Probability	Ang and Buttery Qualitative Descriptor
Certain (almost)	1	
	0.9999	Extremely likely (i.e., almost certain)
	0.9	Very likely
Probable	0.85	
Expected	0.75	
	0.7	Likely
Fifty-fifty	0.5	Indeterminate
Uncertain	0.25	
Improbable	0.15	
	0.1	Probable
	0.01	Unlikely
	0.001	Very unlikely
	0.0001	Extremely unlikely
(Almost) Impossible	0	

Work on eliciting rupture probabilities has been explored by Vo et al. (1991). The experts shared their knowledge of component rupture and then gave their individual assessments of the failure probability of each component (a best estimate and two values, one greater than and one less than the best estimate, that the authors refer to as 'uncertainty estimates', without explicitly saying what they represent) on a paper questionnaire. The experts were also encouraged to write down their reasons for each assessment alongside. The assessments were aggregated across the group, using the best estimates of each expert only. Aggregation was achieved by finding the median and quartiles of the best estimates (for each component), the quartiles being used as the 'uncertainty estimates'. This may imply that the initial 'uncertainty estimates' were also quartiles, but this is not explained directly.

A complete edition of *Radiation Protection Dosimetry* (Volume 90, Number 3) discusses the use of expert opinion in probabilistic nuclear risk assessment. The volume includes examples of the use of a computer program from the European Community (as it was then known) called 'COSYMA – An Accident Consequence Assessment Package'. Additional work on the use of the results from probabilistic nuclear risk assessment can be found in Breeding et al. (1992) and Helton et al. (1992) and further work on elicitation in the nuclear industry can be found in Hofer et al. (1985).

10.5 Veterinary science

McKendrick et al. (2000) describe the creation of a disease diagnosis system, using a Bayesian belief network. The idea behind such a network is to use Bayes' Theorem to assess the probability that an animal has the disease in question given that it is showing a particular sign. Using two consultant veterinarians, the authors created a list of 27 signs which may or may not be present given a disease (the system covers 20 diseases in total). A questionnaire was developed, with a table for each disease. The row headings were the signs, and the expert was asked for an approximate percentage of the cases of the disease she had seen in which the sign had been present, by ticking the appropriate column. The options were: 0%, 1–20%, 21–40%, 41–60%, 61–80%, 81–99%, 100%. The questionnaire was completed by vets, selected for their expertise with each disease (some vets got one form and some got as many as five forms). McKendrick et al. chose to combine the expert opinions mathematically, using the normal distribution to model the variation among the vets. For each disease, the maximum likelihood estimate (MLE) of the probability of the sign being present (and its standard error) were calculated, the MLE being used in the belief network. An important consequence of this work is that the diagnostic software tool that incorporates the elicited probabilities (Cattle Disease Diagnosis System) is available on the Internet[1].

There are several other examples of elicitation methods being used in veterinary science. The aim of van der Fels-Klerx et al. (2002b) is to estimate the effects of bovine respiratory disease on the productivity of dairy heifers and they do this in two stages: gaining consensus on the definitions of the variables involved (using the Delphi method) and then eliciting probability distributions of the variables (using van Lenthe's ELI technique; see van Lenthe, 1993a and 1993b). The density functions of the 21 experts were combined mathematically, using software called EXCALIBR, by Cooke and Solomatine (1992). A description of eliciting from a heterogeneous expert panel is given by van der Fels-Klerx et al. (2002a), which also encourages the use of the Delphi method and the ELI technique of van Lenthe, using bovine respiratory disease as an example.

Horst et al. (1996) provide an introduction to risk modelling in the field of contagious animal diseases, concluding that they favour using subjective probability distributions. Horst et al. (1998) also used the ELI technique of van Lenthe to elicit opinion about contagious animal diseases. In this case, the minimum, maximum and most likely length of time the from first infection to the first detection and from the first detection to the end (i.e., 'when all measures are considered effective') were elicited. Further detail of the elicitation performed for classical swine fever can be found in Horst et al. (1997).

In another study relating to a contagious disease, Forbes et al. (1994) explore the opinions and beliefs of veterinarians about whether New Zealand is likely to have an outbreak of foot-and-mouth disease. Their methods involved using a postal

[1] see http://vie.dis.strath.ac.uk/vie/CaDDiS/

questionnaire and then gathering the experts together for a Delphi conference, at which the experts were encouraged to look at each others' results and discuss the issues before making their own re-assessments. Two methods were used for the original assessments on the postal questionnaire. The first method involved eliciting the minimum and the maximum, dividing the interval into 10 sections of equal length and asking respondents to assign probabilities to each range (i.e., employing the histogram technique). The second method was the variable interval method. The variable interval method alone was used for the re-assessments. In their discussion, Forbes et al. recognise one of the advantages of subjective probability elicitation – that it can be used in rare situations (at the time of their writing there had never been an outbreak of foot-and-mouth disease in New Zealand). They also note the attitude of the experts to the process – some were not sure about the process during the first postal survey but after they had been through the Delphi conference, the experts accepted the methods.

10.6 Agriculture

Smith and Mandac (1995) elicit subjective distributions from farmers in the Philippines about rice crop yields. Two groups of farmers were used, one group that farm on the plains and another group that farm on the plateaux. Their (prior) distributions for the next season's crop yield based on three different levels of nitrogen fertiliser were elicited using the histogram method. First, a minimum and maximum possible yield were elicited from each farmer and then the interval between these values was divided into five equally spaced sub-intervals (or bins). The farmer was then given 25 coins (whose value, in total, was about one day's wage) and was asked to divide them between the bins in such a way as to represent her probability that the yield would fall into that interval. Also, on each farmer's land, three plots were established on which rice was grown, applying one of the three levels of fertiliser to each plot, and the yields at the end of the season were measured.

The elicited distributions were compared to 'objective' estimates produced from an econometric model. The variances of the subjective distributions were smaller than those of the objective distributions, which Smith and Mandac suggest was due to the fact that the farmers only thought about recent weather history and cycles and did not consider the possibility of extreme weather conditions. The variances showed no relationship with the different levels of fertiliser applied, and the variances in yields from the plateaux were smaller than those from the plains. This was contrary to the authors' hypothesis that the distributions of the yields from the plateaux should have larger variances owing to the higher number of droughts that occur there. The following list, proposed by Smith and Mandac (1995, pp. 158) to explain the differences between the subjective and objective distributions, is informative.

1. There are methodological problems in the estimation of objective probabilities: the model specification could be incorrect or an inefficient method of estimation could have been employed.

2. Nitrogen responses may be inherently highly variable from field to field.
3. Farmers lack understanding of the causal structure underlying parameter values (this implies [a] lack of knowledge about [the] effect of nitrogen on yield and [the] effect of stochastic disturbances on the marginal product of nitrogen).
4. There are biases in probability judgements made by farmers.[2]
5. The elicitation procedure fails to capture the farmer's true risk perceptions.

Previously, Grisley and Kellogg (1983) had performed a similar study with farmers in Northern Thailand, eliciting prices and yields for some of the crops that they produced. They also used the histogram technique. In this particular study, the farmers were given 25 one baht coins. The farmers were also given a monetary incentive in that they were told that they would each receive the number of coins that they had placed in the interval in which the correct value, measured at harvest, fell. (However, see the criticism of the linerar scoring rule in Section 8.2.) Unusually for elicitation studies, the farmers were not required to give their opinions straightaway but were given one week to think about them. The study was done in an economic context, with the authors comparing the elicitations of prices and yields. The elicitations were compared with the actual results, achieved later in the year, and the conclusions drawn were that the farmers produced better estimates for rice (the crop they had traditionally grown in the region) than for tobacco, soybeans and peanuts. They report that the variability and uncertainty of the estimates were larger for the higher estimates of yield than for the lower ones. Finally, a regression analysis was performed on the difference between the expected and actual values with various socio-economic variables. Grisley and Kellogg conclude that mathematical and abstract thinking skills (found by testing the farmers) were not significant in explaining the elicitation performance.

10.7 Meteorology

Probabilities have long been used by weather forecasters. Daan and Murphy (1982) discuss the performance of weather forecasters in The Netherlands. They look, for example, at the 12- and 24-hour forecast probabilities of precipitation between 1972 and 1979. The probability elicitation method they used was to divide the 0–1 interval into ten sections of equal length (0.0–0.1, ..., 0.9–1.0) and ask for an integer between 0 and 9, which represented each interval. Their main result was that the calibration curves (which they refer to as *reliability curves*) of the forecasts were close to the line indicating perfect calibration, although there was a tendency to overforecast the 12-hour forecasts and underforecast the 24-hour forecasts. They

[2] Here the authors cite Anderson et al. (1977).

claim that overforecasting (giving higher probabilities than appropriate) is often exhibited by inexperienced probability forecasters and in situations where it is felt that an overforecast is more acceptable to those using the forecast than an underforecast. Further work in this area by Murphy and Daan (1984) compared weather forecasts from two years, where feedback had been given to the forecasters after the first year. (The elicitation technique was the same as before.) They found that the forecasts from the second year (after feedback had been given) were far more reliable (i.e., better calibrated) but there was only a small change in resolution (the ability to distinguish periods when it did rain from those when it did not). There was also a considerable reduction in overforecasting in the second year. The authors suggest that the provision of feedback was a major influence on the improvement in reliability/calibration in the second year, but put the small change in resolution down to the slow improvement in weather forecasting techniques at the time.

Abramson et al. (1996) developed a belief network for modelling severe weather conditions. For the elicitation of the probabilities for the network they used the direct method, asking people to complete tables with conditional probabilities. The authors note an observation they made of someone making his assessments. They felt that the tests for consistency which they had put into the process caused the person to move from thinking that "the numbers being assessed are vague and hard to specify" to "a resentment of the demands for consistency implicit in the checking process." The assessor felt that when the consistency check failed he should rethink all of the values across the distribution, not just a single value.

Other elicitation work in meteorology includes Vescio and Thompson (2001), Coles and Tawn (1996) (using expert opinion in extreme value analysis, with an example of rainfall data), Varis and Kuikka (1997) (using belief networks and assessing correlations between variables) and Morgan and Keith (1995).

10.8 Business studies, economics and finance

As part of a study into modelling and predicting the outcomes of organisational change, Gustafson et al. (2003) used a Bayesian network model and employed a full elicitation process. The experts were nominated after consideration of their knowledge of organisational change, the respect held for them by their colleagues and their ability to work well in a group. At the next stage, the model network was built using two inputs. Firstly, the authors elicited ideas from the experts about what 'implementation success' (of the organisational change) would mean to that individual. Secondly, they investigated which factors each expert felt would best predict the success or failure of the implementation. Once these had been collated, the conditional independence between factors was elicited. Finally, beliefs were elicited by asking questions in the form of likelihood ratios. For example:

> "Think about two healthcare improvement projects. One was successfully implemented and the other was not. Which one is more likely to have the following characteristics."

A factor level was then specified, such as "staff that hate the current situation and believe that change is essential." Both verbal and numerical responses were obtained.

Within the field of economics, Dominitz (1998) elicits opinions about future weekly earnings. The experts were asked to give a sequence of subjective probabilities, $F_1, \ldots, F_4$, that one year later weekly earnings would be less than four specified values, $W_1, \ldots, W_4$. These quantiles from the subjective cumulative distribution function were fitted to a log-normal cumulative distribution function by least squares methods. A similar method is used in Dominitz and Manski (1997), in which they estimate the inter-quartile range (IQR) from the data, finding that it increases as the estimate of the median increases, but less than proportionally, and the estimates of the IQR vary considerably at similar median estimates.

Dhar et al. (1999) investigate the effects of discount sales in the retail industry. Part of their study involved informing their subjects that $x\%$ of a store's stock of a particular item was in the sale and asking for their probability that they would be able to get a discount. This was done with a loose method of asking for either a point estimate or a range without any assistance offered as to how the subject could make the assessment.

Elicited subjective probabilities are used by Pattillo (1998) in her econometric study of Ghanaian manufacturing. She asked the owners of some manufacturing firms about their views of the potential change in demand for their products over the next year and the next three years. The owners were given a list of nine ranges of percentage changes and were asked to assign their probabilities for each range (i.e., the histogram technique). Pattillo used them to gain an understanding of the variance of the expected growth in demand, which was included in her econometric model.

Interest rates and inflation forecasts are studied by Lahiri et al. (1988). They elicit subjective probability distributions of the annual percentage change in the implicit price deflator by the histogram technique. From these distributions the first four moments were calculated, giving their mean, variance, skewness and kurtosis. Owing to the distributions being 'non-normal', the skewness and kurtosis of the distributions are suggested as useful, additional measures of uncertainty or risk.

The use of subjective probabilities has also been considered in accountancy and audit. Solomon (1982) discusses work with 103 auditors to elicit the quantiles of their subjective probability distributions for account balances. The paper explores three different aggregation methods which involved dividing the auditors into three groups. In the first group, 26 auditors worked alone. In the second one, 38 auditors were divided into groups of three (with one of two). Within these groups the auditors determined their own opinions before sharing them with the rest of their group and arriving at a group consensus. In the third group, the remaining 39 auditors were divided into groups of three. Within these groups, the auditors discussed their opinions and formed a group consensus distribution before being asked for their own, individual, quantiles. The conclusions were mixed. Neither the group method nor the individual method was found to be systematically the best. Poor calibration of individuals was a problem and interaction between auditors was found to be more reliable than aggregating the individual auditor assessments (by finding the mean of each quantile elicited). Further work on the calibration problem is discussed in Curtis et al. (1985) in which it is noted that the amount of over- or underconfidence and over- or under-estimation varied across the different types of account under consideration. They also observe that using the distribution of the best group member (rather than the statistical aggregation of the group) resulted in better calibration and smaller error.

Eliciting priors for the error occurrence rate from auditors is explored by Crosby (1980). He elicits from 42 experts their prior distribution for the error rate in an audit using the variable interval method and the equivalent prior sample (EPS) method (the variable interval method used obtained the 95, 75, 50, 25 and 5% quantiles and included checks for consistency). Beta distributions were then fitted to the assessments. As discussed in Section 6.3, the EPS method elicits a sample size for the planned audit directly (considered by Crosby to be the prior sample size, n_{PR}). The sample size implied by the variable interval method is calculated from the fitted beta distribution. After the elicitation, the auditors were asked to plan the extent of attribute sampling they would undertake. They were asked for an expected error rate, a confidence level and an upper precision level based on their judgement of a material error rate (they also gave the sample size that they would choose to test). The first three of these values were used to fit a beta distribution for what Crosby considers to be the auditor's posterior distribution of the error rate. From this beta distribution, the posterior sample size n_{PO} can be determined. Assuming a beta-binomial model, the sample size for the audit implied by the auditor's assessments is then given by $n = n_{PO} - n_{PR}$. For many assessments, the audit sample size was negative. The sample sizes for the audit calculated as described above, using a beta-binomial model and values elicited through both elicitation methods, were generally smaller than those directly elicited from the auditors at the end of the elicitation process, with the sizes from the EPS method being larger than those from the variable interval method.

Crosby also compared the variable interval method with the EPS method for audit error rates in Crosby (1981). His results showed that while the distributions produced by each method led to similar estimates of the mean, the EPS method produced distributions with smaller inter-quartile ranges. There was evidence that the auditors had a mental anchor point, in that the error rate would not be above 10%, which was reflected in the assessments made by the variable interval method. However, the EPS method does not allow for such a cut-off. Having asked the auditors which method they preferred to use, there was a mixed response (19 preferred the variable interval method and 23 preferred the EPS method).

More recently, Laws and O'Hagan (2002) have written about using hierarchical Bayesian models within 'multilocation auditing'. Their model requires the elicitation of several parameters in the prior distributions. The methods they employed include eliciting a 'best' estimate (which is interpreted as the mode) and an 'upper bound' for the average error rate. The 'upper bound' is actually the $100(1-\alpha)\%$

quantile, with α being a variable, but was set to a default value of 0.05. A beta distribution was fitted, with the elicited values being converted into the parameter estimates. Further work on auditing can be found in van Batenberg et al. (1994) and a small amount of elicitation work on financial forecasts can be found in De Bondt (1993).

10.9 Other professions

Walker et al. (2001), working in environmental and public health, give a detailed account of an elicitation of experts' distributions of the level of personal exposure to benzene. The experts were given a choice of the method they used; some choosing to start with a parametric distribution (e.g., the log-normal, giving their estimates of the mean and standard deviation on the log scale), whereas others gave their estimates of the mean and various quantiles (i.e., the variable interval method). The experts were also asked about the distributions of their uncertainty about their estimates of the arithmetic mean and the 90th percentile given in the first stage of the process; specifically, they were asked for their minimum, maximum, 90% credible interval, inter-quartile range and their median for both estimates. The experts had some problems with fully appreciating the idea of a credible interval for the mean. In more recent work, Walker et al. (2003) investigated the calibration of the earlier assessments, using graphical techniques and scoring rules. They found that the experts were "relatively well calibrated although on balance, *underconfident*", but that they were overconfident in their estimates of means.

Asset management within the water industry is addressed by the use of robust modelling in O'Hagan (1994). The method is based on the Bayes linear modelling approach, which requires just the elicitation of means, variances and covariances. O'Hagan suggests that robust results are best obtained in these circumstances by ensuring that the elicitation questions are as simple as possible and directly relate to the elicitee's understanding of the situation.

The use of expert judgement and Bayesian methods in occupational hygiene (hazards in the workplace) is described by Ramachandran et al. (2003). Their elicitation involved 11 experts, each providing assessments of the proportion of (each of) 4 hazard types at 12 different sites via a postal questionnaire. The authors decided to apply a uniform distribution to each proportion, with the experts being asked for their 95% (central) confidence interval for the proportion, that is, their 2.5 and 97.5% quantiles. The study reveals a problem with using uniform distributions in this way, which is that uniform distributions have a fixed range on which they take positive probabilities. The result is that, when the data used to update the distributions have a mean well outside the range of the priors, the updating process produces posteriors with distributions that do not contain the sample mean.

Morgan et al. (2001) elicit opinions from 11 experts about forest ecosystems, particularly about potential climate change. They interviewed each expert and used a range of elicitation techniques. First, they elicited a full probability distribution for a "likely percentage change in ... biomass... at least 500 years after the new climate had taken effect." Second, they used a log scale to elicit the number of species that will go extinct in the next 100 years. Third, they asked the experts to select from three probability ranges ($p \leq 0.2, 0.2 < p < 0.5, p \geq 0.5$) and then draw curves to indicate the change in proportions over time. The results from using these approaches are presented in depth from the ecology perspective but little is done to assess the success of the elicitation techniques. The authors observe that their results show that experts can have differences of opinion and that expert assessments are useful for sensitivity analysis.

Borsuk et al. (2001) elicited the opinions of a physical oceanographer about events in the Neuse estuary (North Carolina, USA), specifically the 'time between mixing events'. They used the fixed interval technique (for which the authors cite Morgan and Henrion, 1990) asking "if you were to observe 100 vertical mixing events, how many do you think would be less than x days apart?" and then fitting an exponential distribution to the results obtained. A similar method was used by Borsuk et al. (2002), using a survival model to investigate a species of estuarine clam (*Macoma balthica*). Borsuk elicits from his co-authors using questions of the type, "Given a *M. balthica* population of 100 individuals and a fixed ambient dissolved oxygen concentration of $1.0\,\text{mg}{\cdot}\text{L}^{-1}$, how much time would you expect to pass before there are x dead individuals?"

To conclude this section, eliciting costs in the US Department of Energy is discussed by Dillon et al. (2002). They used a process similar to that documented in Section 2.2.2 to identify and train experts. The cost was decomposed into 13 cost components and the minimum, maximum, 10, 50 and 90% quantiles for each component were elicited from each expert, after having given them a base cost estimate for each component. The opinions were aggregated using Monte Carlo simulation and then presented to the group of experts for discussion and revision. The estimates that were given for each cost component were generally found to be at the lower end of the elicited distributions, showing that anchoring was not a large problem with the assessments.

10.10 Other examples of the elicitation of subjective probabilities

Other examples of the use of subjective probability include the following.

1. Diagnosis and treatment decisions in medicine

    - Welch et al. (2000) consider eliciting subjective pretest probabilities of cardiac ischemia and compare them with so-called 'objective' pretest probabilities calculated by a score.
    - Eliciting symptoms of mental state, with reference to Bayesian methodology, is discussed by Jacob (2003).

- Saleh et al. (1997) appear to use some probabilities elicited from experts regarding hip surgery. There are no details about how the values were obtained.
- Spiegelhalter et al. (1993) discuss using a Bayesian belief network for congenital heart disease, briefly mentioning eliciting values for the priors.

2. Environment, Conservation and Ecology

- Sherrick (2002) explores the elicitation of subjective probabilities about weather variables from farmers in Illinois. He measures their opinions against past weather data and shows that the farmers overestimated the probabilities of negative events, which has an effect on their attitude to insurance against bad weather.
- Roosen and Hennessy (2001) elicited opinion from farmers about the percentage change in regional apple yield using the histogram technique.
- McDaniels (1995) uses elicitation as part of fisheries resource management. He uses the variable interval method to elicit the distribution of long-term equilibrium productivity.
- Wolfson et al. (1996) elicited opinions about radioactive groundwater contamination and soil lead contamination.
- Christen and Nakamura (2000) elicit the number of species of a particular animal (e.g., bats) that experts expect to observe in a given habitat.
- Whitehead (1992) elicits some probabilities for work on a sea turtle protection programme.
- Stiber et al. (1999) elicit probabilities for a Bayesian belief network that is used for analysing the contamination of groundwater with trichloroethane.
- Diekmann and Featherman (1998) consider probability assessment for costs in 'environmental restoration'.
- Hammond et al. (2001) study 'fish school cluster composition', discussing the elicitation of prior distributions for a Bayesian analysis.
- Qian and Reckhow (1998) elicit opinion about phosphorus contamination in wetlands.
- Further work on elicitation in climate change studies (see also the section 'Other Professions') is reported in Risbey et al. (2000).

3. Veterinary Science

- Sørensen et al. (2002) work on eliciting expert opinion about the control of milk fever.

- Stärk et al. (2002) discuss eliciting opinions about the infectiousness of salmonella in pigs, using the histogram technique.
- Smith and Wilkinson (2002) look at models for disease spread. They elicit the number of contacts that an animal might make in a given time period, from which contact probabilities are calculated.

4. Psychology

   - Normand et al. (2002) investigate depression by asking experts for the probability that a patient's depression status would be in each of four categories, namely full remission, mild depression, moderate depression and major depression, after a treatment programme.
   - De Jong and Muris (2002) consider elicitation of information about anxiety disorders.
   - Huber (1995) explores human experiences in ambiguous situations and compares this to situations where control exits.

5. Business Studies, Economics and Finance

   - Önkal and Muradoğlu (1996) explore the forecasting of stock prices. They use the histogram technique on the percentage change of a stock price over one week.
   - Bajari and Ye (2003) describe procurement auctions, particularly the detection of competition and collusion. Their complex statistical modelling includes a small amount of elicitation, obtaining expert opinions of the 25th, 50th, 75th and 99th percentiles of auction mark-ups.

6. Engineering

   - Sexsmith (1999) considers elicitation from experts in structural safety.
   - Coolen et al. (1992) use the histogram technique to assess priors in Bayesian reliability models.
   - Stephens et al. (1993) discuss deriving and using probability density functions for model parameters, giving a brief description of the variable interval method.
   - Hodge et al. (2001) consider the elicitation of engineering knowledge about reliability.

7. Law – Lindley (1991) discusses the use of subjective probability for making legal judgements.

8. Sport – Baseball, e.g., Steffey (1992), Basketball, e.g., Paese (1993), Hockey (field), e.g., Vertinsky et al. (1986).

9. Archaeology – Buck et al. (1996) and Buck and Millard (2004) provide a wide range of illustrations of the use of subjective prior information in archaeology and chronology building.

10. Game Theory – Nau (1992) discusses eliciting probabilities of the other players in a game.

11. Ophthalmology – Aspinall and Hill (1984) give an introduction to subjective probability and descriptions of betting methods and the variable interval method for ophthalmology.

12. Demography – Daponte et al. (1997) explore the Iraqi Kurdish population.

13. Emergency Services – Hirsch et al. (1998) conduct an elicitation using experts in firefighting.

14. Maps – Green and Strawderman (1994) elicit probabilities of pixel classification for thematic maps.

Chapter 11

Guidance on Best Practice

From Chapter 1 – Fundamentals of Probability and Judgement

- The distinction between aleatory and epistemic uncertainty is important for elicitation practice. Elicitation usually focuses on uncertainties that are either purely epistemic or have an epistemic component. However, people are most familiar with the concepts of probability in the context of aleatory uncertainties.
- It is important to remember that elicited statements of probability are judgements made in response to the facilitator's questions, not pre-formed quantifications of pre-analysed beliefs. The psychophysics literature suggests that all such judgements are intrinsically relative.
- The range-frequency compromise suggests that in some situations experts will tend to distribute their elicited probabilities evenly over (the whole or part of) the probability scale.
- Elicited probabilities may suffer from biases and non-coherence in practice, but the goal of elicitation is to represent the expert's knowledge and beliefs as accurately as possible.

From Chapter 2 – The Elicitation Context

- Elicitation is best conducted as a face-to-face interaction between the expert(s) and the facilitator.
- The elicitation process has several stages, as in Section 2.2.2, all of which are important. The actual elicitation comes at the end, and its success depends on the foundations built in the preceding stages.

Uncertain Judgements – Eliciting Experts' Probabilities A. O'Hagan, C. E. Buck, A. Daneshkhah, J. R. Eiser, P. H. Garthwaite, D. J. Jenkinson, J. E. Oakley and T. Rakow

From Chapter 3 – The Psychology of Judgement under Uncertainty

- Research shows that humans cannot be guaranteed to act as rational agents who follow the prescriptions of probability and decision theories. In many situations where people are called upon to provide judgements under uncertainty, they are liable to choose from a selection of easy-to-use strategies (heuristics). These strategies can be effective, especially when time or information is scarce, but they are not always optimal.
- Substantive expertise in a specialist area is no guarantee of normative expertise in providing coherent probability assessment. Careful thought needs to be given to the training in probability and statistics that the expert should receive at the beginning of the elicitation exercise. Furthermore, assistance and coaching as the elicitation proceeds will often be desirable.
- The facilitator and, quite often, the expert ought to be aware of the main biases that can occur in judgement under uncertainty and the reasons proposed for their occurrence. The elicitation process ought to be structured with this in mind.
- The facilitator should recognise the potential for inadvertently introducing bias into the elicitation process (e.g., anchoring effects induced by particular orders of questions) and should seek ways to reduce or avoid such problems.
- The level of detail with which an event is described can alter the probability assessments that are provided for that event. Particularly when eliciting assessments for a series of mutually exclusive events, it will be necessary to assess the coherence of the set of assessments provided by the expert.
- Several errors and biases can be attributed to adopting too narrow a focus: focussing too closely on a single instance of some broader class of events, or examining only one hypothesis as opposed to considering competing alternatives. The expert may need to be actively encouraged to avoid this.
- It can be helpful to introduce procedures that explicitly encourage an expert to think analytically. For instance, the facilitator should explore whether aids to judgement are available that help the expert to analyse all the substantive expertise that is relevant to a particular assessment. Or, the facilitator should consider whether and how the expert could be involved in the process of checking her own set of assessments for coherence.
- The facilitator and expert ought to consider how the expert's particular experience (her access to samples of data) may influence or bias the assessment of uncertain quantities. A full consideration of the uncertainty associated with a quantity ought to take into account how well the expert's past experience informs the specific elicitation tasks.

- Where the 'outputs' of an elicitation exercise become 'inputs' for decision analytic models, the sensitivity analyses on such models can usefully be informed by research on judgement biases. The statistician or decision analyst can use what is known about the likely direction and magnitude of biases to help him set the range of possible parameter values to examine within the model.

From Chapter 4 – The Elicitation of Probabilities

- It is important that feedback is given on training exercises to help the expert learn how to make better probability assessments.
- Subjective probabilities can be well calibrated, but often they are not. This highlights the importance of training and ought to be taken account of in sensitivity analyses (if expert assessments are subsequently used in decision analysis).
- The interpretation of verbal expressions of uncertainty varies considerably across individuals and situations. Attempts to impute specific values to them that do not take account of this are fraught with danger.
- Alternative ways of describing uncertain events or eliciting a response using different response scales can prompt experts to change their information-processing strategies. These can result in alternative assessments being made. Facilitators ought to be aware of how sensitive probability judgements can be to changes in language and response format.
- Aids and procedures for debiasing judgement ought to be considered. However, people can be resistant to their implementation, and facilitators ought to be aware of the dangers of imposing debiasing measures upon the expert without consultation.

From Chapter 5 – Eliciting Distributions – General

- Eliciting a probability distribution in practice entails eliciting a (relatively small) number of summaries from the expert and then fitting a suitable distribution that conforms to those elicited judgements. It is important to recognise that the fitted distribution implies many more statements that the expert did not make, and feedback may be used to check that these implications are acceptable to the expert.
- To elicit a univariate distribution, summaries based on probabilities are most widely used, including individual probabilities, quantiles, credible intervals and (in some cases) ratios of probabilities. There is conflicting evidence about which summaries are elicited most accurately, but in general it seems that eliciting credible intervals with high probabilities (e.g., 90% or higher) should be avoided.

- Although there is some evidence that people find it easier to think of proportions than probabilities, it is impractical to elicit epistemic uncertainties in this way. However, training in the assessment of probability can use proportions to give an interpretation of the probability scale.

- The task of eliciting a multivariate distribution is substantially more complex, and it is important to begin by structuring the task, for instance, in terms of independent variables (whether the direct variables of interest, functions of them or latent variables).

- Approaches to eliciting association in a joint distribution include joint and conditional probabilities and regression relationships.

- Any fitted distribution is an approximation to the expert's true underlying beliefs and will differ from the ideal distribution because (a) there are many possible distributions that would fit the elicited summaries and (b) the expert cannot, in practice, specify the requested summaries accurately and reliably. It is important to recognise in some way the resulting uncertainty or imprecision in the fitted distribution.

From Chapter 6 – Eliciting and Fitting a Parametric Distribution

- In complex situations, structure must be imposed on the probability distribution used to represent the expert's opinion for the elicitation problem to be manageable. Usually, it is assumed that her opinion can be modelled by some specified parametric distribution, when the elicitation problem reduces to estimating the parameters of the distribution.

- For multivariate problems, it is typically assumed that opinion can be modelled by a conjugate prior distribution. This assumption is also commonly made with univariate problems, especially if it is envisaged that sample data will become available.

- Assumptions about the prior distribution should be checked with the expert, as the purpose of the distribution is to represent her opinions. Feedback and overfitting are recommended and can help identify an inappropriate distribution.

- Feedback and overfitting can also highlight assessments that are out of line. Inconsistencies in an expert's assessments may be reconciled through reassessment by the expert or through some form of averaging.

- The most widely used tasks in elicitation methods are the assessment of a central measure (a mean, median or mode), the assessment of quantiles and the revision of opinion when given hypothetical sample data. To elicit quantiles, the variable interval method is most generally used. The most commonly elicited quantiles are the lower and upper quartiles, which can be assessed using the method of bisection, but there is some evidence that the assessment of 0.33 and 0.67 quantiles leads to distributions that are better calibrated.

- Many elicitation methods reported in the literature have only been devised theoretically and have never been used to elicit anyone's opinions. In contrast, other methods have been developed with careful consideration given to practical details and they have been tested and used in real problems by genuine experts. When there is a choice of elicitation method, in practice it is clearly better to use one of the latter methods.

- Many distributions are available to statisticians that are more flexible than the commonly studied conjugate families, but in most cases, no elicitation methods for these have been proposed in the literature.

- Interactive computing is almost essential if an elicitation method requires a sequence of questions in which some questions are determined by an expert's earlier answers. Benefits from interactive computing include the facility to provide feedback to the expert and to identify apparent inconsistencies in her assessments as they are made. Interactive graphics have also been used in various problems and seem to be particularly useful in regression problems.

- The development of a decent elicitation method requires the knowledge of certain areas of the psychology literature and, especially in multivariate problems, it also requires a degree of statistical expertise. In addition, implementing an elicitation method can require computing skills. Thus, interdisciplinary teams are best suited to their development.

From Chapter 7 – Eliciting Distributions – Uncertainty and Imprecision

- The accuracy of any fitted distribution as a representation of the expert's opinion is compromised by (1) imprecision in the expert's stated summaries and (2) the fact that only a limited number of summaries can be elicited in practice.

- While there is as yet no consensus on how to express the implied uncertainty in the elicited distribution (e.g., probabilistically or using upper and lower bounds), it is important to acknowledge it explicitly when asserting a fitted distribution.

From Chapter 8 – Evaluating Elicitation

- One form of accuracy for an elicited probability distribution is that it should accurately reflect the assessor's opinion. Another is that it should accurately reflect reality. The latter is easier to measure but the former should be paramount, as it is the defining quality of a subjective distribution.

- Poor calibration of an expert's judgements, or poor performance on scoring rules, may reflect inaccurate knowledge rather than poor elicitation. Although most of the psychological literature is based on such measures, to the extent that consistent biases are found across a variety of experts and knowledge areas, they point to sources of poor elicitation.
- Proper scoring rules are useful for comparing assessors and assessment methods. They also provide an incentive for experts to record their opinions well and can play a part in training experts to quantify their opinions accurately.
- On the basis of the frequency of use, the probability scoring rule (also called the *quadratic score* or the *Brier score)* is the preferred scoring rule for judging subjectively assessed *discrete* distributions and the logarithmic scoring rule is the preferred rule for continuous distributions.
- In principle, various decompositions of the probability scoring rule can be used to focus on different features of a set of assessments (such as calibration), but their efficacy is unclear in practice.
- There has been only limited empirical success at relating coherence to better calibration.
- The most natural way of improving the accuracy with which a subjective distribution represents an expert's opinion is through feedback.
- Where possible, overfitting should be coupled with feedback to avoid 'leading questions'.
- Interactive software is almost essential for the effective use of feedback.

From Chapter 9 – Multiple Experts

- The simple average (equal-weighted linear opinion pool) of distributions from a number of experts provides a simple, robust, general method for aggregating expert knowledge.
- More complex mathematical aggregation, such as the method of Cooke (1991), has the potential to perform better but success depends on having a well-structured elicitation process. Experts should be chosen to have a broad range of substantive knowledge without undue overlap.
- Group elicitation probably has even greater potential, since it can bring better synthesis and analysis of knowledge through the group interaction. However, success depends even more on the abilities of the facilitator, who must encourage (a) the sharing of knowledge (as opposed to opinions), (b) the recognition of expertise and (c) the study of feedback, but must avoid (d) the group being dominated by shared knowledge or over-strong opinions, (e) the kinds of biases found in individual assessments and particularly (d) the tendency of groups towards overconfidence.

Chapter 12

Areas for Research

From Chapter 1 – Fundamentals of Probability and Judgement

- To what extent do the existence of an absolute scale (0 to 1) for probabilities and the way that training usually gives the expert other anchors or landmarks on the scale allow absolute (rather than relative) judgements?
- What are the implications of the range-frequency compromise in the context of probability elicitation?
- Does elicitation using proper scoring schemes (as propounded by Savage and others) lead to less accurate assessments than the direct elicitation of probabilities?

From Chapter 2 – The Elicitation Context

- The value of many aspects of the overall elicitation process has not been formally studied. For instance, does keeping a transcript of the interview help the expert(s) commit more fully to the exercise or might it inhibit them?

From Chapter 3 – The Psychology of Judgement under Uncertainty

- Is it possible, or useful, to screen experts for those with sufficient normative expertise to take part in an elicitation exercise?
- In the context of elicitation, what kinds of training and aids to judgement are effective in reducing biases in probability judgement?

Uncertain Judgements – Eliciting Experts' Probabilities A. O'Hagan, C. E. Buck, A. Daneshkhah, J. R. Eiser, P. H. Garthwaite, D. J. Jenkinson, J. E. Oakley and T. Rakow

- It is known that different methods have a tendency to yield different assessments, possibly ones that are biased in different directions. Does this mean that more accurate or more coherent assessments can be achieved by using multiple methods when an individual expert provides an assessment of an uncertain quantity?

From Chapter 4 – The Elicitation of Probabilities

- What are the best ways to provide the substantive expert with the normative expertise that they need to contribute fully to an elicitation exercise?

- The calibration of subjective probabilities is generally better for predictions of future events than it is for the truth of almanac items. Is this because the uncertainty associated with future events is *both* epistemic *and* aleatory, rather than solely epistemic? Or, is this because the typical study populations for these kinds of study differ in the degree of substantive expertise that they can bring to bear on the task? Determining the implications that calibration research has for practical elicitation partly depends upon the answers to these questions.

- How should the *expert's* preferences for carrying out the elicitation be taken into account? Sometimes, there are different ways in which an elicitation task can be structured, for instance, when different numbers of response categories or different types of response scale are feasible. In such cases, how helpful is it to ask the expert how they would like to proceed?

- Sometimes the effects of different forms of language or response formats are known, but one cannot say which alternative is the best. How does one decide which forms of language or response formats are the best in a given situation? Could it be useful to use multiple methods with the same expert as a means of checking for coherence?

- When it is possible to obtain probability assessments from different sources (multiple experts and/or methods), how should one proceed? Should one average across several sources or should one seek to identify the single best method that one believes will be most accurate?

- How can the facilitator best work with the expert to determine what debiasing methods can be applied and to decide how they should be employed?

From Chapter 5 – Eliciting Distributions – General

- The distinction between eliciting epistemic and aleatory uncertainty has not been clear in much of the research literature. There is a need for study into the extent to which research on people's ability to assess proportions, means, variances and correlations in random samples can provide guidance on eliciting epistemic uncertainty.

- Although univariate summaries have been well studied, there has been little research on eliciting multivariate distributions. It is not clear what kinds of probabilities or other summaries are most useful or how many summaries are needed to elicit a joint distribution effectively in two or more dimensions. In particular, the conjectures in Section 5.3.3 concerning possible biases in the assessment of joint and conditional probabilities should be tested empirically.

From Chapter 6 – Eliciting and Fitting a Parametric Distribution

- Empirical research is needed to examine and compare some of the many elicitation methods that have been proposed but never tested.
- Effective elicitation methods need to be developed for several standard problems. For instance, methods are needed for quantifying opinion about a positive scalar, such as the parameter of a Poisson distribution, the parameters of a univariate normal distribution and those of a gamma distribution.
- Flexible user-friendly software for eliciting distributions needs to be developed by inter-disciplinary collaboration between experts in statistics, psychology and programming.

From Chapter 7 – Eliciting Distributions – Uncertainty and Imprecision

- Much more research is needed into formal expressions of imprecision in elicited probability distributions.
- To what extent is it possible to determine a limit on the achievable accuracy of any elicited distribution? Is it lower when eliciting joint distributions than marginal distributions?

From Chapter 8 – Evaluating Elicitation

- The score given to an assessed distribution is largely determined by the accuracy of the mean of the distribution, with only a minor role being played by the variances, covariances and degrees of freedom parameters. Methods of decomposing scores so as to focus on these parameters are needed so that ways of eliciting them can be evaluated effectively.
- Work on whether agreement between an expert's probabilities and reality related to coherence has been limited and should be extended. In the same vein, work is needed that examines whether coherence relates to the accuracy with which an expert's opinions are represented.
- Using feedback in elicitation methods appears to work well in practice. However, its benefits should be demonstrated rigorously by experiment and the usefulness of different forms of feedback should be compared. Also, the presumed superiority of overfitting over simple feedback should be tested empirically.

From Chapter 9 – Multiple Experts

- There is a need for more research into behavioural aggregation of probability judgements and probability distributions (as opposed to decisions or estimates). As with individual elicitation, the use of substantive experts in such experiments is much preferred.
- There is still much to be done to identify what characteristics of the experts, the facilitation and other aspects of the elicitation process are conducive to success with the more complex mathematical aggregation methods or group elicitation.

Glossary

A

Aleatory uncertainty: Two general types of uncertainty exist: aleatory and **epistemic**. Aleatory uncertainties can be characterised as due to randomness. They are inherent, irreducible and unpredictable in nature.

Anchoring-and-adjustment: A cognitive **heuristic** used for quantitative judgements, where a final judgement is made by adjusting from an initial value ('the anchor'). The **heuristic** is typically an efficient way of making a judgement; but is prone to produce biased judgements because people often make insufficient adjustment from the initial anchor. This is particularly problematic when people are provided with an 'arbitrary' anchor (e.g., a randomly chosen value) – as final judgements will then lie too close to the anchor. The heuristic implies that the order in which a series of values are elicited will influence later values in the sequence.

Arithmetic mean: See **mean**.

Association: Two uncertain variables (or observables) are said to be associated if the value taken by one tends to be linked to the value of the other. It is usual to think of association in terms of random samples, where, for instance, if high values of one variable tend to occur with high/low values of the other, we say that the variables have positive/negative (monotone) association. (More complex, non-monotone forms of association are also possible, such as when a very high *or* very low value of one variable tends to occur with a high value of the other.) When eliciting an expert's beliefs about two **parameters**, association should be interpreted as saying that if the expert were to learn the true value of one **parameter**, then this would affect the expert's beliefs about the other **parameter**. For instance, in the case of positive monotone association, the larger the first **parameter** is found to be, the larger the expert will believe the second one likely to be.

Uncertain Judgements – Eliciting Experts' Probabilities A. O'Hagan, C. E. Buck, A. Daneshkhah, J. R. Eiser, P. H. Garthwaite, D. J. Jenkinson, J. E. Oakley and T. Rakow

Availability: A cognitive **heuristic** used for judging the **probability** or **frequency** of events, or the size of classes, where judgements are made according to how easily examples of the event or members of the class come to mind. The **heuristic** is typically effective, as examples of frequent events or large classes are called to mind more readily than those infrequent events or small classes. However, the **heuristic** can result in biased judgements, as features unrelated to event probability/frequency or class size will also influence how easily examples can be generated. For instance, low **probability** events that attract disproportionate press attention will often be judged to be more probable than is actually the case.

B

Base-rate effect (base-rate neglect): A tendency for people to ignore or underweight the average **frequency** or a priori **probability** of an event and to base **probability** estimates only upon information about the individual instance currently being assessed – even when this 'individuating information' has limited diagnostic power. Thus, a particular event, known to be rare in the physical world, may be assigned a reasonably high **probability** on the basis of imperfect indicators if these indicators are judged to be **representative** of the event.

Bayes' theorem: This states how to update one's probabilities after receiving new data. For an uncertain event E and data D, Bayes theorem states that

$$P(E|D) = \frac{P(E)P(D|E)}{P(D)}.$$

Note that for a **repeatable event** E, *all* statisticians, regardless of whether they would classify themselves as Bayesian or frequentist, would use Bayes' theorem in this context. For a continuous uncertain variable θ, the equivalent version of Bayes' theorem states that

$$f(\theta|D) = \frac{f(\theta)f(D|\theta)}{f(D)},$$

where $f(\theta|D)$ is known as *the posterior density*, $f(\theta)$ is the prior density and $f(D|\theta)$ is the likelihood function. The denominator $f(D)$ simply ensures that the posterior density integrates to 1.

Bayesian reasoning: The application of the principles of **Bayesian statistics** to scientific reasoning in general.

Bayesian statistics: The fundamental principle of Bayesian statistics is that it is appropriate to describe uncertainty about unknown **parameters** in terms of a probability distribution. Almost all followers of the Bayesian approach would then accept that subjective probability is the correct tool for describing one's uncertainty about an uncertain quantity or event. Operationally, a Bayesian will formulate a joint **prior** probability distribution for the unknown **parameters** and update this in the light of new data by deriving the **posterior distribution** for the **parameters** using **Bayes' theorem**. This approach is quite different from that of **frequentist statistics** in which **parameters** are not permitted to have probability distributions.

Behavioural aggregation: The process of generating a single consensus judgement from an interacting group. The group's judgement is arrived at after some process of discussion, information sharing and/or negotiation (and not solely by the application of a mechanical rule for aggregating judgement).

Bernoulli distribution: A random variable representing a Bernoulli trial has a Bernoulli distribution. It has one **parameter** usually denoted by p, which is the probability that the trial results in a 'success', rather than a 'failure'.

Bernoulli trials/Bernoulli process: A Bernoulli trial is a single trial of an event that has two possible outcomes, usually encoded 'success' and 'failure'. A series of **independent** Bernoulli trials is called a *Bernoulli process*.

Beta distribution: The Beta distribution is a continuous probability distribution over the range 0 to 1. It has two **parameters**, usually encoded α and β, which define its shape. If X is a random variable with a beta distribution, we write $X \sim Beta(\alpha, \beta)$ and its probability density function (pdf) is

$$f(x) = \frac{1}{B(\alpha, \beta)} x^{(\alpha-1)}(1-x)^{(\beta-1)}$$

where $B(\alpha, \beta)$ is called the *beta function* and is defined by $B(a, b) = \frac{\Gamma(a)\Gamma(b)}{\Gamma(a+b)}$, where $\Gamma(a)$ is the gamma function.

The shape of the distribution is affected by whether the **parameter** values are greater than or less than 1. The range of the distribution, between 0 and 1, and its variety of shapes make the Beta distribution a good distribution for representing beliefs about a probability (for some examples, see Figure G.1).

Bias: In psychology, any process or influence that makes one form of response or behaviour more or less likely to occur than would be expected from the content of the input or presented stimulus considered by itself. In frequentist statistics, the difference between the expectation of an estimator and the true value of the **parameter** being estimated.

Bimodal: See **mode**.

C

Calibration of subjective probabilities: The process of comparing **subjective probabilities** for event outcomes with the observed **relative frequency** of

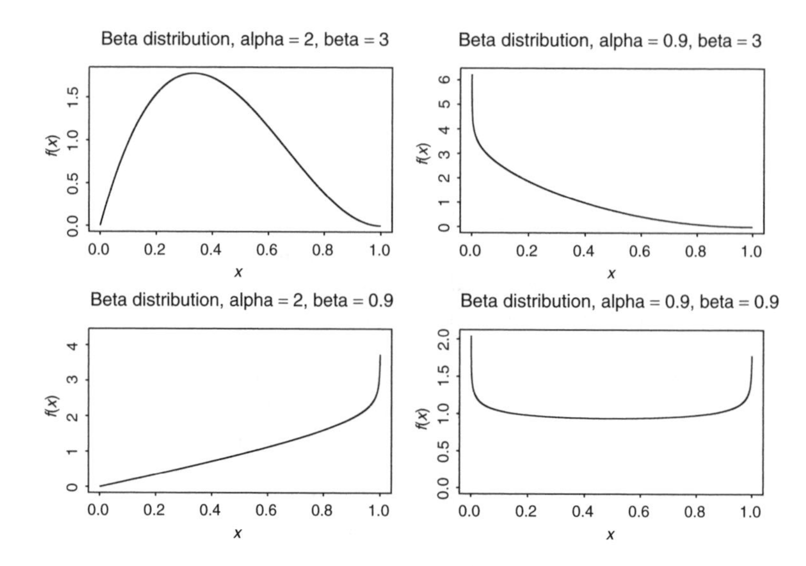

Figure G.1: Beta distribution of various shapes.

these events. For perfectly calibrated judgement, for all sets of judgements of identical subjective probability, π, the observed relative frequency of the specified events should be π. The calibration of subjective probabilities can be represented graphically by means of a calibration curve (Figures G.2 and G.3). In psychological studies, one of two task formats is commonly employed: (a) full range (see Figure G.2), or (b) half range (see Figure G.3). Full-range task formats allow response across the entire probability range, $0 < p < 1$; for example, "What is the probability of rain in Sheffield tomorrow?" Half-range tasks require (implicitly or explicitly) specifying which of two options is more likely to occur, before specifying a response in the upper half of the range, $0.5 < p < 1$, for example, "Do you think it will rain in Sheffield tomorrow?" then, "How confident are you?" (Most, but not all, experimenters have not allowed responses below 0.5, as an answer to the first question implies $p > 0.5$ for that response.) Poorly calibrated probability judgement most commonly exhibits **overconfidence** or poor discrimination.

Central tendency measures (or central tendency parameter): Measures of central tendency are numerical summaries (typically a single number) of an empirical (observed) or theoretical distribution. Such measures are usually devised or selected in an attempt to convey the central or most likely values of distributions. The most common measures of central tendency include the **arithmetic mean**, the **mode** and the **median**. See also **geometric mean** and **location parameter**.

Classical statistics: Synonymous with **frequentist statistics**.

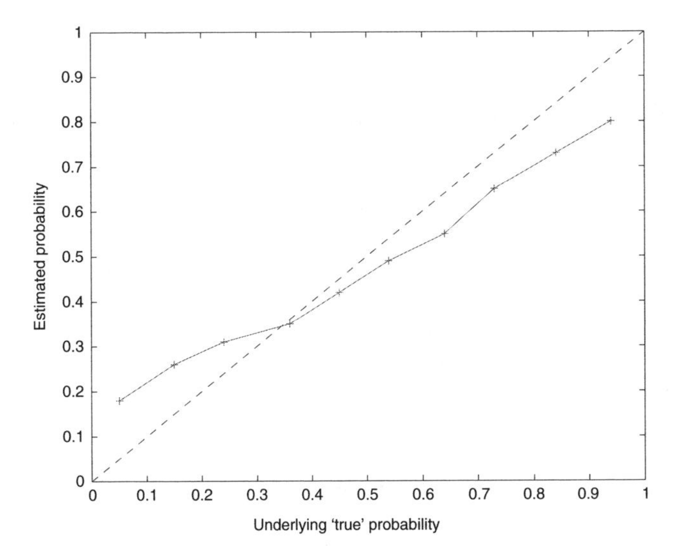

Figure G.2: Calibration curve (solid line) which represents full-range calibration data. Perfect calibration would be represented by the dashed diagonal line.

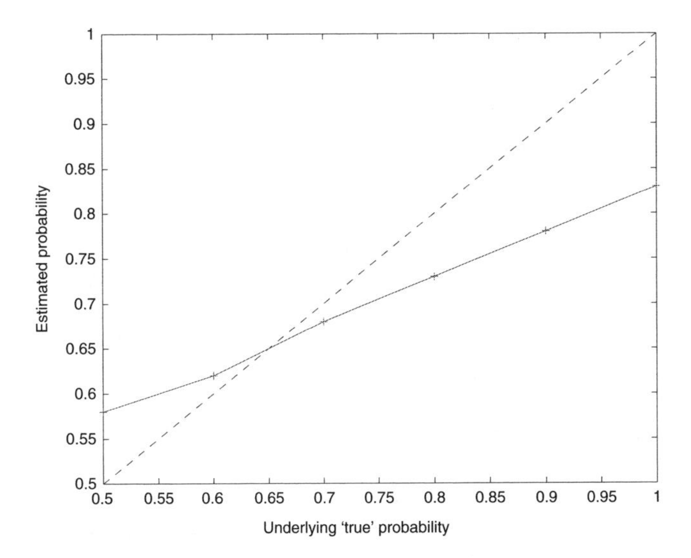

Figure G.3: Calibration curve (solid line) which represents half-range calibration data. Perfect calibration would be represented by the dashed diagonal line.

Coefficient of variation: A random variable's coefficient of variation is its **standard deviation** divided by its **mean**.

Cognitive factors: Factors pertaining to cognition (the use or handling of knowledge). Cognitive approaches to psychology consider human behaviour as the outcome of several information-processing tasks, including memory, attention, associative learning, reasoning, judgement and decision-making.

Coherent: A set of probability judgements are said to be coherent if they do not contradict each other according to the laws of probability. As an example of incoherent probability judgements, suppose England are to play France at football. I state $P(\text{England win}) = 0.3$, $P(\text{France win}) = 0.4$ and $P(\text{draw}) = 0.4$. But from the first two probabilities, $P(\text{draw}) = 1 - P(\text{England win}) - P(\text{France win}) = 1 - 0.3 - 0.4 = 0.3$, contradicting my assertion that $P(\text{draw}) = 0.4$.

Conditional distribution (probability): Conditional probability refers to the probability of an event, say B, given that another event, say A, occurs. This is written as $P(B|A)$, and satisfies the equation $P(B|A) = \frac{P(B \text{ and } A)}{P(A)}$. The conditional distribution is, by extension, the distribution of a random variable or vector, say Y, given the value of another random variable or vector, say X. The conditional distribution function is written as $F(y|x)$ and the conditional density function is written as $f(y|x)$. If the value of X does not effect the distribution of Y, that is, $f(y|X = x) = f(y)$, then Y and X are said to be (statistically) independent (see also **independence**).

Conditional independence: Suppose X_1 and X_2 are the heights of two randomly selected adult males and that the heights of adult males are described by a normal distribution with mean m and variance v. If I am uncertain about m and/or v then, according to the Bayesian approach to statistics, X_1 and X_2 would not be independent; learning the value of X_1 would give me new information about m and v, and so change my beliefs about the likely value of X_2. However, if I already know the values of m and v, then X_1 and X_2 are *conditionally independent* given m and v; learning the value of X_1 tells me nothing I did not know before about m and v, and so my beliefs about X_2 are unchanged.

Confusion of the inverse: A frequently noted tendency for people to confuse a **conditional probability** with its inverse – namely, to state, to infer, or to act as if $P(A|B) = P(B|A)$. For instance, doctors are often seen to confuse test sensitivity, $P(\text{positive test}|\text{no disease})$, with the predictive value of a positive test, $P(\text{disease} \mid \text{positive test})$ or to confuse test specificity, $P(\text{negative test} \mid \text{disease})$, with the predictive value of a negative test, $P(\text{no disease} \mid \text{negative test})$.

Conjugate prior: Given data x about an uncertain **parameter** θ, a conjugate prior $f(\theta)$ is a prior distribution with the property that when combined with the likelihood $f(x|\theta)$, the posterior distribution $f(\theta|x)$ has the same form as the prior distribution. For example, suppose we have binomial data; x 'successes' observed in n trials where the population proportion θ of successes is uncertain. The conjugate prior for θ is the **beta distribution**; when this is combined with the **likelihood function** $f(x|\theta)$, the **posterior distribution** is also a beta distribution. Conjugate priors are mathematically very convenient, as the resulting posterior distributions are very easy to derive.

Conjunction fallacy: The tendency to assess the probability of the conjunction of two events or propositions, $P(A \bigcap B)$ as larger than either of its components $P(A)$ or $P(B)$. For instance, detailed descriptions of a person listing several (plausibly associated) attributes may be judged as more probable than descriptions mentioning a single attribute.

Conjunctive (and disjunctive) propositions: In logic, conjunction and disjunction combine propositions in the same way that intersection and union (respectively) combine events. (It may be noted that **subjective probability** is more naturally formulated for propositions than events.)
Disjunctive propositions are of the form 'Either A or B'; such a proposition is 'True' if and only if at least one of these propositions are 'True'.
Conjunctive propositions, or conjunctions, sometimes called *joint assertions*, have the form, 'A and B'; such a proposition is 'True' if and only if both propositions are 'True'.

Conservatism: The tendency to revise judgement on the acquisition of sample data by less than that which would be prescribed by **Bayes Theorem**. The tendency is a common one in experimental studies of belief revision, but stands in apparent contradiction to the **overconfidence** that is often seen in **calibration** studies – as overconfidence implies that assessors assume that the information they have removes more uncertainty than is warranted by the data.

Continuous random variable: A random variable that can take any value within a range, as opposed to a **discrete random variable** which must take one of a set of separate values. For example, a person's height is a continuous variable because it can take any positive value and, in principle, one's height might be 1.8769302286 m (or even take any value between this and 1.8769302287 m).

Copula: A copula is a function that constructs a joint distribution for a set of variables from the univariate marginal distributions of the individual variables. There are various forms of copula available, and in many cases they are equipped with parameters that may be chosen to introduce varying degrees of correlation between the variables.

Correlation coefficient and correlation matrix: A correlation coefficient is a measure of **association** between two variables. It is a statistical measure of the extent to which variations in one variable are related to variations in another. There exist a number of different formulae (or formal definitions) for correlation coefficients. The usual (Pearson's) coefficient measures the strength of linear association, while others (e.g., Spearman's coefficient) convey the strength of monotonic relation and are suitable for summarising the strengths of a range of linear and non-linear relationships between variables. Whatever definition is used, correlation is reported on the scale -1 to $+1$, where -1 represents perfect negative correlation and $+1$ represents perfect positive correlation. The value 0 represents no correlation. **Independence** between two variables implies that they have zero correlation, but the converse is not true. Where we need to consider more than two variables, we can compute a correlation for each pair of variables. Such correlations can be reported in a square matrix known as a *correlation matrix*.

Covariance and covariance matrix: Covariance is a measure of **association** between two variables. The Pearson **correlation coefficient** is the result of scaling the covariance by dividing it by the product of the **standard deviations** of the two variables; this must then lie between -1 and $+1$. In the case of a random sample of values x_i and y_i for two variables X and Y, the *sample* covariance is

$$\mathrm{Cov}(X, Y) = \frac{\sum_{i=1}^{n}(x_i - \overline{x})(y_i - \overline{y})}{n-1},$$

where $\overline{x}$ is the sample mean of the x values and $\overline{y}$ is the sample mean of the y values. For two uncertain random variables or parameters X and Y, the (population) covariance is defined in terms of **expectations** as

$$\mathrm{Cov}(X, Y) = E\{(X - E(X))(Y - E(Y))\}.$$

Thus, covariance is a statistical value that measures the simultaneous deviations of X and Y variables from their means and its definition is a natural extension of the definition of the **variance** for a single variable. Where we need to consider more than two variables, we can compute a variance for each variable and a covariance for each pair of variables. Such variances and covariances can be reported in a square matrix (which is symmetric since, from the formulae above, the covariance between X and Y is the same as the covariance between Y and X). Such a matrix is known as a *variance–covariance matrix*, a *covariance matrix* or a *variance matrix*.

Credible interval or credible region: A credible interval is a range in which the value of a particular random variable or uncertain parameter has a specified probability (called the *coverage probability*) of lying. For instance, the interval from a to b is a 95% credible interval for a parameter X if the probability that X lies between a and b is 0.95. Such intervals can be constructed in several different ways and the method selected for a particular purpose will depend on the nature of the distribution to be summarised and on the use to which it is to be put. An equal-tailed credible interval has a and b chosen such that there are equal probabilities that X will lie below a or above b. Another form of credible interval or credible region is a highest posterior density (HPD) interval or region. The interval from a to b is an HPD interval if all values of the parameter that lie between a and b have a higher probability density than those that lie outside this range. For unimodal (i.e., single peaked) distributions, there is an HPD interval for any required coverage probability, but for multi-modal distributions, certain coverage probabilities will lead to an HPD *region* made up of two or more disjoint intervals. HPD intervals or regions can be used to summarise any probability density and, as such, the terms highest prior density region and highest posterior density region are also common. Credible intervals/regions are used in **Bayesian statistics** instead of the **frequentist** confidence intervals.

Cues and cue weights: Variables that inform (social, cognitive or perceptual) judgement are conventionally referred to as *cues* in several psychological paradigms. If cues exert differential influence upon a judgement, they are described as being weighted differently – hence they are ascribed different cue weights. For instance, if the relation between multiple cues and a judgement is analysed by regressing judgement onto cues (across multiple cases), the resulting regression coefficients are often taken to be indicators of the weight given to each cue.

Cumulative distribution function (CDF): The CDF of a random variable X is $F(x) = P(X \leq x)$, the probability that X is less than or equal to x. The CDF is an increasing function: if $y \geq x$, then $F(y) \geq F(x)$. If X is an absolutely continuous random variable, then $F(x)$ is a smooth function. For example, the graph of the CDF of the standard normal distribution is shown in Figure G.4. It is a smooth curve and its 'S shape' is typical of a unimodal distribution. For discrete random variables, the cumulative distribution is a step function. For example, the CDF of the single roll of a fair, six-faced die (with the numbers one to six, one on each face) is shown in Figure G.5.

D

Decision-making: Selecting a course of action. If a course of action is selected from a prescribed set of options, psychologists refer to this as choice, whereas more general decision-making may be made without first defining a set of options.

Decision maker: The agent who makes the decision. In psychology, it is not assumed that the agent is necessarily rational.

Decision theory: A body of knowledge and related analytical techniques of different degrees of formality designed to help a **decision maker** choose among a set of alternatives in the light of their possible consequences. This theory can

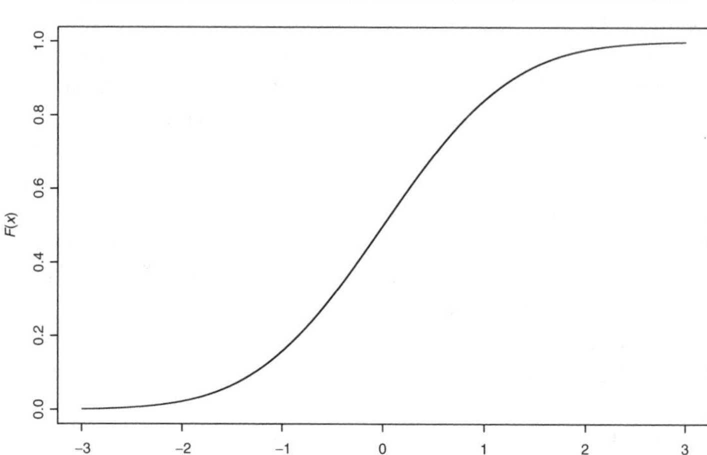

Figure G.4: CDF of the standard normal distribution.

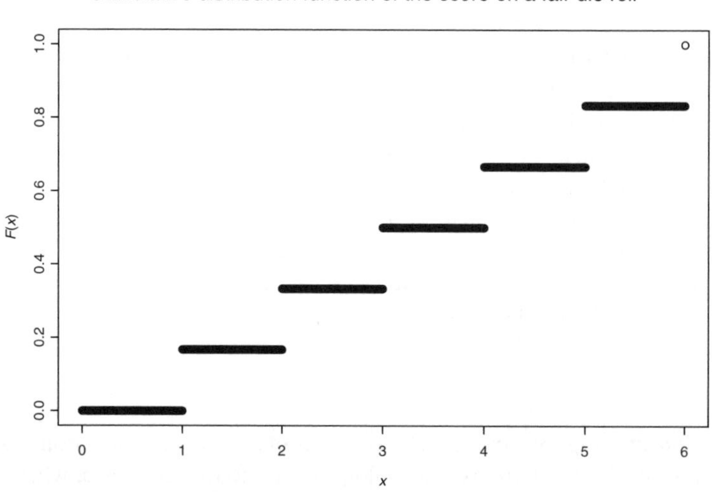

Figure G.5: CDF of the single roll of a fair cubic die.

apply to conditions of certainty, risk, or uncertainty. Decision under certainty means that each alternative leads to one and only one consequence and a choice among alternatives is equivalent to a choice among consequences. But, in decision under risk, each alternative will have one of several possible consequences, and the probability of occurrence for each consequence is known. Therefore, each alternative is associated with a probability distribution and a choice among probability distributions.

When the (frequency) probability distributions are unknown, one speaks about *decision under uncertainty*. Decision theory recognises that the ranking produced by using a criterion has to be consistent with the **decision maker's** objectives and preferences. The theory offers a rich collection of techniques and procedures to reveal preferences and to introduce them into models of decision. It is not concerned with defining objectives, designing the alternatives or assessing the consequences; it usually considers them as given from outside or previously determined. Given a set of alternatives, a set of consequences and a correspondence between those sets, decision theory offers conceptually simple procedures for choice. In a decision situation under certainty, the **decision maker's** preferences are simulated by a single-attribute or multi-attribute value function that introduces ordering on the set of consequences and, thus, also ranks the alternatives. Decision theory for risk conditions is based on the concept of **utility**. The **decision maker's** preferences for the mutually exclusive consequences of an alternative are described by a utility function that permits the calculation of the **expected utility** for each alternative. The alternative with the highest **expected utility** is considered the most preferable. For the case of uncertainty, decision theory offers two main approaches. The first exploits criteria of choice developed in a broader context by game theory, as for example the MAX-MIN rule, where we choose the alternative such that the worst possible consequence of the chosen alternative is better than (or equal to) the best possible consequence of any other alternative. The second approach is to reduce the uncertainty case to the case of risk by using **subjective probabilities** on the basis of expert assessments or on analysing previous decisions made in similar circumstances.

Default Bayesian methods: synonymous with **objective Bayesian** methods.

Delphi method: A group communication structure used to facilitate communication on a specific task. The method usually involves anonymity of responses, feedback to the group as a whole of individual and/or collective views and the opportunity for any respondent to modify an earlier judgement. In fact, this method can be considered as a series of repeated examinations, usually by means of questionnaires, of a group of individuals whose opinions or judgements are of interest. After the initial examination of each individual, each subsequent examination is accompanied by information regarding the preceding round of replies, usually presented anonymously. The individual is thus encouraged to reconsider and, if appropriate, to change her previous reply in the light of the replies of other members of the group. After two or three rounds, the group position may be determined by averaging. The method is usually conducted via paper and mail but can be executed within a computerised conferencing environment.

Dempster–Shafer theory: Dempster–Shafer theory, also known as the *theory of belief functions*, is a generalisation of the Bayesian theory of **subjective**

probability. Whereas the Bayesian theory requires probabilities for each question of interest, in Dempster–Shafer theory, belief functions allow us to base degrees of belief for one question on probabilities for a related question. These degrees of belief may or may not follow coherent axioms of probabilities; how much they differ from probabilities will depend on how closely the two questions are related. It should be noted that belief functions are not probability distributions because they do not obey the laws of probability. Their meaning and interpretation are therefore less clear than for probabilities and are defined only implicitly within the theory.

Density function: Synonymous with **probability density function**.

Discrete random variable: A random variable restricted to take one of a set of distinct values, usually integers. For example, the number of people attending a lecture would be a discrete variable. A person's height in centimetres is also discrete, whereas a person's height is, in principle, a **continuous random variable** if measured to arbitrary accuracy.

Distribution function: Synonymous with **cumulative distribution function**.

Dutch Book: A combination of bets that guarantees a profit, regardless of the outcome. A Dutch book can be constructed whenever a set of probability judgements are made that are not **coherent**.

E

Environment(al): Psychologists use the terms 'environment' and 'environmental' in a variety of contexts that are only minimally recognisable from their use in everyday language (which is perhaps more akin to how these terms are employed in climatology or in environmental and ecological science). The use of these terms in psychology is not easily defined, but, roughly speaking, the terms are applied to refer to the space, location, activity or information in which or with which an organism interacts. Consider the following examples.

- In developmental psychology, 'environmental factors' or 'environmental influences' in child development can refer to all factors/influences upon behaviour that are not hereditary (e.g., upbringing, schooling, interaction with other children, nutrition, accidental injury, infectious diseases, etc.).
- Experimental cognitive psychologists may use the phrase 'task environment' as a shorthand for 'all the features of the task which the participant must learn (or use, or respond to)'.
- In judgement research, the phrase 'probabilistic environment' can be used to denote a set of imperfect relationships between variables. (A participant in a study would then learn, or make use of, the information in some way.)

This use of the term 'environment' to refer to a set of variables and/or to the relationships between them is not uncommon in the psychological studies that are of relevance to this book.

Epistemic uncertainty: Two general types of uncertainty exist: **aleatory** and epistemic. Epistemic uncertainty is subjective in nature and arises primarily from limited or imperfect knowledge. It is, in principle, reducible by obtaining more or better information. Uncertainty about **parameters** of a **statistical model** is generally epistemic.

Equivalent prior sample: When eliciting prior probability distributions, in certain cases it is possible to equate the information in the elicited distribution with that which would be obtained from a particular sample of data. For example, if you represent your beliefs about an uncertain proportion θ by a $Beta(3, 7)$ distribution, this can be thought of as equivalent to the beliefs you would have about θ if you had observed three 'successes' in a sample of ten trials (and had no other information). The stronger your prior beliefs, the greater would be the equivalent prior sample size.

Error: This is a widely used term in the behavioural sciences. It is used differently in different contexts, sometimes synonymously with 'imprecision', though it most commonly refers to 'any and all variability that we do not apportion to a specified factor'. It is probably fair to say that all techniques for investigating and analysing behaviour employed by psychologists assume that measurement of any psychological variable is subject to error. Almost every research method will discuss the following equation (and will assume that the error is never truly and precisely zero): measured value = true value + error.

Expectation (expected value): For a random variable X with density function $f_X(x)$, the expectation (or expected value or mean) of any function $g(X)$ is defined as

$$E\{g(X)\} = \int g(x) f_X(x) dx.$$

Informally, when X is a random variable corresponding to a repeatable event, for example, the height of a randomly selected adult male, $E(X)$ can be thought of as the sample mean value that would be obtained in a very large sample of heights.

Expected utility: See **utility**.

Expert: Studies of decision-making use several different criteria for defining expertise: peer assessment, length of experience and demonstrable ability. Cognitive scientists ascribe qualitatively different processes of judgement or reasoning to experts (in comparison to novices and intermediates), and estimate the time required to acquire expertise within a professional domain to be of the order of 10 years.

Expert system: An expert system is a computer program that uses a knowledge base of human expertise to aid in solving problems or **decision-making**. The degree of problem solving is based on choosing suitable tools to examine the data (mostly by using statistical techniques) and rules obtained from the human expert. Expert systems are designed to perform at a human expert level.

Extensional reasoning: A form of reasoning based on the assumption that the probability assigned to classes of events depends on their extension, that is, the entire set of events included within that class. This assumption is challenged by **support theory** in that alternative descriptions of the same event often produce different probability judgements.

F

Facilitator: The individual in an expert elicitation who interviews the experts and constructs the probability distribution for their beliefs based on their responses.

Fitting a distribution: Fitting a distribution to a dataset involves identifying the distribution in some chosen class that best fits the data. Similarly, fitting a distribution to an expert's expressed opinions involves identifying the distribution that reflects an expert's elicited assessments. This can be done by assuming that a specific family of parametric distributions adequately represents the expert's opinions (e.g., the normal distribution) and calculating the values of the parameters of the distribution from the assessments. It can also be done without making distributional assumptions by various techniques, including histograms and drawing smooth curves.

Frequency probability (frequentist probability): The definition of probability that is usually taught in elementary work is the frequency definition. According to this formulation, the probability of an event is the frequency with which events occur in an infinitely long sequence of events containing the event of interest. The usual illustration is with simple, repeatable chance outcomes like tossing Heads with a coin. Thus, the probability of Heads in a coin toss is the limiting frequency with which Heads occurs in an infinitely long sequence of tosses. The probability of an event can only be defined using frequency probability if it is repeatable and is then only defined in relation to the particular sequence in which it is embedded. The latter point causes no trouble if the sequence is understood unambiguously or is clearly defined, but can be a source of difficulty for **frequentist statistics**. The fact that events must be repeatable means that frequency probability is essentially only usable to quantify **aleatory** uncertainty and cannot be used where the uncertainty is **epistemic**. The most widely used alternative to frequency probability is **subjective probability**.

Frequentist statistics: Frequentist statistics is most clearly distinguished from **Bayesian statistics** by the fact that in the frequentist framework, parameters are fixed but unknown quantities, and are not permitted to have probability distributions. This distinction is closely linked to the fact that frequentist statistics is based on **frequency probability**, which is not able to assign probabilities to one-off, non-repeatable quantities like parameters. Inference in the frequentist paradigm is based on estimators, confidence intervals and significance tests, whose justification, interpretation and properties come through embedding the observed sample of data in a notional infinite sequence of repetitions. Thus, for instance, if the range from a to b is a 95% confidence interval for a parameter X, this means that if the rule used to compute the interval a to b were repeated on each of the notional infinite sequence of data samples, then 95% of such intervals would contain X. This interpretation should be contrasted with that of a Bayesian 95% **credible interval**, which says that there is a 95% probability that the true value of X is in the *particular* interval a to b.

Fuzzy Set Theory: A radically different approach to dealing with uncertainty compared to **subjective probabilities** or other statistical methods. The essential feature of a fuzzy set is a membership function that assigns a grade of membership between 0 and 1 to each member of a set A. Mathematically, a membership function of a fuzzy set A is a mapping from a space $\mathcal{X}$ to the unit interval $m_A : \mathcal{X} \to [0, 1]$. In other words, fuzzy set includes 0 and 1 as extreme cases of truth (or 'the state of matters', 'fact' or 'uncertainty') but also includes the various states of truth in between so that, for example, the result of a comparison between two things could be not 'tall' or 'short' but '0.38 of tallness'.
Because memberships take their values in the unit interval, it is tempting to think of them as probabilities; however, memberships do not follow the laws of probability and it is possible to allow an object to simultaneously hold non-zero degrees of membership in sets traditionally considered mutually exclusive.
Methods derived from the theory have been proposed as alternatives to **subjective probability** in different areas, although they have not met with universal acceptance and a number of statisticians have commented that they have found no solution using such an approach that could not have been achieved at least as effectively using **subjective probability** and other relevant statistical approaches.

G

Gamma distribution The Gamma distribution is a continuous probability distribution on the range of 0 to ∞. If the distribution of the random variable X is gamma with parameters α and β, we write $X \sim Gamma(\alpha, \beta)$ and its pdf is

$$f(x) = \frac{\beta^\alpha}{\Gamma(\alpha)} x^{(\alpha-1)} e^{-\beta x}, x \geq 0$$

where $\Gamma(x) = \int_0^\infty t^{x-1} e^{-t} dt$ is known as the *Gamma function*. The first parameter α determines the shape (particularly the degree of skewness) of the distribution and β is a scale parameter. If events occur randomly in time, at a constant average rate λ, then the time that one would wait for k events to occur has a $Gamma(k, \lambda)$ distribution.

Geometric mean: The geometric mean is a measure of **central tendency**. It is sometimes used in place of the **arithmetic mean** when a variable necessarily takes positive values and is particularly useful for providing a **robust** measure of **central tendency** for positively **skewed distributions**. The sample geometric mean is the nth root of the product of the data values (assuming we observe n values) that is, it is $(x_1 \times x_2 \times x_3 \times \cdots \times x_n)^{\frac{1}{n}}$. It is worth noting that the geometric mean is always less than or equal to the **arithmetic mean** and that its logarithm equals the **arithmetic mean** of the logarithms of the data. For a positive-valued random variable X, its geometric mean is $\exp\{E(\ln X)\}$, where as usual 'exp' is the exponential function and 'ln' denotes the natural logarithm.

Graphical models (Bayesian networks): One use of the word 'graph' in mathematics is to describe a set of points (called *nodes*) connected by lines (called *edges*). Such a graph can be used to describe some statistical models by using a node to represent a parameter and edges to represent the relationships between parameters. Graphs are often useful for representing Bayesian hierarchical models and such graphs are usually termed *Bayesian networks*.

H

Heuristics: Strategies or implicit rules used by decision makers to reach decisions under uncertainty, usually on the basis of selective or incomplete evidence.

(Bayesian) hierarchical model: In a parametric model, the values of the parameters are unknown and are estimated from data (in classical statistics) or from data and expert opinion (in Bayesian statistics). Within the Bayesian methodology, quantities that are unknown are given probability distributions. These probability distributions have their own parameters (see **parameters (hyperparameters)**) which, in turn, may be treated as unknown and are then given a probability distribution. Thus, there is a hierarchy of parameters in a Bayesian model, hence hierarchical models. The process generally stops after two or three levels of the hierarchy, as greater levels of complexity are seldom needed.

Histogram: A histogram is a graphical representation of an observed or theoretical continuous distribution. It is created by dividing the range of the distribution into non-overlapping intervals and counting the number of observed (or proportion of theoretical) values which fall into each interval. The intervals are also known as *bins or classes*. The graphical representation is then created by drawing rectangles for each bin with the width of the rectangle proportional to the bin width and the area proportional to the number (or proportion) recorded in that bin. Histograms in which all bins are of the same width are most easily drawn since, in this situation, the heights of the rectangles (or bars) are proportional to the number (or proportion) in each interval (such graphs are sometimes known as *bar charts*).

Hyperparameters: See **parameters**.

Hypothetical future sample: In elicitation, the expert may be asked what they would believe after observing a hypothetical future sample of data. This process can, for example, give information about the strength of an expert's beliefs; an expert with strong opinions would not change their views significantly in light of a small sample of data.

I

Independence: In probability theory, if two events are independent, knowing whether one of them occurred makes it neither more probable nor less probable that the other occurred. Similarly, if two random variables are independent, knowing something about the value of one of them does not yield any information about the value of the other. If two variables are independent, then both their **covariance** and their **correlation coefficient** are 0. (However, the converse is not true.) If two variables are not independent, they are said to be dependent.

Intuitive reasoning: A general term describing forms of reasoning that do not involve explicit calculations or logical analysis.

J

Joint (probability) distribution: See **multivariate distribution**.

K

Kendall's τ: This is a measure of correlation (cf **correlation coefficient**) and so measures the strength of the relationship between two variables. Kendall's τ is carried out on the ranks of the data; that is, for each variable separately, the values are put in order and numbered, 1 for the lowest value, 2 for the next lowest, and so on. Then the usual (Pearson) correlation coefficient is calculated between the rank values. In common with other measures of correlation, Kendall's τ will take values between -1 and $+1$, with a positive correlation indicating that the variables increase together, while a negative correlation indicates that as one variable increases, the other decreases. However, the degree of correlation does not relate only to the extent to which there is a *linear* relationship between the variables.

Kurtosis: See **tails of a distribution**.

L

Law of large numbers: The law of large numbers says that in repeated, independent trials with the same probability, p, of success at each trial, the proportion of observed successes is increasingly likely to be close to the probability of success as the number of trials increases. Formally, the probability that the proportion of successes differs from the probability, p, by more than a fixed positive amount, $\epsilon > 0$, converges to 0 as the number of trials, n, goes to ∞. This is sometimes known as *Bernoulli's law of large numbers* and is fully formalised via the **Bernoulli distribution** and the **Bernoulli process**.

Law of small numbers: A cognitive **heuristic** implied where judges show over-reliance on small samples and assume that sample parameters, sequences and trends must match those of the population.

Lens model: The lens model framework of Brunswik (1952) for investigating multiple cue judgements in a probabilistic environment (Figure G.6) has been highly influential among psychologists who investigate judgement under uncertainty. One side of the lens (the 'external world') models the relationship between cues and an (objective) outcome criterion. In medicine, statistical risk factor analyses provide practitioners with information about this side of the lens. The other side of the lens ('psychological processes') models the relationship between cues and clinical judgement(s) *at the level of the individual judge* – most typically using regression (regressing cues onto judgement over a set of judgements to produce a linear model of the judge).

The process of building a model of a judge is called *judgement analysis* or *policy capturing*. In particular contexts, the names *clinical judgement analysis* and *social judgement analysis* are also used to describe this process (e.g., in medical judgement and social policy judgements, respectively). Judgement analysis is used to infer the 'weight' given to each cue by each judge – that is, which cues exert more, or less, influence upon a judgement that relies upon multiple (fallible) indicators (Holzworth, 2001). The reliability of judgement processes can

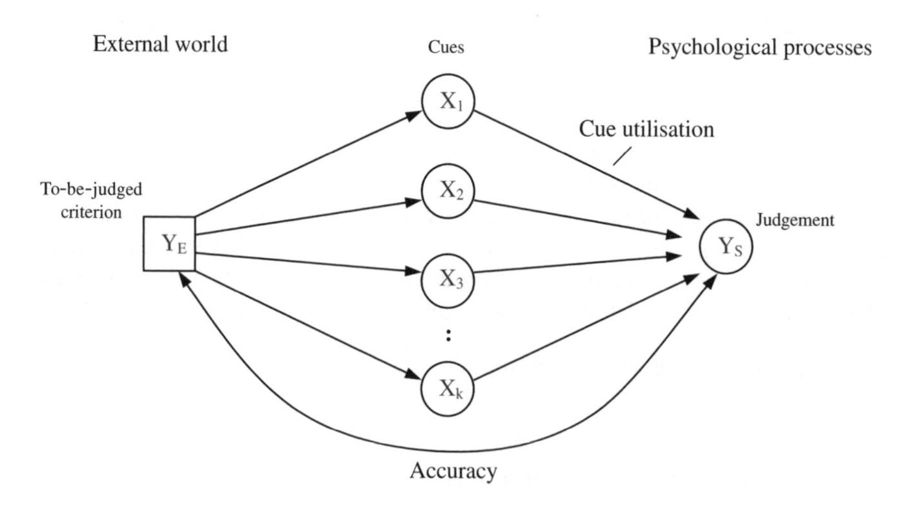

Figure G.6: The illustrations of the lens model.

be assessed by how consistently each judge uses cues across a set of judgements (referred to as the degree of *cognitive control* they possess, Hammond and Summers, 1972). Furthermore, judgement analysis provides the means to explain disagreement between different judges in terms of their differential weighting of cues (Schilling and Hogge, 2001). Owing to its idiographic nature (i.e., analysis at the level of the individual), a small number of participants are often used in this type of research. For example, Stewart et al., (1989) studied seven meteorologists, and, in one of the most frequently cited lens model studies, Einhorn (1974) studied three pathologists.

Judgement analysis has been applied in contexts as diverse as meteorology, educational attainment, medicine and financial management (for details see Hammond and Stewart, 2001). It has been applied to different kinds of judgement, including the judgement of probabilities. Typically, linear models of judgement based on only three or four cues predict most of the variance in judgement – implying that people tend to rely on relatively few cues and that their influence is largely linear (Hastie and Dawes, 2001). A linear model of the judge usually generates more accurate judgements than the judge herself – attributable to removing 'noise' from judgement (Kleindorfer et al., 1993). Hence, even when criterion information is unavailable, a model of an individual's judgement could still be useful by eliminating the random error and thereby increasing the validity of prediction.

Judgement analysis has been performed upon sets of judgements made in response to hypothetical or actual cases. Hypothetical cases increase the investigators control of the study and enhance statistical power. By employing factorial designs that remove correlations between cues, a relatively large number of cue-criterion relationships can be examined using a relatively small number of cases. Rare case features can be examined efficiently by inflating the frequency with which they occur. In contrast, if real cases are randomly or systematically selected, inter-cue correlations that exist in the 'real world' can be preserved. Arguably, inter-cue correlations can play an important role in 'real' judgements such as when two highly correlated cues can be used interchangeably, rendering the simultaneous use of both cues unnecessary (Brunswik, 1957/2001). However, using a sample of real cases creates difficulties for the investigator (or for the statistician in an elicitation), such as limiting the statistical power to examine the impact of rare case characteristics, unless extremely large numbers of cases are considered. Thus, each approach has strengths and weaknesses.

In studies where values of the criterion to be judged can be determined, a full analysis of both sides of the lens can be conducted. In other words, it is possible to construct one model that maps cues onto judgement and a second model that maps cues onto the criterion. These two models may then be compared. Studies which do this contribute to the body of literature on the comparison of clinical judgement and statistical models for prediction and throw considerable light upon when and why these two approaches differ.

Likelihood function: Suppose the probability distribution of a random variable/vector **X** depends on a parameter (vector) θ. We write this probability distribution as $f(\mathbf{x}|\theta)$ to show its dependence on θ. If we look upon $f(\mathbf{x}|\theta)$ as a function of θ (rather than **x**), then we refer to $f(\mathbf{x}|\theta)$ as the likelihood function and write it as $L(\theta|\mathbf{x})$.

Location parameter or location measure: Location parameters (or measures) are summaries of distributions (typically a single parameter for any given distribution) that can be used to represent or summarise the location of the most likely or most central value taken by that distribution. The most common location parameter in statistics is the arithmetic **mean**, but the **geometric mean**, **mode** and the **median** are also important in the elicitation of prior information. See also **central tendency measure**.

Log-likelihood function: Often the logarithm of the likelihood function is used instead of the likelihood function because this gives a function that is easier to maximise; the parameter values that maximise the log-likelihood also maximise the likelihood.

M

Main effects: In the design of an experiment, there are often several variables, say $X_1, \ldots, X_k$, that can be used to define covariates in the model. A covariate can be simply one of the variables, a product of variables (an *interaction* term) or some other functions of variables, for example, X_3^2. A covariate that is an original variable on its own is called a *main effect*.

Markov Chain Monte Carlo (MCMC): A powerful computational tool used to a great extent in applied Bayesian statistics. MCMC is a means of generating a sample of random values from any probability distribution. The sample of random values can then be used, for example, to estimate the mean and variance of the distribution, when it is not possible to identify these quantities analytically. This procedure is typically used whenever it is not possible to identify a suitable **conjugate prior** distribution.

Mathematical aggregation: Mathematical aggregation methods consist of processes or analytical models that seek to combine the judgements of several experts by operating on the individual probability distributions to produce a single 'combined' probability distribution. These methods range from simple summary measures such as arithmetic or geometric means of probabilities (judgements) to procedures based on axiomatic approaches or on various models of the information-aggregation process requiring inputs regarding characteristics such as the quality of, and dependence among, the experts' probabilities. In contrast, *behavioural aggregation* approaches attempt to generate agreement among the experts by having them interact in some way.

Maximise (minimise) a function: Finding the value(s) of the arguments of a function that give the function its maximum (minimum) value.

Mean: There are several forms of 'mean' that are measures of **central tendency** for a distribution. When the phrase 'the mean' is used, it invariably refers to the **arithmetic mean**, but statisticians also sometimes use the **geometric mean** or harmonic mean. For a sample, the **arithmetic mean** is simply the sum of all observed values divided by the number observed and it is also often called the *average*. The presence of a few extreme values can result in an **arithmetic mean** that is not a very good description of the **central tendency** of the distribution as a whole and in such situations other measures of **central tendency** such as the **median**, **mode** and/or **geometric mean** may be more appropriate/informative.

Mean absolute deviation The mean absolute deviation is a measure of how spread out a distribution is (but the **variance** is much more widely used for this purpose). For a sample, x_i, with $\overline{x}$ being its arithmetic mean, its mean absolute deviation is calculated by

$$\text{MAD}(x) = \frac{1}{n}\sum_{i=1}^{n} |x_i - \overline{x}|$$

where $|x|$ is the absolute value of x. Thus, it is the arithmetic mean of the absolute values of the differences between each component in the sample and the sample mean. Like the **standard deviation**, it is measured in the same units as the sample.

Mean squared deviation: See **variance**.

Mechanistic aggregation: This term is equivalent to *mathematical aggregation*.

Mechanistic model: A representation of reality whose components are selected to describe and/or explain physical phenomena observed in the real world. Put simply, they are devised to help represent or explain mechanisms. In statistical papers, the term is usually used to refer to mathematical models (rather than physical ones) that lack any stochastic components. In some statistical research, mechanistic models are an important starting point onto which stochastic components are added in order to allow us to represent the inherent uncertainty in the production and observation processes in the real world.

Median: The median is the point in the range of a distribution at which 50% of the distribution lies above it and 50% lies below it. The median is therefore a measure of **central tendency**. In symmetric distributions, like the **normal distribution**, the **mean** and the **median** take the same value. In **skewed distributions**, however, the **median** is often a more **robust** measure of **central tendency** and is used in preference to the **mean**.

Mode: The mode of a sample is the value that occurs most frequently in that sample. The mode of a random variable is the value for which its density function is maximised. A distribution is called *unimodal* if the probability density rises steadily (monotonically) to a single peak at the mode and then decreases monotonically. It is called *multi-modal* if the density has two or more separate peaks. Figure G.7 provides illustrations of densities that are (a) unimodal, (b) bimodal and (c) multimodal. In the case of two modes, the distribution is called *bimodal*. However, with two or more modes, *the* mode refers to the location of the largest of the modes. So the mode is, loosely speaking, the most likely (or most common) value in a distribution and as such is a measure of **central tendency**. In symmetric, unimodal distributions, like the **normal distribution**, the **mean**, the **median** and the mode take the same value. Although multi-modal distributions arise frequently in statistics, an expert's probability distribution for an uncertain parameter will usually be unimodal.

Monotone or monotonic (monotonicity): When mathematicians describe a relationship between two variables as monotone or monotonic, they mean that as values of one variable increase, values of the other either systematically increase (i.e., there is a positive trend) or systematically decrease (i.e., there is a negative trend).

Multivariate distribution: The majority of the distributions described here are the distributions of one variable and are technically called *univariate* distributions. It is possible to consider the distribution of two or more variables together. The distribution of two variables is a *bivariate* distribution, that of three variables a *trivariate* distribution, and so on. Any distribution of more than one variable is called a *multivariate* distribution.

The combined distribution of more than one variable is also called their *joint* distribution. The distribution of one of the variables alone from the joint distribution is then called its *marginal* distribution. For example, if X and Y are two random variables, their joint distribution is described by a joint density function $f(x, y)$, while $f(x)$ and $f(y)$ would represent their marginal density functions. From the definition of conditional probability, their joint distribution is the product of the conditional distribution of Y given X and the marginal distribution of X, that is,

$$f(x, y) = f(y|x)f(x) = f(x|y)f(y).$$

In the case of independence between X and Y, we have $f(x, y) = f(x)f(y)$. These results extend to three or more variables, although the general formula involves increasingly complex conditional distributions.

Multivariate normal distribution: The most common **multivariate distribution** is the multivariate normal distribution. Its density function has two parameters: a mean vector, μ, and a covariance matrix, Σ. We write $\mathbf{X} \sim N(\mu, \Sigma)$ or $\mathbf{X} \sim MVN(\mu, \Sigma)$, and its probability density function is

$$f(\mathbf{x}) = \frac{1}{\sqrt{2\pi}} \mid \Sigma \mid^{-1/2} \exp\left[-\frac{1}{2}(\mathbf{x} - \mu)'\Sigma^{-1}(\mathbf{x} - \mu)\right]$$

The marginal distribution of each variable is the univariate normal distribution.

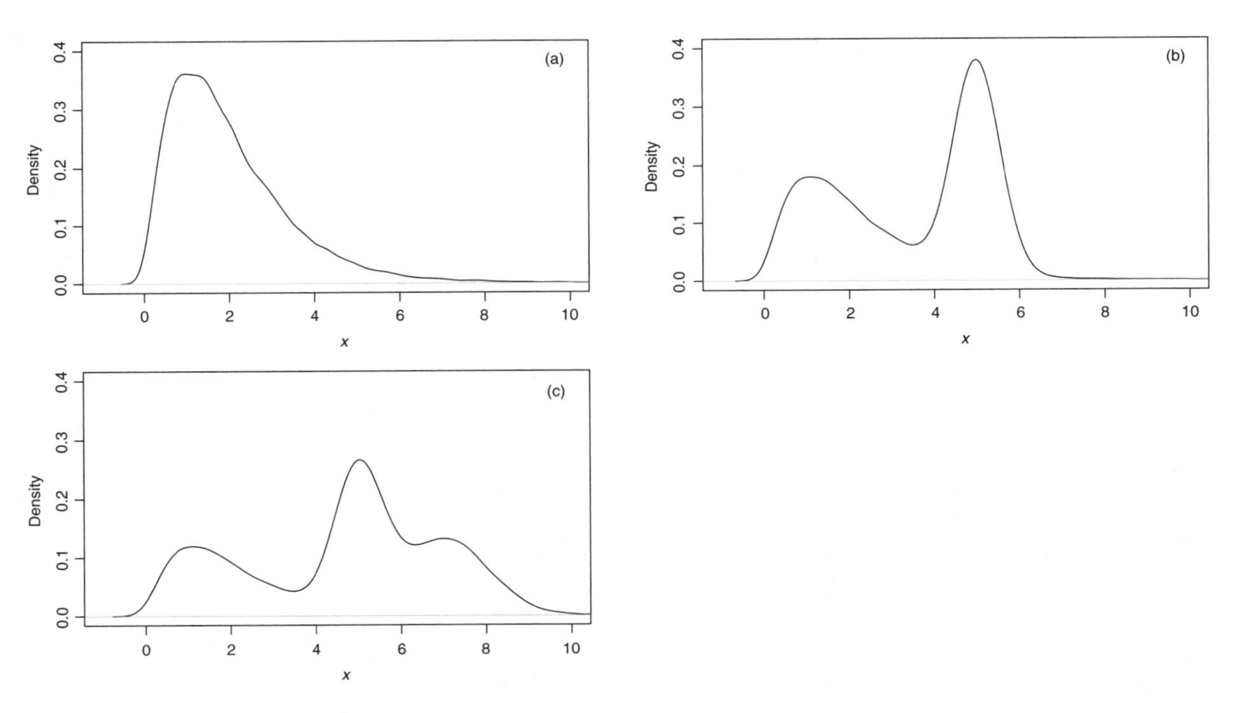

Figure G.7: The illustrations of (a) unimodal, (b) bimodal and (c) multi-modal densities.

N

Non-negative definite: See **positive-definite**.

Non-parametric model (see also **parametric model**): A non-parametric model is a statistical model in which the data are not assumed to come from any specified parametric family of distributions. Using a non-parametric model has the advantage that it does not impose the constraints of a particular distribution family, such as a particular distributional shape, onto the data. A disadvantage is that non-parametric models are harder to interpret.

Normal distribution: This is the most used distribution in statistics. It occurs in many natural situations and is characterised by the 'bell-shaped' curve of its density function. It has two parameters: the **mean** and the **variance**. If a random variable X has a normal distribution with mean μ and variance σ^2, we write $X \sim N(\mu, \sigma^2)$. The density function of the normal distribution is given by

$$f(x) = \frac{1}{\sqrt{2\pi}\,\sigma} \exp\left(-\frac{1}{2}\frac{(x-\mu)^2}{\sigma^2}\right) - \infty < x < \infty.$$

The prominence of the normal distribution is partially caused by the Central Limit Theorem, which states that for a set of independent random variables (each with finite mean and variance) the mean value of the random variables tends towards having a normal distribution as the number of random variables tends to infinity. Calculations on binomial and Poisson random variables are often made by approximation to the normal distribution.

The normal distribution with a mean of 0 and a variance of 1, $N(0, 1)$, is known as the *standard normal* distribution (illustrated in Figure G.8). Any normally distributed random variable can be transformed to have the standard normal distribution by subtracting its mean and then dividing by its **standard deviation**, that is, if $X \sim N(\mu, \sigma^2)$, then $\frac{X-\mu}{\sigma} \sim N(0, 1)$.

The normal distribution is also known as *the Gaussian distribution*.

Normalisation: By definition, a probability density function must integrate across its range to 1. A function which needs to be a density function but does not integrate over its range to 1 can be multiplied by an appropriate constant so that it does. This is known as *normalisation*.

Normative expert(ise): In the context of the elicitation of expert probabilities, normative expertise refers to adherence to the axioms of probability theory, and other prescribed normative principles such as the revision of probabilities consistent with Bayes theorem, or quoting probabilities that are well calibrated. (We should remember that normative expertise does not guarantee the expression of 'useful' or 'valuable' information, such as when a set of identical probabilities are trivially

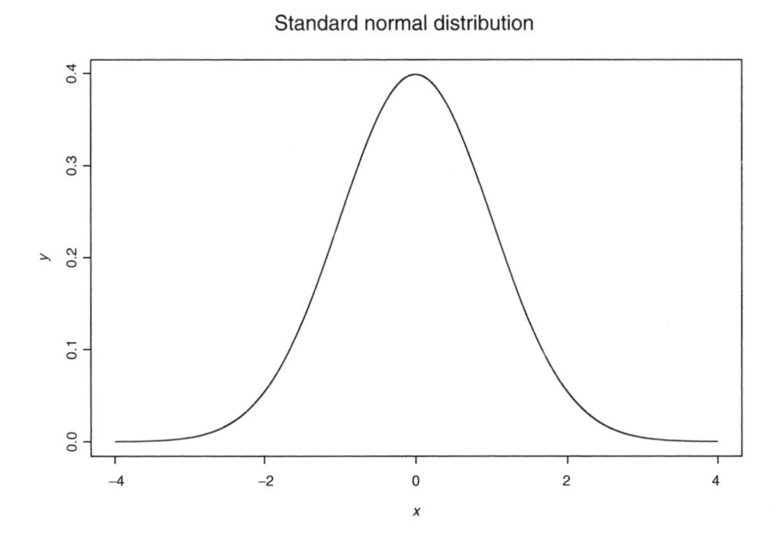

Figure G.8: Density of the standard normal distribution.

well calibrated because they match the long-run relative frequency of the occurrence of the judged event – cf **substantive expertise**.)

O

Objective and subjective Bayesians: Our definition of **Bayesian statistics** emphasises that its chief distinguishing feature is the willingness to express uncertainty about parameters in terms of probability distributions. This is usually taken to mean that such probabilities must be subjective, since **frequency probability** does not apply to non-repeatable events. However, the implied subjective nature of Bayesian statistics has been a source of criticism of the Bayesian approach. Those who believe that an element of subjectivity is unavoidable are sometimes called *subjective Bayesians*. This is not to say that subjective Bayesians make a virtue of subjectivity, but rather that while they view the scientific method as a process that strives towards objectivity, it is never possible to remove (expert) judgement completely and that the subjective nature of judgements should not be hidden or denied.

The *objective Bayesian* approach tries to eliminate subjectivity from the Bayesian method by not using subjective **prior distributions**. The prior distribution is the most obvious place where subjectivity enters into Bayesian statistics, representing the scientist's own prior information (although it is hardly defensible to claim that the **likelihood** is objective). Objective Bayesians replace the prior distribution by a conventional form that is variously interpreted as representing ignorance (i.e., the absence of prior information), vague prior information or a reference point against which genuine prior information may be compared. Such formal prior distributions are correspondingly known as *ignorance priors*, *non-informative priors*, *reference priors*, *default priors* or *objective priors*. However, there are many competing theories for such prior formulations, and so there is no uniquely agreed objective prior for any situation. Non-material changes to the problem can lead to different prior distributions being advocated.

By denying the use of the investigator's prior knowledge, the objective Bayesian approach can be accused of failing to use all the available information, in the same way as **frequentist statistics**.

Observables: An observable quantity is one whose value may be determined by a suitable experiment. Sample data are observable (although the term would only be used for data that have not yet been observed). Parameters are generally not observable, except in the limiting sense that they can be evaluated to any desired accuracy from sufficiently large samples.

Opinion pool: A mechanistic rule for combining probabilities or distributions elicited from two or more sources into a single probability or distribution.

Optimal decision: The decision which results in the best possible result: maximum gain or minimum loss. Since a statistical model is modelling outcomes that are, usually, unknown, optimal decisions are those that minimise the *expected* loss.

Orthogonal: The word orthogonal has many meanings in mathematics. The most straightforward meaning is 'pertaining to or composed of right angles', but this is not always what is meant in statistical reports, so some other definitions follow.

> Two vectors v and w (each of length n) are orthogonal if the sum of the n pairwise products of the two vectors is 0.
>
> A matrix, M, is orthogonal if $M \times M^T$ gives the identity matrix (i.e., a square matrix with ones down the main diagonal and zeroes everywhere else).
>
> Two curves are orthogonal at a point P at which they cross if the tangents at P are perpendicular (i.e., at right angles).

Overconfidence (see also **calibration of subjective probabilities**): This is a common observation in judgement under uncertainty that can be seen in the following.

1. (a) Half-range calibration tasks when subjective probabilities are (on average) too high (when assessed against the corresponding observed relative frequencies).

   (b) Full-range tasks when subjective probabilities close to 0 and to 1 are used too frequently, for example, events judged to have probability 0.05

are observed to occur with relative frequency 0.2, whereas events judged to have probability 0.95 are observed to occur with relative frequency 0.8. (Some prefer the term 'over-extremity' to overconfidence in this the context of full-range tasks).

2. The tendency of subjective credible intervals to be too narrow. Subjective 90%, 95% and 98% intervals are often found to capture the target value on around 40–60% of occasions.

Overfitting: In general statistical modelling, to say that a model is 'overfitted' is a criticism, implying that a more complex model has been used, and more parameters estimated, than can be justified from the data.

In elicitation, the term more often refers to the practice of eliciting from the expert more judgements than are needed to fit the parametric distribution that the facilitator believes will represent the expert's beliefs adequately. In this context, overfitting is a useful practice, allowing the adequacy of the assumed parametric distributional form to be assessed.

P

Parameters (hyperparameters): The word *parameter* has several, related, meanings within mathematics and statistics. The Oxford English Dictionary gives, amongst others, the following two definitions which are relevant to statistics:

> A quantity which is constant (as distinct from the ordinary variables) in a particular case considered, but which varies in different cases; especially a constant occurring in the equation of a curve or surface, by the variation of which the equation is made to represent a family of such curves or surfaces.
>
> A numerical characteristic of a population (as distinguished from a 'statistic', which relates to a sample).

The standard probability distributions discussed are parameterised, in the sense of the first definition. For example, any data or variable considered to be 'normally distributed' has a distribution belonging to the family of normal distributions (cf 'family of such curves or surfaces') which are usually parameterised with two parameters, the mean and the variance (other parameterisations are possible, but using the mean and the variance is the most useful). While these satisfy the first definition – they are constant in, say, a $N(0, 1)$ distribution and vary in the general $N(\mu, \sigma^2)$ distribution – they are also 'characteristics of a population', as in the second definition.

In a statistical model, the unknown quantities that we wish to estimate are also called *parameters*. For example, in a simple linear model, $Y = X\beta + \varepsilon$, $\varepsilon \sim N(0, \sigma^2)$, where Y is the response variable, X is the design matrix and β and σ^2 are the parameters.

The Bayesian approach to estimating the values of the parameters of a model is to consider them to be unknown and, therefore, to apply probability distributions to them. These distributions have parameters, which are sometimes called *hyperparameters* to distinguish them from the parameters of the sampling model.

In elicitation, we typically ask for expert knowledge about parameters and, depending on the context, these may be parameters in either of the dictionary senses. Thus, Bayesian statisticians frequently wish to elicit prior knowledge from an expert about parameters in a statistical model. The elicited **prior distribution** will be combined with the **likelihood** (representing the information in the data) to yield a **posterior distribution**. In other contexts, we seek expert knowledge about physical quantities that are either unobservable or for which observation is impractical. These quantities are often required as input parameters to a risk or decision model.

Parametric distribution (family, model): A parametric distribution is a probability distribution that is fully specified once values are assigned to a finite number of **parameters**. The collection of distributions that are obtained by varying the parameters of a particular parametric form is called a *parametric family*. For example, a **normal distribution** is a parametric distribution because it is completely specified once we assign values to its two parameters, the mean and variance. The standard normal distribution is a parametric distribution and is the member of a normal family of distributions obtained by setting the mean to 0 and the variance to 1.

A parametric model is a **statistical model** in which the joint probability distribution of the data is specified in terms of a finite number of unknown parameters.

Pareto distribution: A Pareto distribution is a continuous distribution whose density function is given by

$$f(x) = \frac{ab^a}{x^{a+1}}, x > b.$$

Since $f(x) = 0$ when $x < b$, the Pareto distribution is useful as a threshold model and is often a suitable choice as the prior distribution for the upper endpoint of a uniform distribution. An illustration of the Pareto distribution is given in Figure G.9.

Percentile: A percentile of a probability distribution is a value such that a specified percentage of the distribution lies at or below this value. The 50th percentile is the **median** and the 25th, 50th and 75th percentiles are known as *quartiles* (because they divide the distribution into quarters). The 25th percentile is known as the *lower quartile* and the 75th percentile is the *upper quartile*. Percentiles are also **quantiles**; the rth percentile is the $\frac{r}{100}$ quantile.

The interval from the rth to the sth percentile of a random variable's distribution is a $(s - r)\%$ **credible interval** for that variable, where $0 \leq r < s \leq 100$.

Personal probability: Synonymous with **subjective probability**.

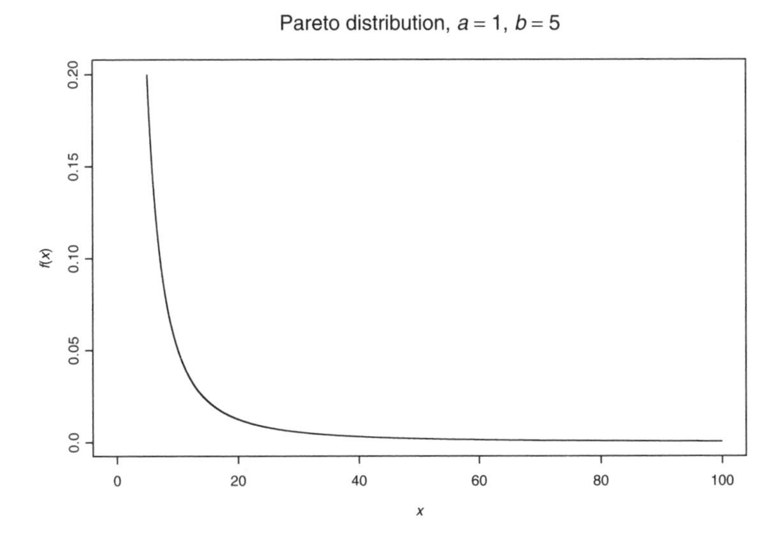

Figure G.9: A Pareto density function.

Positive-definite: By definition, in order to be a covariance matrix, a matrix must satisfy several conditions. In addition to being symmetric and containing only positive entries on its main diagonal, it must also be what is called *positive-definite*. This property is difficult to explain intuitively but vital for much of the multivariate statistical theory that is used to work with covariance matrices. It is not necessary for the non-statistician to know or understand the theory, but it is important that anyone seeking to, for example, elicit prior covariance matrices understands the definition that follows.

A symmetric matrix A is non-negative definite if for any vector x the result of pre- and post-multiplying A by x, written mathematically as $x^T Ax$, is non-negative. This is a rather abstract definition, and there is a large amount of mathematical theory relating to such matrices – for instance, an equivalent definition is that all the eigenvalues of the matrix must be non-negative. A is called *positive-definite* if $x^T Ax$ is always strictly positive except when x is a zero vector; equivalently, all its eigenvalues are positive. Finally, A is positive semi-definite if it is non-negative definite but not positive-definite, in which case it has at least one zero eigenvalue or there exists at least one non-zero x for which $x^T Ax = 0$.

A **covariance matrix** for a set of random variables is, by definition, symmetric and it can be shown that it is also necessarily non-negative definite. Indeed, a covariance matrix will be strictly positive-definite *unless* there exists a linear combination of its variables (represented by the vector x) which has zero **variance**. (The existence of such a linear combination means that the set of random variables contains a redundancy because that linear combination has no uncertainty/variability.)

Positive semi-definite: See **positive-definite**.

Posterior (distribution): A term used within Bayesian statistics to describe the probability distribution of the parameters *after* data are observed. When the **prior distribution** is updated to take account of observing data, the updated distribution is known as the *posterior distribution*, often abbreviated to 'the posterior'. **Bayes' Theorem** implies

$$\text{posterior} \propto \text{prior} \times \text{likelihood}$$

where $\propto$ means 'is proportional to' and 'likelihood' refers to the **likelihood function**.

Prior (distribution): Used within Bayesian statistics to describe the probability distribution of the parameters *before* data are observed. It represents the knowledge, experience, beliefs or opinions of the experimenter. The prior distribution, often abbreviated to 'the prior', may be obtained formally through elicitation. In so-called **objective Bayesian** methods, the prior distribution is a conventional representation of prior ignorance.

Probability: A measure of the chance of occurrence of a particular event. There are various alternative formal definitions of probability, of which the leading two are known as **frequency probability** and **subjective probability**. The latter is sometimes called *personal probability* and in some formulations replaces the idea of occurrence of an event with truth of a proposition.

The probability of event (or proposition) E is denoted by $P(E)$ (but see also **conditional probability**). For any E, $0 \leq P(E) \leq 1$, with $P(E) = 0$ implying that E is certain not to occur (or to be false) and $P(E) = 1$, that it is certain to occur (or to be true).

Probability density function (pdf): A pdf describes the probability distribution of a **continuous random variable**, say X. The pdf, which is never negative, is defined for all values in the range of X, describing a curve. Formally, the pdf is the differential of the **cumulative distribution function**. The probability, $P(x_1 < X \leq x_2)$, that X lies between the limits x_1 and x_2 is thus the area under the pdf curve between $x = x_1$ and $x = x_2$. The total area under the curve defined by the pdf is equal to 1, since X is certain to take a value somewhere in its range.

Although the equivalent for a **discrete random variable** is the **probability mass function**, the term *pdf* is also used in situations where it is either unknown or irrelevant whether the variable is continuous or discrete.

Probability mass function (pmf): The pmf for a **discrete random variable**, say X, is a function which gives the probability that the variable takes each of the individual values in its range. The probability, $P(x_1 \leq X \leq x_2)$, that X lies between the limits x_1 and x_2 (including the limits themselves) is therefore the sum of all the values of the pmf for all possible X values between (and including) those limits. Since X is certain to take a value somewhere in its range, the sum of the pmf over all possible values of x is 1.

Probability wheel: The probability wheel is often seen as the most successful device that has been used to elicit prior probabilities from an expert, in practice. In its simplest form, the wheel is divided into two regions with different sizes and colours, for example, blue and orange. In the elicitation process, the expert is asked to adjust the proportion of the wheel that is, for example, blue until the probability that a spinning pointer attached to the wheel will end up on blue is judged by the expert to be the same as the probability that the specific event will occur.

Prospect theory: Proposed by Kahneman and Tversky (1979, Econometrica) as a general theory of decision-making under uncertainty. Departures from subjective **expected utility** theory include the notions that (a) preferences involve comparisons of outcomes in terms of gains or losses relative to a previously established reference point, rather than as absolute end-states, and (b) judges tend to show a preference for certain over uncertain prospects of the same expected utility when comparing gains (i.e., they are 'risk-averse for gains'), whereas they tend to prefer uncertain to certain losses (i.e., they are 'risk-seeking for losses'). Like **support theory**, prospect theory is primarily a descriptive theory of how people actually assess probabilities, whereas probability theory is normative, describing how people's probability judgements should, ideally and logically, behave.

Q

Quantile: The q quantile of a probability distribution (where q is a number between 0 and 1) is the value such that the probability in the distribution that is less than or equal to this value is q. In terms of the **cumulative distribution function**, $F(x)$, the q quantile x_q satisfies $F(x_q) = q$. The 0.5 quantile is the **median** and the 0.25, 0.5 and 0.75 quantiles are known as *quartiles* (because they divide the distribution into quarters). The 0.25 quantile is known as the *lower quartile* and the 0.75 quantile is the *upper quartile*. See also **percentile**.

Quartile: See **quantile** and **percentile**.

R

Repeatable event: An important distinction exists between repeatable and non-repeatable events. If we take the event 'a randomly selected adult in the United Kingdom wears glasses', this is repeatable in the sense that different adults could be selected on different occasions, with different outcomes observed in each case. A frequentist statistician would declare the proportion of adults wearing glasses in the population as 'the' probability of this event occurring. If we now consider the event 'the proportion of adults in the United Kingdom wearing glasses is greater than 40%', this event is non-repeatable; there is only one population of UK adults. Uncertainty about whether this statement is true or not can only (quantitatively) be expressed using subjective probability, and there is no 'true value' for this probability. Statistical inference (Bayesian or frequentist) is primarily concerned with inference for non-repeatable events.

Representativeness: A cognitive **heuristic** used for judging the **probability** of class membership or the **probability** that an outcome has been generated by a specified process. Judgements about a case are made according to how similar to, or how typical of, the class or process this target case is. The **heuristic** is typically efficient, but prone to result in biases, as features that bear little relation to **probability** may have considerable bearing upon similarity or typicality (and vice versa). Representativeness is held to be responsible for **conjunction fallacies**, as adding detail to a description increases the similarity to the prototype or stereotype of a class, yet decreases the class size and hence decreases the **probability** of class membership. Judgements by representativeness are also held to be insufficiently influenced by the prior **probability** for the event (as this has no bearing upon similarity or typicality). The terms '**base-rate neglect**' or the under-weighting of 'base-rates' are often associated with such judgements.

Robust Bayesian methods: A major criticism of **Bayesian statistics** has focused on the need to specify a single, often subjective, **prior distribution** for the parameters of interest. One way to address this criticism is to use methods for assessing the robustness of the posterior distribution to the specification of the prior. The robust Bayesian approach to data analysis replaces the prior distribution (or the **likelihood function** or both of them) with a class of prior distributions and investigates how the inferences might change as the prior varies over this class. Inferences which remain unchanged with respect to these changes are called *robust*.

S

Scale parameter: Scale parameters are properties of distributions that can be used to represent or summarise the range of values that will probably be taken by a random variable with that distribution. The most common scale parameter in statistics is the **variance** (or its square root, the **standard deviation**).

Scoring Rule: A formula that is used to provide a measure of the accuracy of a set of judgements (often forecasts). Most of these are measures of the distance (or closeness) between the judged and observed values. Examples include the Skill Score and the Brier Score (or mean probability score), both of which can be decomposed into several component measures that assess different aspects of performance in judgement.

Sensitivity analysis: The process of investigating the impact of a particular assumption on the final conclusion. For example, suppose a decision is to be made on the basis of an elicited **prior distribution** and some data. A sensitivity analysis might involve investigating whether changing the prior distribution to represent differing opinions changes the resulting **optimal decision**.

Skew or skewed distribution: A skew (or skewed) distribution is one that is not symmetric or that shows distortion in a positive or negative direction. There is zero skew (or skewness) when the shape of the distribution is symmetrical. It is skewed when a higher proportion of the values it takes are at one end of the range covered by the distribution than the other end. The skew is described as positive when there is greater density at the right-hand end of the distribution than the left, and negative when the there is greater density at the left-hand end of the distribution than the right (for examples, see Figure G.10).

Smooth function (rough function): A smooth function is a function whose graph is a smooth curve, that is, one without any breaks (discontinuities) or spikes. A smooth function is shown in Figure G.11 and an example of a spike and a break, in Figure G.12.

Standard deviation: See **variance**.

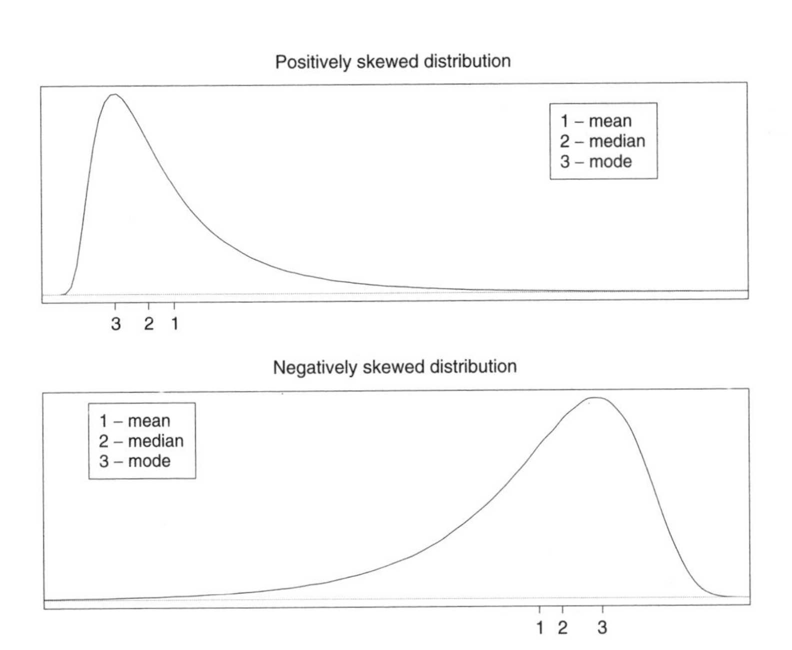

Figure G.10: Two examples of skewed distributions.

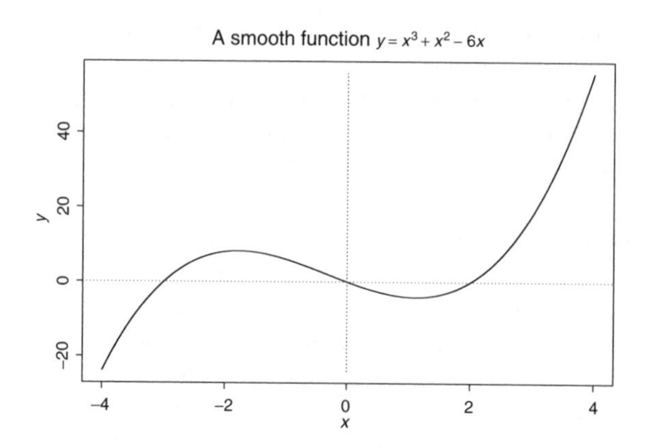

Figure G.11: An example of a smooth function.

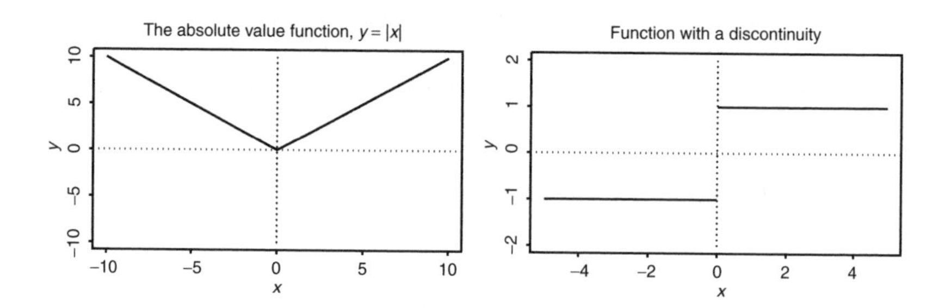

Figure G.12: Examples of non-smooth functions: one with a spike (the absolute value function) and one with a discontinuity.

Statistical model: A description of the assumed structure of a set of observations that can range from a fairly imprecise verbal account to, more usually, a formalised mathematical expression of the process assumed to have generated the observed data. The purpose of such a description is to aid in understanding the data, and, in particular, to enable the **likelihood function** to be derived.

Statistical summary: Statistical summaries are tools used to report important features of both observed and theoretical distributions. Sometimes, we use graphical summaries like **histograms**, scatter plots or line graphs. More often, particularly in formal statistical work, we use numerical summaries including measures of **central tendency** (like **means**, **medians**, **modes**), measures of range, variability or spread (like **variances**, **standard deviations** and **credible intervals**) and individual probabilities, **quantiles** or **percentiles**.

Student *t* distribution: See ***t* distribution**.

Subjective Bayesians: See **objective and subjective Bayesians**.

Subjective probability: One of the two most widely used definitions of **probability**, the other being **frequency probability**. A subjective probability (also known as *personal probability*) represents a degree of belief in the truth of a particular proposition, based on all the information available to a person. Two people will, in general, have different knowledge and so may assign different probabilities to the same proposition. The only constraint is that a single person's probabilities should be **coherent**.

Frequency probability is always defined in terms of probabilities of events, rather than propositions, with an event being understood to be something **repeatable**. It is easy to define the subjective probability for an event by considering the proposition that the event occurs. Although subjective probability is apparently a very different formulation from frequency probability, the two are consistent. It can be shown that if a person observes a long sequence of repeatable events, then that person's subjective probability for another event in the sequence will be the limiting frequency with which events occurred in the observed sequence; that is, subjective and frequency probabilities will coincide whenever the frequency probability is available. However, subjective probability is more general.

Substantive expert(ise): A substantive expert is an individual with knowledge or skill pertaining to a particular domain, discipline, area of knowledge or task. (They may not necessarily have insight into the extent or value of their own knowledge or be able to provide coherent or well-calibrated probability assessments – cf **normative expertise**.)

Summary: See **statistical summary**.

Support theory: Proposed by Tversky and Koehler (1994) to account for the fact that subjective probability judgements can be influenced by how easily judges may call to mind instances or examples that support one hypothesis (e.g., about the cause of some event) rather than another. In particular, adding details by listing subcategories of a specific class of a cause can make that cause seem more probable (cf **availability**), whereas 'catch-all' categories, such as 'all other causes' are more difficult to imagine. Such findings are inconsistent with **extensional reasoning**, since a given class of events or causes should have the same probability, regardless of how it is subcategorised. Similar to **prospect theory**, support theory is primarily a descriptive theory of how people actually assess probabilities, whereas probability theory is normative, describing how people's probability judgements should, ideally and logically, behave.

T

***t* distribution**: The t distribution is similar to the normal distribution in shape, but it has heavier (thicker) **tails**. It is **unimodal** with its mode at zero and is symmetrical about its mode (zero skew). The usual Student t distribution has just one parameter, known as the *degrees of freedom*, which primarily controls the heaviness of the tails. As the degrees of freedom tends to infinity, the Student t distribution tends to the standard **normal distribution**. Examples of t distributions with different degrees of freedom are shown in Figure G.13.

In **Bayesian statistics**, the t distribution is often given a location parameter and a scale parameter, so that the Student t distribution becomes the special case where the location parameter is 0 and the scale parameter, 1.

Tails of a distribution (heavy and light tails): The tails of a distribution refer to the parts of the distribution towards the ends of the range of the random variable. A heavy-tailed distribution has more of its probability in the tails of the distribution (a light-tailed distribution has less probability in its tails), the comparison usually being made with the normal distribution. One measure of the 'weight' of the tails is *kurtosis*: the normal distribution has a kurtosis of three, a heavy tailed distribution will have kurtosis more than three and a light-tailed distribution will have kurtosis less than three.

Transitive ranking: A transitive ranking is one in which whenever x is greater than y and y is greater than z, then x is greater than z.

Triangular distribution: This distribution has a triangular pdf; examples are given in Figure G.14. The distribution has a finite range between two points, say

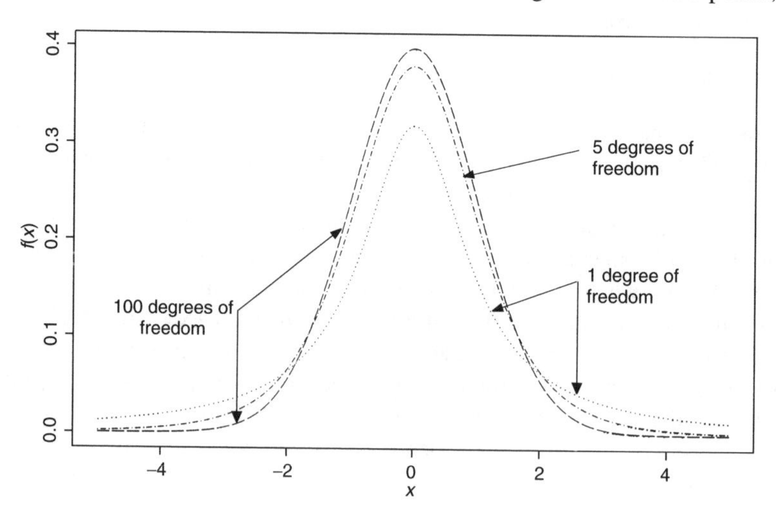

Figure G.13: t distributions with various degrees of freedom.

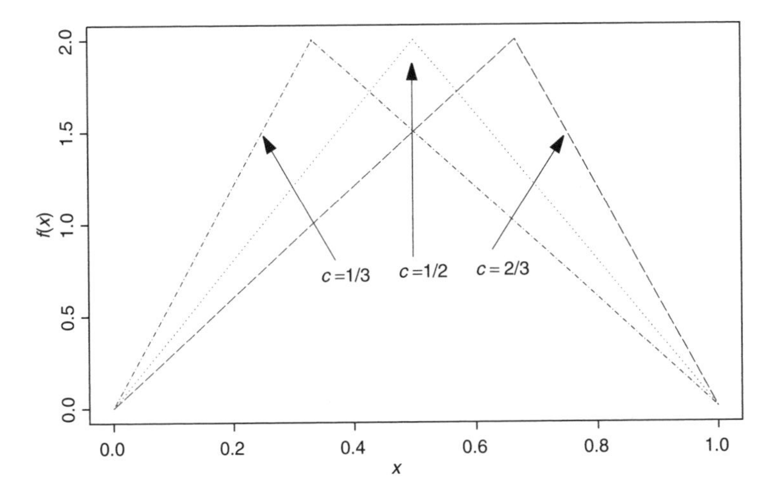

Figure G.14: Triangular distributions, with $a = 0, b = 1$.

a and b. If c is the mode of the distribution, $a \leq c \leq b$, then the density function is

$$f(x) = \begin{cases} \frac{2(x-a)}{(b-a)(c-a)} & a \leq x \leq c \\ \frac{2(b-x)}{(b-a)(b-c)} & c \leq x \leq b \end{cases}$$

If $c = \frac{1}{2}(a + b)$, then the distribution is symmetric.

U

Uniform distribution: The uniform distribution comes in both continuous and discrete forms. The continuous case requires a finite range, $x \in [a, b]$, and its density function is

$$f(x) = \begin{cases} 0 & x < a \\ \frac{1}{(b-a)} & a \leq x \leq b \\ 0 & x > b \end{cases}$$

The distribution has mean $\frac{1}{2}(a + b)$ and variance $\frac{1}{12}(b - a)^2$.

The discrete case refers to a distribution where X is equally likely to take any one of a finite set of values. If there are n possible values it could take, $x_1, x_2, \ldots, x_n$, then $P(x_i) = \frac{1}{n}$ for each x_i.

Unimodal: See **mode**.

Utility: A numerical measure of your preference for a particular outcome or situation. Unlike probability, there is no absolute scale for utility; a utility scale only measures *relative* preferences for different outcomes. A key requirement is that your utility scale should be linear; two units of utility should genuinely be 'twice as good' as one for you. In **decision theory**, the **optimal decision** is that which has maximal utility. More generally, a decision may have uncertain consequences, depending on some uncertain quantity X. Then, the utility of a decision d conditional on X taking the value x is written as $U(d, x)$ and it can be proven that the optimal decision will maximise the expected utility $E\{U(d, X)\}$, where the **expectation** is taken with respect to the probability distribution of X.

V

Vague prior: A prior distribution that attempts to describe (almost) complete ignorance about a parameter, so that posterior inferences are based on the data only. See **objective Bayesians**.

Variance: The variance is a **summary** which measures how spread out or dispersed a distribution is. For a sample of data, the sample variance is calculated as the sum of the squares of the differences between each observation and the **mean**, divided by one less than the total number of observations. Consequently, it is also known as the *mean squared deviation* of the data from the **mean**. (Strictly, the divisor for the variance should simply be the number of observations, and dividing by one less than this number instead yields the unbiased estimator of the population variance; however, we accord here with common usage.) In mathematical notation, the sample variance is

$$s^2 = \frac{\Sigma_{i=1}^{n}(x_i - \overline{x})^2}{n-1},$$

where $x_1, x_2, \ldots, x_n$ are the data values and $\overline{x}$ is the sample mean of those data. The value calculated will always be greater than or equal to 0, with larger values corresponding to data that are more spread out. If all data values are identical, the variance is equal to 0.

For a probability distribution, the variance of a random variable X is analogously defined as the **expectation** of the squared difference of X from its mean, that is,

$$var(X) = E\{(X - E(X))^2\}.$$

The square root of the variance is called *the standard deviation*. Since the standard deviation is measured in the same units as X, it is more frequently reported than the variance. However, a great deal of statistical theory is based on the variance and so it is used in preference in more technical papers and/or in situations in which careful definition of properties of distributions is important.

Variance–covariance matrix: See **covariance matrix**.

W

Weighted average: The weighted average is a measure of **central tendency** of a set of observations $x_1, x_2, \ldots, x_n$, calculated using a set of weights $w_1, w_2, \ldots, w_n$ to adjust the importance of the information conveyed by each observation. For example, in some methods for assessing students, each piece of submitted work carries equal weight for the final assessment, but in others some pieces of work carry more weight than others. In the latter case, we would compute a weighted average of the exam marks (rather than a simple **arithmetic mean** as we would for the former case). The weighted average is obtained as

$$\frac{\Sigma_{i=1}^{n}(x_i \times w_i)}{\Sigma_{i=1}^{n} w_i}.$$

Bibliography

Abdolmohammadi, M. and Shanteau, J. (1992). Personal attributes of expert auditors. *Organizational Behavior and Human Decision Processes*, **53**, 158–172.

Abramson, B., Brown, J., Edwards, W., Murphy, A. and Winkler, R.L. (1996). Hailfinder: A Bayesian system for forecasting severe weather. *International Journal of Forecasting*, **12**, 57–71.

Adelman, L. and Bresnick, T.A. (1992). Examining the effect of information sequence on Patriot air defence officer judgments. *Organizational Behavior and Human Decision Processes*, **53**, 204–228.

Adelman, L., Tollcott, M.A. and Bresnick, T.A. (1993). Examining the effect of information order on expert judgement. *Organizational Behavior and Human Decision Processes*, **56**, 348–369.

Al-Awadhi, S.A. (1997). Elicitation of Prior Distributions for a Multivariate Normal Distribution. Thesis (PhD), University of Aberdeen.

Al-Awadhi, S.A. and Garthwaite, P.H. (1998). An elicitation method for multivariate normal distributions. *Communications in Statistics – Theory and Methods*, **27**, 1123–1142.

Al-Awadhi, S.A. and Garthwaite, P.H. (2001). Prior distribution assessment for a multivariate normal distribution: An experimental study. *Journal of Applied Statistics*, **28**, 5–23.

Alicke, M.D., Klotz, M.L., Breitenbecher, D.L., Yurak, T.J. and Vredenburg, D.S. (1995). Personal contact, individuation, and the better-than-average effect. *Journal of Personality and Social Psychology*, **68**, 804–825.

Allen, G. (1982). Probability and judgement in weather forecasting. *Ninth Conference on Weather Forecasting and Analysis (pp. 1–6). Boston, MA: American Meteorological Society.*

Alpert M. and Raiffa H. (1969). A Progress Report on the Training of Probability Assessors. In *Judgment Under Uncertainty: Heuristics and Biases*, D. Kahneman,

Uncertain Judgements – Eliciting Experts' Probabilities A. O'Hagan, C. E. Buck, A. Daneshkhah, J. R. Eiser, P. H. Garthwaite, D. J. Jenkinson, J. E. Oakley and T. Rakow

P. Slovic, and A. Tversky (eds.), 294–305. Cambridge: Cambridge University Press.

Alpert, M. and Raiffa, H. (1982). A progress report on the training of probability assessors. In D. Kahneman, P. Slovic and A. Tversky (Eds.), *Judgment Under Uncertainty: Heuristics and Biases* (pp. 294–305). Cambridge: Cambridge University Press.

Anderson, J.R., Dillon, J.L. and Hardaker, J.B. (1977). *Agricultural Decision Analysis*. Ames, IA: Iowa State University Press.

Anderson, M. and Sunder, S. (1995). Professional traders as intuitive bayesians. *Organizational Behavior and Human Decision Processes*, **64(2)**, 185–202.

Ang, M.L. and Buttery, N.E. (1997). An approach to the application of subjective probabilities in level 2 PSAs. *Reliability Engineering and System Safety*, **58**, 145–156.

Anzai, Y. (1991). Learning and use of representations for physics expertise. In K.A. Ericsson and J. Smith (Ed.), *Toward a General Theory of Expertise*, Cambridge: Cambridge University Press.

Arkes, H.R. (1991). Costs and benefits of judgment errors: Implications for debiasing. *Psychological Bulletin*, **110**, 486–498.

Arkes, H.R., Dawson, N.V., Speroff, T., Harrell, F.E., Alzola, C., Phillips, R., Desbiens, N., Oye, R.K., Knaus, W. and Connors, A.F. (1995). The covariance decomposition of the probability score and its use in evaluating prognostic estimates. *Medical Decision Making*, **15**, 120–131.

Arkes, H.R., Saville, P.D., Wortman, R.L. and Harkness, A.R. (1981). Hindsight bias among physicians weighing the likelihood of diagnosis. *Journal of Applied Psychology*, **66**, 252–254.

Armor, D.A. and Taylor, S.E. (1998). Situated optimism: Specific outcome expectancies and self-regulation. In M.P. Zanna (Ed.), *Advances in Experimental Social Psychology* (Vol. 30, pp. 309–379). San Diego, CA: Academic Press.

Aspinall, P. and Hill, A.R. (1984). Clinical inferences and decisions – II. Decision trees, receiver operator curves and subjective probability. *Ophthalmic and Physiological Optics*, **4**, 31–38.

Ayton, P. (1992). On the competence and incompetence of experts. In G. Wright and F. Bolger (Eds.), *Expertise and Decision Support*. New York: Plenum Press.

Babad, E. (1987). Wishful thinking and objectivity among sports fans. *Social Behaviour*, **4**, 231–240.

Bajari, P. and Ye, L. (2003). Deciding between competition and collusion. *The Review of Economics and Statistics*, **85**, 971–989.

Barclay, S. and Peterson, C.R. (1973). Two Methods for Assessing Probability Distributions. Technical Report 73-1, Decisions and Designs, McLean.

Bargh, J.A., Chen, M. and Burrows, L. (1996). Automaticity of social behavior: Direct effects of trait construct and stereotype activation on action. *Journal of Personality and Social Psychology*, **71**, 230–244.

Bargh, J.A. and Ferguson, M.J. (2000). Beyond behaviorism: On the automaticity of higher mental processes. *Psychological Bulletin*, **126**, 925–945.

Bar-Hillel, Y. (1973). Primary truth bearers. *Dialectics*, **27**, 303–312.

Bartlett, F.C. (1932). *Remembering: A study in experimental and social psychology*. Cambridge: Cambridge University Press.

Basak, I. (1998). Probabilistic judgments specified partially in the analytic hierarchy process *European Journal of Operational Research*, **108**, 153–164.

Beach, L.R. (1992). Epistemic strategies: Causal thinking in expert and nonexpert judgment. In G. Wright and F., Bolger, (Eds.), *Expertise and Decision Support*. New York: Plenum Press.

Beach, L.R. and Braun, G.P. (1994). Laboratory studies of subjective probability: A status report. In G. Wright and P. Ayton (Eds.), *Subjective Probability*. Chichester: John Wiley.

Beach, L.R. and Scopp, T.S. (1968). *Intuitive Statistical Inferences about Variances. Organizational Behaviour and Human Performance*, **3**, 109–123.

Beach, L.R. and Swenson, R.G. (1966). Intuitive estimation of means. *Psychonomic Science*, **5**, 161–162.

Bedrick, E.J., Christensen, R. and Johnson, W. (1996). A new perspective on priors for generalized linear models. *Journal of the American Statistical Association*, **91**, 1450–1460.

Benson, P.G. and Önkal, D. (1992). The effects of feedback and training on the performance of probability forecasters. *International Journal of Forecasting*, **8**, 559–573.

Benson, P.G. and Whitcomb, K.M. (1993). The effectiveness of imprecise probability forecasts. *Journal of Forecasting*, **12**, 139–159.

Berbaum, K.S., Dorfman, D.D., Franken, E.A. and Caldwell, R.T. (2002). An empirical comparison of discrete ratings and subjective probability ratings. *Academic Radiology*, **9**, 756–763.

Berger, J.O. (1994). An overview of robust Bayesian analysis (with discussion). *Test*, **3**, 5–124.

Berger, J.O. and Berliner, M. (1986). Robust Bayes and empirical Bayes analysis with ϵ-contaminated priors. *Ann Statist*, **14**, 461–486.

Berger, J.O. and O'Hagan, A. (1988). Ranges of posterior probabilities for unimodal priors with specified quantiles. In J.M. Bernardo, M.H. DeGroot, D.V. Lindley and A.F.M. Smith (Eds.), *Bayesian Statistics 3* (pp. 45–66). Oxford University Press.

Bernardo, J.M. (1979). Expected information as expected utility. *Annals of Statistics*, **7**, 686–690.

Beyth-Marom, R. (1982). How probable is probable? A numerical translation of verbal probability expressions. *Journal of Forecasting*, **1**, 257–269.

Bigaroni, A., Perrier, A. and Bounameaux, H., (2000). Is clinical probability assessment of deep vein thrombosis by a score really standardized? *Thrombosis and Haemostasis*, **83**, 788–789.

Blanning, R.W. and Reining, B.A. (1999). Cross-impact analysis using group decision support systems: An application to the future of Hong Kong. *Futures*, **31**, 39–56.

Blattberg, R.C. and Hoch, S.J. (1990). Database models and managerial intuition: 50% model + 50% manager. *Management Science*, **36**, 887–899.

Bloom, B.S. (1985). *Developing Talent in Young People*. New York: Ballentine Books.

Bobbio, M., Detrano, R., Shandling, A.H., Ellestad, M.H., Clark, J., Brezden, O., Abecia, A. and Martinez-Caro, D. (1992). Clinical assessment of the probability of coronary artery disease: Judgmental bias from personal knowledge. *Medical Decision Making*, **12(3)**, 197–203.

Bolger, F. and Wright, G. (1993). Coherence and calibration in expert probability judgment. *OMEGA International Journal of Management Science*, **21**, 629–644.

Bolger, F. and Wright, G. (1994). Assessing the quality of expert judgment: Issues and analysis. *Decision Support Systems*, **11**, 1–24.

Borsuk, M., Clemen, R., Maguire, L. and Reckhow, K. (2001). Stakeholder values and scientific modeling in the Neuse river watershed. *Group Decision and Negotiation*, **10**, 355–373.

Borsuk, M.E., Powers, S.P. and Peterson, C.H. (2002). A survival model of the effects of bottom-water hypoxia on the population density of an estuarine clam *(Macoma balthica). Canadian Journal of Fisheries and Aquatic Sciences*, **59**, 1266–1274.

Bower, G.H. (1981). Mood and memory. *American Psychologist*, **36**, 129–148.

Braun, P.A. and Yaniv, I. (1992). A case study of expert judgment: Economists' probabilities versus base-rate model forecasts. *Journal of Behavioral Decision Making*, **5**, 217–231.

Breeding, R.J., Helton, J.C., Gorham, E.D. and Harper, F.T. (1992). Summary description of the methods used in the probabilistic risk assessments for Nureg-1150. *Nuclear Engineering and Design*, **135**, 1–27.

Brenner, L., Koehler, D., Liberman, V. and Tversky, A. (1996). Overconfidence in probability and frequency judgments: A critical examination. *Organizational Behavior and Human Decision Processes*, **56**, 212–219.

Brenner, L.A. and Koehler, D.J. (1999). Subjective probability of disjunctive hypotheses: Local-weight models for decomposition of evidential support. *Cognitive Psychology*, **38**, 16–47.

Brenner, L.A., Koehler, D.J. and Rottenstreich, Y. (2002). Remarks on Support Theory: Recent advances and future directions. In T. Gilovich, D. Griffin and D. Kahneman (Eds.), *Heuristics and Biases: The Psychology of Intuitive Judgment* (pp. 489–509). Cambridge: Cambridge University Press.

Brier, G.W. (1950). Verification of forecasts expressed in terms of probability. *Monthly Weather Review*, **78**, 1–3.

Broniarczyk, S.M. (1994). Theory versus data in prediction and correlation tasks. *Organizational Behavior and Human Decision Processes*, **57**, 117–139.

Browne, G.J., Curley, S.P. and Benson, P.G. (1999). The effects of subject-defined categories on judgmental accuracy in confidence assessment tasks. *Organizational Behavior and Human Decision Processes*, **80**, 134–154.

Bruine de Bruin, W., Fischbeck, P.S., Stiber, N.A. and Fischhoff, B. (2002). What number is "fifty-fifty"? Redistributing excess 50% responses in risk perception studies. *Risk Analysis*, **22**, 725–735.

Brun, W. and Teigen, K.H. (1988). Verbal probabilities: Ambiguous, context-dependent, or both? *Organizational Behavior and Human Decision Processes*, **41**, 390–404.

Brunswik, E. (1952). The conceptual framework of psychology. Chicago: University of Chicago Press.

Brunswik, E. (1957/2001). Scope and aspects of the cognitive problem. In K.R. Hammond and T.R. Stewart (Eds.), The Essential Brunswik: Beginnings, Explications, Applications (pp. 300–312). New York: Oxford University Press.

Buck, C.E., Cavanagh, W.G. and Litton, C.D. (1996). *Bayesian Approach to Interpreting Archaeological Data*, Wiley, Chichester.

Buck, C.E. and Millard, A.R. (Eds.) (2004). *Tools for constructing chronologies: crossing disciplinary boundaries*, Lecture Notes in Statistics (Vol. 177). London: Springer-Verlag.

Budescu, D.V. and Wallsten, T.S. (1985). Consistency in interpretation of probabilistic phrases. *Organizational Behaviour and Human Decision Processes,* **36**, 391–405.

Budescu, D.V., Weinberg, S. and Wallsten, T.S. (1988). Decisions based on numerically abd verbally expressed uncertainties. *Journal of Experimental Psychology: Human Perception and Performance*, **14**, 281–294.

Buehler, R.J. (1971). Measuring information and uncertainty. In V.P. Godambe and D.A. Sprott (Eds.) *Foundations of Statistical Inference.* Toronto: Holt, Rinehart and Winston, 330–341.

Burgus, G.R., Chapman, G.B., Levy, B.T., Ely, J.W. and Oppliger, R.A. (1998). Clinical diagnosis and the order of information. *Medical Decision Making*, **18**, 412–417.

Burton, P.R. (1994). Helping doctors to draw appropriate inferences from the analysis of medical studies. *Statistics in Medicine*, **13**, 1699–1713.

Burton, P.R., Gurrin, L.C. and Campbell, M.J. (1998). Clinical significance not statistical significance: A simple Bayesian alternative to p values. *Journal of Epidemiology and Community Health*, **52**, 318–323.

Cagno, E., Caron, F., Mancini, M. and Ruggeri, F. (2000). Using AHP in determining the prior distributions on gas pipeline failures in a robust Bayesian approach. *Reliability Engineering and System Safety*, **67**, 275–284.

Cairns, J. and Shackley, P. (1999). What price information? Modelling threshold probabilities of fetal loss. *Social Science and Medicine*, **49**, 823–830.

Campodónico, S. and Singpurwalla, N.D. (1994). A Bayesian analysis of the logarithmic-poisson execution time model based on expert opinion and failure data. *IEEE Transaction on Software Engineering*, **20**, 677–683.

Carlson, B.W. (1993). The accuracy of future forecasts and past judgments. *Organizational Behavior and Human Decision Processes*, **54**, 245–276.

Centor, R.M., Dalton, H.P. and Yates, J.F. (1984). Are physicians probability estimates better or worse than regression model estimates? In *Sixth Annual Meeting of the Society for Medical Decision Making*, Bethesda, MD.

Chaloner, K. (1996). The elicitation of prior distributions. In D. Berry and D. Stangl (Eds.), *Case Studies in Bayesian Biostatistics* (pp. 141–156). New York: Dekker.

Chaloner, K., Church, T., Louis, T.A. and Matts, J.P. (1993). Graphical elicitation of a prior distribution for a clinical trial. *The Statistician*, **42**, 341–353.

Chaloner, K.M. and Duncan, G.T. (1983). Assessment of a beta prior distribution: PM elicitation. *The Statistician*, **32**, 174–180.

Chaloner, K. and Duncan, G.T. (1987). Some properties of the Dirichlet-multinomial distribution and its use in prior elicitation. *Communications in Statistics – Theory and Methods*, **16**, 511–523.

Chaloner, K. and Rhame, F.S. (2001). Quantifying and documenting prior beliefs in clinical trials. *Statistics in Medicine*, **20**, 581–600.

Chapman, L.J. and Chapman, J. (1971). Test results are what you think they are. *Psychology Today*, November, pp. 18–22 and 106–110.

Chen, M.H., Ibrahim, J.G. and Shao, Q.M. (2000). Power prior distributions for generalized linear models. *Journal of Statistical Planning and Inference*, **84**, 121–137.

Chen, M.H., Ibrahim, J.G., Shao, Q.M. and Weiss, R.E. (2003). Prior elicitation for model selection and estimation in generalized linear mixed models. *Journal of Statistical Planning and Inference*, **111**, 57–76.

Chen, M.H., Ibrahim, J.G. and Yiannoutsos, C. (1999). Prior elicitation, variable selection and Bayesian computation for logistic regression models. *Journal of the Royal Statistical Society, Series B*, **61**, 223–242.

Chen, M.H. and Shao, Q.M. (1999). Properties of prior and posterior distributions for multivariate categorical response data models. *Journal of Multivariate Analysis*, **71**, 277–296.

Christen, J.A. and Nakamura, M. (2000). On the analysis of accumulation curves. *Biometrics*, **56**, 748–754.

Christiansen, F., Nilsson, T., Måre, K. and Carlsson, A. (1997). Adding a visual linear scale probability to the PIOPED probability of pulmonary embolism. *Acta Radiologica*, **38**, 458–463.

Christensen-Szalanski, J.J.J. and Beach, L. (1984). The citation bias: Fad and fashion in the judgement and decision literature. *American Psychologist*, **39**, 75–78.

Christensen-Szalanski, J.J.J., Beck, D.E., Christensen-Szalanski, C.M. and Koepsell, T.D. (1983). Effects of expertise and experience on risk judgements. *Journal of Applied Psychology*, **68**, 278–284.

Christensen-Szalanski, J.J.J. and Busheyhead, J.B. (1981). Physicians' use of probabilistic information in a real clinical setting. *Journal of Experimental Psychology: Human Perception and Performance*, **7(4)**, 928–935.

Clark, D.A. (1990). Verbal uncertainty expressions: A critical review of two decades of research. *Current Psychology: Research and Reviews*, **9**, 203–235.

Clemen, R.T. (1989). Combining forecasts: A review and annotated bibliography. *International Journal of Forecasting*, **5**, 559–583.

Clemen, R.T., Fischer, G.W. and Winkler, L.R. (2000). Assessing dependence: Some experimental results. *Management Science*, **46(8)**, 1100–1115.

Clemen, R.T., Jones, S.K. and Winkler, R.L. (1996). Aggregating forecasts: An empirical evaluation of some Bayesian methods. In Donald A. Berry, Kathryn M. Chaloner, and John K. Geweke (Eds.), *Bayesian Statistics and Econometrics: Essays in Honor of Arnold Zellner* (pp. 3–13). New York: Wiley.

Clemen, R.T. and T. Reilly (2001). *Making Hard Decisions with Decision Tools*. Pacific Grove, CA: Duxbury Press.

Clemen, R.T. and Winkler, R.L. (1987). Calibrating and combining precipitation probability forecasts. In R. Viertl (Ed.), *Probability and Bayesian Statistics* (pp. 97–110). New York: Plenum.

Clemen, R.T. and Winkler, R.L. (1990). Unanimity and compromise among probability forecasters. *Management Science*, **36**, 767–779.

Clemen, R.T. and Winkler, R.L. (1999). Combining probability distributions from experts in risk analysis. *Risk Analysis*, **19**, 187–203.

Cohen, L.J. (1981). Can human irrationality be experimentally determined? *Behavioral and Brain Sciences*, **4**, 317–331.

Coles, S.G. and Powell, E.A. (1996). Bayesian methods in extreme value modelling: A review and new developments. *International Statistical Review*, **64**, 119–136.

Coles, S.G. and Tawn, J.A. (1996). A Bayesian analysis of extreme rainfall data. *Applied Statistics*, **45**, 463–478.

Connolly, T. and Thorn, B.K. (1987). Predecisional information acquisition: Effects of task of variables on suboptimal search strategies. *Organizational Behaviour and Human Decision Processes*. **39**, 397–416.

Cooke, R.M. (1991). *Experts in Uncertainty: Opinion and Subjective Probability in Science*. New York: Oxford University Press.

Cooke, R.M. and Goossens, L.H.J. (2000). Procedures guide for structured expert judgement in accident consequence modelling. *Radiation Protection Dosimetry*, **90**, 303–09.

Cooke, R.M., Goossens, L.H.J. and Kraan, B.C.P. (2001). *Probabilistic Accident Consequence Uncertainty Assessment Procedures Guide For Structured Expert Judgement EUR 18820.* European commission, Directorate-General for Research.

Cooke, R.M. and Goossens, L.H.J. (2006). TU Delft expert judgement data base. *Technical report, Delft University of Technology*. http://www.rff.org/expert judgementdocuments/relevantdocs/CookeGoossens-EJDatabase.pdf

Cooke, R.M. and Solomatine, D. (Eds.) (1992). *EXCALIBR – Integrated System for Processing Expert Judgements, User's Manual, Version 3.0.* Delft University of Technology and SoLogic Delft, Delft, the Netherlands.

Cooksey, R. (1996). *Judgment Analysis: Theory, Methods and Applications.* London: Academic Press.

Coolen, F.P.A., Mertens, P.R. and Newby, M.J. (1992). A Bayes-competing risk model for the use of expert judgment in reliability estimation. *Reliability Engineering and System Safety*, **35**, 23–30.

Coolen, F.P.A. (1996). On Bayesian reliability analysis with informative priors and censoring. *Reliability Engineering and System Safety*, **53**, 91–98.

Craig, P.S., Goldstein, M., Seheult, A.H. and Smith, J.A. (1996). Bayes linear strategies for matching hydrocarbon reservoir history. In J.O. Berger, J.M. Bernardo, A.P. David, and A.F.M. Smith (Eds.), *Bayesian Statistics 5* (pp. 69–95). Oxford University Press.

Craig, P.S., Goldstein, M., Seheult, A.H. and Smith, J.A. (1998). Constructing partial prior specifications for models of complex physical systems. *The Statistician*, **48**, 37–53.

Crosby, M.A. (1980). Implications of prior probability elicitation on auditor sample size decisions. *Journal of Accounting Research*, **18**, 585–593.

Crosby, M.A. (1981). Bayesian statistics in auditing: A comparison of probability elicitation techniques. *The Accounting Review*, **66**, 355–365.

Curtis, P.E., Ferrell, W.R. and Solomon, I. (1985) A Model of Calibration for Subjective Probability Distributions and its Application to Aggregation of Judgements. *Journal of Forecasting*, **4**, 349–361.

Daan, H. and Murphy, A.H. (1982). Subjective probability forecasting in The Netherlands: Some operational and experimental results. *Meteorologische Rundschau*, **35**, 99–112.

Dalgleish, T., Taghavi, R., Neshat-Doost, H., Moradi, A., Yule, W. and Canterbury, R. (1997). Information Processing in Clinically Depressed and Anxious Children and Adolescents. *Journal of Child Psychology and Psychiatry and Allied Disciplines*, **38**, 535–541.

Dalgleish, T., Neshat-Doost, H., Taghavi, R., Moradi, A., Yule, W., Canterbury, R. and Vostanis, P. (1998). Information processing in recovered depressed children and adolescents. *Journal of Child Psychology and Psychiatry and Allied Disciplines*, **39**, 1031–1035.

DallAglio, G., Kotz, S. and Salinetti, G. (1991). *Advances in Probability Distributions with Given Marginals*. Dordrecht: Kluwer.

Daponte, B.O., Kadane, J.B. and Wolfson, L.J. (1997) Bayesian demography: Projecting the Iraqi Kurdish population, 1977–1990. *Journal of the American Statistical Association*, **92**, 1256–1267.

Dawson, N.V., Connors, A.F., Speroff, T., Kemka, A., Shaw, P. and Arkes, H.R. (1993). Hemodynamic assessment in managing the critically ill: Is physician confidence warranted? *Medical Decision Making*, **13**, 258–266.

de Bondt, W.F.M. (1993). Betting on trends: Intuitive forecasts of financial risk and return. *International Journal of Forecasting*, **9**, 355–371.

de Finetti, B. (1962). Does it make sense to speak of 'good probability appraisers'? In I.J. Good (Ed.), *The Scientist Speculates – An Anthology of Partly-Baked Ideas*. London: Heineman, 356–364.

de Finetti, B. (1974). *Theory of Probability* (Vol. 2). Translated by A. Machi and A.F.M. Smith. Chichester: Wiley.

DeGroot, M.H. (1970). *Optimal Statistical Decisions*. McGraw-Hill.

DeGroot, M.H. (1974). Reaching a consensus. *Journal of the American Statistical Association*, **69**, 118–121.

DeGroot, A.D. (1978). *Thought and Choice in Chess*. The Hague: Mouton.

DeGroot, M.H. and Fienberg, S. (1983). The comparison and evaluation of forecasters. *The Statistician*, **32**, 12–22.

de Jong, P.J. and Muris, P. (2002) Spider phobia. Interaction of disgust and perceived likelihood of involuntary physical contact. *Journal of Anxiety Disorders*, **16**, 51–65.

Delbecq, A.L., van de Ven, A.H. and Gustafson, D.H. (1975). *Group Techniques for Program Planning*. Glenview, IL: Scott, Foresman and Company.

de Smet, A.A., Fryback, D.G. and Thornbury, J.R. (1979). A second look at the utility of radiographic skull examinations for trauma. *American Journal of Radiology*, **132**, 95–99.

de Wispelare, A.R., Herren, L.T. and Clemen, R.T. (1995). The use of probability elicitation in the high-level nuclear waste regulation program. *International Journal of Forecasting*, **11**, 5–24.

Dhar, S.K., González-Vallejo, C. and Soman, D. (1999). Modeling the effects of advertised price claims: Tensile versus precise claims? *Marketing Science*, **18**, 154–177.

Dickey, J.M., Dawid, A.P. and Kadane, J.B. (1986). Subjective probability assessment methods for multivariate-t and matrix-t models. In P. Goel and A. Zellner (Eds.), *Bayesian Inference and Decision Techniques: Essays in Honor of Bruno de Finetti* (pp. 177.195). New York: Elsevier.

Dickey, J.M., Jiang, J.M. and Kadane, J.B. (1983). Bayesian methods for multinomial sampling with noninformatively missing data. *Research Report 6/83 – #15*, State University of New York at Albany, Department of Mathematics and Statistics.

Dickey, J.M., Lindley, D.V. and Press, S.J. (1985). Bayesian estimation of the dispersion matrix of a multivariate normal distribution. *Communications in Statistics – Theory and Methods*, **14**, 1019–1034.

Diekmann, J.E. and Featherman, W.D. (1998). Assessing cost uncertainty: lessons from environmental restoration projects. *Journal of Construction Engineering and Management*, **124**, 445–451.

Dillon, R.L., John, R. and von Winterfeldt, D. (2002). Assessment of cost uncertainties for large technology projects: A methodology and an application. *Interfaces*, **32**, 52–66.

Dolan, J.G., Bordley, D.R. and Mushlin, A.I. (1986). An evaluation of clinicians' subjective prior probability estimates. *Medical Decision Making*, **6**, 216–223.

Dominitz, J. (1998). Earnings expectations, revisions, and realisations. *The Review of Economics and Statistics*, **80**, 374–388.

Dominitz, J. and Manski, C.F. (1997). Using expectations data to study subjective income expectations. *Journal of the American Statistical Association*, **92**, 855–867.

Dowie, J. (1976). On the efficiency and equity of betting markets. *Economica*, **43**, 139–150.

Dreyfus, H.L. and Dreyfus, S.E. (1986). *Mind Over Machine*. New York: The Free Press.

Druzdzel M.J. and van de Gaag, L.C. (1995). Elicitation of probabilities for belief networks: Combining qualitative and quantitative information. In: *Proceedings of the Eleventh Conference on Uncertainty in Artificial Intelligence*, 141–148.

Dunn, R.A. and Ramsing, K.D. (1981). *Management Science*. Macmillan: New York.

Duran, B.S. and Booker, J.M. (1988). A Bayes sensitivity analysis when using beta distribution as a prior. *IEEE Transactions on Reliability*, **37**, 239–247.

Echternacht, G.J. (1972). The use of confidence testing in objective tests. *Review of Educational Research*, **42**, 217–36.

Eddy, D.M. (1982). Probabilistic reasoning in clinical medicine: Problems and opportunities. In D. Kahneman, P. Slovic and Tversky, A. (Eds.), *Judgement under Uncertainty: Heuristics and Biases*. Cambridge: Cambridge University Press.

Edwards, W. (1965). Optimal strategies for seeking information. *Journal of Mathematical Psychology*, **2**, 312–329.

Edwards, W. (1982). Conservatism in human information processing. In D. Kahneman, P. Slovic, and A. Tversky, (Eds.), *Judgment under Uncertainty: Heuristics and Biases* (pp. 359–369). Cambridge: Cambridge University Press.

Einhorn, H.J. (1971). Use of nonlinear, noncompensatory models as a function of task and amount of information. *Organizational Behaviour and Human Performance*, **6**, 1–27.

Einhorn, H. (1974). Expert judgement: Some necessary conditions and an example. *Journal of Applied Psychology*, **59**, 562–571.

Eisenberg, R.L., Heineken, P., Hedgcock, M.W., Federle, M. and Goldberg, H.I. (1982). Evaluation of plain abnormal radiographs in the diagnosis of abdominal pain. *Annals of Internal Medicine*, **97**, 257–261.

Eiser, J.R. (1990). *Social Judgement*. Buckingham: Open University Press.

Eiser, J.R. and Hoepfner, F. (1991). Accidents, disease, and the greenhouse effect: Effects of response categories on estimates of risk. *Basic and Applied Social Psychology*, **12**, 195–210.

Eiser, J.R., Pahl, S. and Prins, Y.R.A. (2001). Optimism, pessimism and the direction of self-other comparisons. *Journal of Experimental Social Psychology*, **37**, 77–84.

Eiser, J.R. and Stroebe, W. (1972). Categorization and Social Judgement. *European Monographs in Social Psychology*(Vol. 3). London: Academic Press.

Ellsberg, D. (1961). Risk, ambiguity, and the savage axioms. *Quarterly Journal of Economics*, **75**, 643–669.

Elstein, A.S., Shulman, L.S. and Sprafka, S.A. (1978). *Medical Problem Solving: An Analysis of Clinical Reasoning*. Cambridge, MA: Harvard University Press.

Epstein, E.S. (1969). A scoring rule system for probability forecasts of ranked categories. *Journal of Applied Meteorology*, **8**, 985–987.

Erev, I. and Cohen, B.L. (1990). Verbal versus numerical probabilities: Efficiency, biases, and the preference paradox. *Organizational Behavior and Human Decision Processes*, **4**, 1–18.

Erev, I., Wallsten, T.S. and Budescu, D.V. (1994). Simultaneous over- and underconfidence: The role of error in judgment processes. *Psychological Review*, **101**, 519–527.

Ericsson, K.A. and Crutcher, R.J. (1990). The nature of exceptional performance. In P.B. Baltes and D. Featherman and R.M.Lerner (Eds.), Life-span development and behaviour (pp. 188–218). Hillsdale, NJ: Lawrence Erlbaum.

Erlick, D.E. (1964). Absolute judgement of discrete quantities randomly distributed over time. *Journal of Experimental Psychology*, **67**, 475–482.

Ettenson, R., Shanteau, J. and Krogstad, J. (1987). Expert judgement: Is more information better? *Psychological Reports*, **60**, 227–238.

Fazio, R.H. (2001). On the automatic activation of associated evaluations: An overview. *Cognition and Emotion*, **15**, 115–141.

Fechner, G.T. (1860). *Elmente der Psychophsik*. Leipzig: Breitkopf and Hartel.

Ferrell, W.R. (1985). Combining individual judgements. In G. Wright (Ed.), *Behavioral Decision Making*, 111–145. New York: Plenum.

Ferrell, W.R. (1994). Discrete subjective probabilities and decision analysis: Elicitation, calibration and combination. In G. Wright and P. Ayton (Eds.), *Subjective Probability* (pp. 411–451). Chichester: John Wiley.

Ferrell, W.R. and McGoey, P.J. (1980). A model of calibration for subjective probabilities. *Organizational Behavior and Human Decision Processes*, **26**, 32–53.

Fiedler, K. (1988). The dependence of the conjunction fallacy on subtle linguistic factors. *Psychological Research*, **50**, 123–129.

Fiedler, K. (2000). Beware of samples! A cognitive-ecological sampling approach to judgment biases. *Psychological Review*, **107**, 659–676.

Fiedler, K. and Juslin, P. (2006). *Information sampling and adaptive cognition*. New York: Cambridge University Press.

Finucane, M. L., Alhakami, A., Slovic, P., and Johnson, S. M. (2000). The affect heuristic in judgments of risks and benefits. *Journal of Behavioural Decision Making*, **13**, 1–17.

Fischer, G.W. (1982). Scoring-rule feedback and the overconfidence syndrome in subjective probability forecasting. *Organizational Behavior and Human Performance*, **29**, 352–369.

Fischhoff, B. (1982). Debiasing. In D. Kahneman, P. Slovic and A. Tversky (Eds.), *Judgment Under Uncertainty: Heuristics and Biases* (pp. 422–444). Cambridge: Cambridge University Press.

Fischhoff, B. (1989). Eliciting knowledge for analytical representation. *IEEE Transactions on Systems, Man and Cybernetics*, **19**, 448–461.

Fischhoff, B. (2000). Informed consent in eliciting environmental values. *Environmental Science and Technology*, **38**, 1439–1444.

Fischhoff, B. and Beyth-Marom, R. (1975). I knew it would happen: Remembered probabilities of once-future things. *Organizational Behaviour and Human Decision Processes*, **13**, 1–16.

Fischhoff, B. and Beyth-Marom, R. (1983). Hypothesis evaluation from a Bayesian perspective. *Psychological Review*, **90**, 239–260.

Fischhoff, B. and Bruine de Bruin, W. (1999). Fifty-fifty = 50%? *Journal of Behavioral Decision Making*, **12**, 149–196.

Fischhoff, B., Slovic, P. and Lichtenstein, S. (1977). Knowing with certainty: The appropriateness of extreme confidence. *Journal of Experimental Psychology: Human Perception and Performance*, **3**, 552–564.

Fischhoff, B., Slovic, P. and Lichtenstein, S. (1978a). Fault trees: Sensitivity of estimated failure probabilities to problem representation. *Journal of Experimental Psychology: Human Perception and Performance*, **4**, 330–344.

Fischhoff, B., Slovic, P., Lichtenstein, S., Reid, S. and Coombs, B. (1978b). How safe is safe enough? A psychometric study of attitudes towards technological risks and benefits. *Policy Sciences*, **9**, 127–152.

Fisk, J.E. (2002). Judgments under uncertainty: Representativeness or potential surprise? *British Journal of Psychology*, **93**, 431–449.

Fisk, J.E. (2004). Conjunction fallacy. In R.F. Pohl (Ed.), *Cognitive Illusions: A Handbook on Fallacies and Biases in Thinking, Judgement and Memory* (pp. 23–42). Hove, UK: Psychology press.

Fisk, J.E. and Pidgeon, N. (1998). Conditional probabilities , potential surprise, and the conjunction fallacy. *Quarterly Journal of Experimental Psychology*, **51A**, 655–681.

Forbes, R.N., Sanson, R.L. and Morris, R.S. (1994). Application of subjective methods to the determination of the likelihood and consequences of the entry of foot-and-mouth disease into New Zealand. *New Zealand Veterinary Journal*, **42**, 81–88.

Fox, B.L. (1966). A Bayesian approach to reliability assessment. *Memorandum RM-5084-NASA* (p. 23). Santa Monica, CA: The Rand Corporation.

Fox, C.R., Rogers, A.A. and Tversky, A. (1996). Options traders exhibit subadditive decision weights. *Journal of Risk and Uncertainty*, **13**, 5–17.

Fraser, J.M. and Smith, P.J. (1992). A catalog of errors. *International Journal of Man-Machine Studies*, **37**, 265–307.

French, S. (1980). Updating of belief in the light of someone else's opinion. *Journal of the Royal Statistical Society, Series A*, **143**, 43–48.

French, S. (1985). Group consensus probability distributions: A critical survey (with discussion). In J.M. Bernardo, M.H. DeGroot, D.V. Lindley and A.F.M. Smith (Eds.), *Bayesian Statistics 2* (pp. 183–197). Amsterdam: North-Holland.

Friedman, D. (1983). Effective scoring rules for probabilistic forecasts. *Management Science*, **29**, 447–454.

Gambara, H. and León, O.G. (1996). Evidence of data and confidence in clinical judgments. *European Journal of Psychological Assessment*, **12**, 193–201.

Garthwaite, P.H. (1989). Fractile assessments for a linear regression model: An experimental study. *Organizational Behavior and Human Performance*, **43**, 188–206.

Garthwaite, P.H. (1992). Preposterior expected loss as a scoring rule for prior distributions. *Communications in Statistics – Theory and Methods*, **21**, 3601–3619.

Garthwaite, P.H. (1983). Assessment of Prior distributions for normal linear models. PhD Thesis, Aberystwyth: University College of Wales.

Garthwaite, P.H. (1994). Assessment of prior distributions for regression models: An experimental study. *Communications in Statistics – Simulation and Computation*, **23**, 871–895.

Garthwaite, P.H. and Al-Awadhi, S.A. (2001). Non-conjugate prior distribution assessment for multivariate normal sampling. *Journal of the Royal Statistical Society, Series B*, **63**, 95–110.

Garthwaite, P.H. and Al-Awadhi, S.A. (2006). Quantifying opinion about a logistic regression using interactive graphics. *Technical Report of the Statistics Department, Open University*.

Garthwaite, P.H. and Dickey, J.M. (1985). Double- and single-bisection methods for subjective probability assessment in a location-scale family. *Journal of Econometrics*, **29**, 149–163.

Garthwaite, P.H. and Dickey, J.M. (1988). Quantifying expert opinion in linear regression problems. *Journal of the Royal Statistical Society, Series B*, **50**, 462–474.

Garthwaite, P.H. and Dickey, J.M. (1991). An elicitation method for multiple linear regression models. *Journal of Behavioral Decision Making*, **4**, 17–31.

Garthwaite, P.H. and Dickey, J.M. (1992). Elicitation of prior distributions for variable-selection problems in regression. *The Annals of Statistics*, **20**, 1697–1719.

Garthwaite, P.H. and Dickey, J.M. (1996). Quantifying and using expert opinion for variable-selection problems in regression. *Chemometrics and Intelligent Laboratory Systems*, **35**, 1–26.

Garthwaite, P.H., Kadane, J.B. and O'Hagan, A. (2005). Statistical methods for eliciting probability distributions. *Journal of the American Statistical Association*, **100**, 680–701.

Garthwaite, P.H. and O'Hagan, A. (2000). Quantifying expert opinion in the UK water industry: An experimental study. *The Statistician*, **49**, 455–477.

Garthwaite, P.H., Kadane, J.B. and O'Hagan, A. (2005). Statistical methods for eliciting prior distributions. *Journal of the American Statistical Association*, **100**, 680–701.

Gavasakar, U. (1988). A comparison of two elicitation methods for a prior distribution for a binomial parameter. *Management Science*, **34**, 784–790.

Gelfand, A.E., Mallick, B.K. and Dey, D.K. (1995) Modeling expert opinion arising as a partial probabilistic specification. *Journal of the American Statistical Association*, **90**, 598–604.

Genest, C. and McConway, K.J. (1990). Allocating the weights in the linear opinion pool. *Journal of Forecasting*, **9**, 53–73.

Genest, C. and Schervish, M.J. (1985). Modelling expert judgments for Bayesian updating. *Annals of Statistics*, **13**, 1198–1212.

Genest, C. and Zidek, J.V. (1986). Combining probability distributions: A critique and an annotated bibliography (with discussion). *Statistical Science*, **1**, 114–148.

Gigerenzer, G. (1994). Why the distinction between single-event probabilities and frequencies is important for psychology (and vice versa). In G. Wright and P. Ayton (Eds.), *Subjective Probability* Chichester: John Wiley.

Gigerenzer, G. (1996a). On narrow norms and vague heuristics: A reply to Kahneman and Tversky. *Psychological Review*, **103**, 592–596.

Gigerenzer, G. (1996b). The psychology of good judgment: Frequency formats and simple algorithms. *Medical Decision Making*, **16**, 273–280.

Gigerenzer, G. (2002). *Reckoning With Risk.* London: Allen Lane The Penguin Press.

Gigerenzer, G., Hoffrage, U. and Kleinbölting, H. (1991). Probabilistic mental models: A Brunswikian theory of confidence. *Psychological Review*, **98**, 506–528.

Gigone, D. and Hastie, R. (1993). The common knowledge effect: Information sharing and group judgement. *Journal of Personality and Social Psychology*, **65**, 959–974.

Gilless, J.K. and Fried, J.S. (2000). Generating beta random rate variables from probabilistic estimates of fireline production times. *Annals of Operations Research*, **95**, 205–215.

Gilovich, T. (1991). *How we know what isn't so: The fallibility of human reason in everyday life*. New York: The Free Press.

Gilovich, T., Griffin, D. and Kahneman, D. (2002). *Heuristics and Biases: The Psychology of Intuitive Judgment.* Cambridge: Cambridge University Press.

Gokhale, D.V. and Press, S.J. (1982). Assessment of a prior distribution for the correlation coefficient in a bivariate normal distribution. *Journal of the Royal Statistical Society A*, **145**, 237–249.

Gore, S.M. (1987). Biostatistics and the medical Research Council. *Medical Research Council News*, **35**, 19–21.

Gottschalk, A., Juni, J.E., Sostman, H.D., Coleman, R.E., Thrall, J., McKusick, K.A., Froelich, J.W. and Alavi, A. (1993). Ventilation-Perfusion Scintigraphy in the PIOPED Study. Part I. Data Collection and Tabulation. *The Journal of Nuclear Medicine*, **34**, 1109–1118.

Green, E.J. and Strawderman, W.E. (1994). Determining accuracy of thematic maps. *The Statistician*, **43**, 77–85.

Griffin, D. and Brenner, L. (2004). Perspectives on probability judgment calibration. In D.J. Koehler and N. Harvey (Eds.), *Blackwell Handbook of Judgment and Decision Making* (pp. 177–199). Oxford: Blackwell.

Griffin, D. and Buehler, R. (1999). Frequency, probability, and prediction: Easy solutions to cognitive illusions? *Cognitive Psychology*, **38**, 48–78.

Griffin, D. and Tversky, A. (1992). The weighting of evidence and the determinants of confidence. *Cognitive Psychology*, **24**, 411–435.

Grisley, W. and Kellogg, E.D. (1983). Farmers' subjective probabilities in Northern Thailand: An Elicitation Analysis. *American Journal of Agricultural Economics*, **65**, 74–82.

Gross, A.J. (1971). The application of exponential smoothing to reliability assessment. *Technometrics*, **13**, 877–883.

Gustafson, D.H., Sainfort, F., Eichler, M., Adams, L., Bisognano, M. and Steudel, H. (2003). Developing and testing a model to predict outcomes of organizational change. *Health Services Research*, **38**, 751–776.

Hamm, R.M. (1988). Clinical intuition and clinical analysis: Expertise and the cognitive continuum. In J. Dowie and A. Elstein (Eds.), *Professional Judgment: A Reader in Clinical Decision Making* (pp. 78–105). Cambridge: Cambridge University Press.

Hammersley, J.S., Kadous, K. and Magro, A.M. (1997). Cognitive and strategic components of the explanation effect. *Organizational Behavior and Human Decision Processes*, **70**, 149–158.

Hammond, K.R. and Summers, D.A. (1972). Cognitive control. *Psychological Review*, **79**, 58–67.

Hammond, T.R., Swartzman, G.L. and Richardson, T.S. (2001). Bayesian estimation of fish school cluster composition applied to a Bering Sea acoustic survey. *ICES Journal of Marine Science*, **58**, 1133–1149.

Harmanec, D., Leong, T.Y., Sundaresh, S., Poh, K.L., Yeo, T.T., Ng, I. and Lew, T.W.K. (1999). Decision analytic approach to severe head injury management. *Proceedings of the 1999 AMIA Annual Symposium*, 271–275.

Harris, P. and Middleton, W. (1994). The illusion of control and optimism about health: On being less at risk but no more in control than others. *British Journal of Social Psychology*, **33**, 369–386.

Harvey, N. (1997). Confidence in judgment. *Trends in Cognitive Science*, **1**, 78–82.

Hastie, R. and Dawes, R.M. (2001). *Rational Choice In An Uncertain World: The Psychology of Judgment and Decision Making* (2nd edition). Thousand Oaks, CA: Sage Publications.

Hayes, J.R. (1981). *The Complete Problem Solver.* Philadelphia, PA: Franklin Institute Press.

Heath, C. and Gonzalez, R. (1995). Interaction with others increases decision confidence but not decision quality: Evidence against information collection views of interactive decision making. *Organizational Behaviour and Human Decision Processes*, **61**, 305–326.

Helton, J.C., Breeding, R.J. and Hora, S.C. (1992). Probability of containment failure mode for fast pressure rise. *Reliability Engineering and System Safety*, **35**, 91–106.

Helton, J.C. and Oberkampf, W.L. (2004). Special issue: Alternative representations of epistemic uncertainty. *Reliability Engineering and System Safety*, **95**, 1–3, 39–72.

Hershman, R.L. and Levine, J.R. (1970). Deviations from optimal purchase strategies in human decision making. *Organizational Behaviour and Human Performance*, **5**, 313–329.

Hill, G.W. (1982). Group versus individual performance: are N + 1 heads better than one? *Psychological Bulletin*, **91**, 517–539.

Hirsch, K.G., Corey, P.N. and Martell, D.L. (1998). Using expert judgement to model initial attack fire crew effectiveness. *Forest Science*, **44**, 539–549.

Ho, C.H. and Smith, E.I. (1997). Volcanic hazard assessment incorporating expert knowledge: application to the Yucca mountain region, Nevada, USA. *Mathematical Geology*, **29**, 615–627.

Hodge, R., Evans, M., Marshall, J., Quigley, J. and Walls, L. (2001). Eliciting engineering knowledge about reliability during design-lessons learnt from implementation. *Quality and Reliability Engineering International*, **17**, 169–179.

Hofer, E., Javeri, V., Löffler, H. and Struwe, D.F. (1985). A survey of expert opinion and its probabilistic evaluation for specific aspects of the SNR-300 risk study. *Nuclear Technology*, **68**, 180–225.

Hofstatter, P.R. (1939). Uber die Schatzung von Gruppeneigenschaften. *Zeitschrift fur Psychologie*, **145**, 1–44.

Hogarth, R.M. (1975). Cognitive processes and the assessment of subjective probability distributions. *JASA*, **70**, 271–294.

Hogarth, R.M. and Einhorn, H. J. (1992). Order effects in belief updating: The Belief Adjustment Model. *Cognitive Psychology*, **24**: 1–55.

Holzworth, J.R. (2001). Multiple cue probability learning. In K.R. Hammond and T.R. Stewart (Eds.), The Essential Brunswik: Beginnings, Explications, Applications. (pp. 348–350). New York: Oxford University Press.

Hora, S.C., Hora, J.A. and Dodd, N.G. (1992). Assessment of probability distributions for continuous random variables: A comparison of the bisection and fixed value methods. *Organizational behavior and human decision processes*, **51**, 133–155.

Hora, S.C. and von Winterfeldt, D. (1997). Nuclear Waste and Future Societies: A look into the deep future. *Technological Forecasting and Social Change*, **56**, 155–170.

Horst, H.S., Dijkhuizen, A.A. and Huirne, R.B.M. (1996). Outline for an integrated modelling approach concerning risks and economic consequences of contagious animal diseases. *Netherlands Journal of Agricultural Science*, **44**, 89–102.

Horst, H.S., Dijkhuizen, A.A., Huirne, R.B.M. and De Leeuw, P.W. (1998). Introduction of contagious animal diseases into The Netherlands: Elicitation of expert opinions. *Livestock Production Science*, **53**, 253–264.

Horst, H.S., Huirne, R.B.M. and Dijkhuizen, A.A. (1997). Risks and economic consequences of introducing classical swine fever into the Netherlands by feeding swill to swine. *Revue Scientifique Et Technique De L'Office International Des Epizooties*, **16**, 207–214.

Huber, O. (1995). Ambiguity and perceived control. *Swiss Journal of Psychology*, **54**, 200–210.

Hughes, W.R. (1993). Consistent utility and probability assessment using AHP Methodology. *Mathematical and Computer Modelling*, **17**, 171–177.

Hughes, G. and Madden, L.V. (2002). Some methods for eliciting expert knowledge of plant disease epidemics and their application in cluster sampling for disease incidence. *Crop Protection*, **21**, 203–215.

Hurd, M.D. and McGarry, K. (2002). The predictive validity of subjective probabilities of survival. *The Economic Journal*, **112**, 966–985.

Ibrahim, J.G. and Laud, P.W. (1994). A predictive approach to the analysis of designed experiments. *Journal of the American Statistical Association*, **89**, 309–319.

Inhelder, B. and Piaget, J. (1958). *The Growth of Logical Thinking from Childhood to Adolescence*. New York: Basic Books.

Jacob, K.S. (2003). Mental state examination: The elicitation of symptoms. *Psychopathology*, **36**, 1–5.

Jeffreys, H. (1967). *Theory of Probability*. Oxford: Oxford University Press.

Jenkins, H.M. and Ward, W.C. (1965). The judgement of contingency between responses and outcomes. *Psychological Monographs*, **79**(1, Whole No. 594).

Jensen, F.A. and Peterson, C.R. (1973). Psychological effects of proper scoring rules. *Organizational Behavior and Human Performance*, **9**, 307–317.

Johnson, T.R., Budescu, D.V. and Wallsten, T.S. (2001). Averaging probability judgments: Monte Carlo analyses of asymptotic diagnostic values. *Journal of Behavioral Decision Making*, **14**, 123–140.

Johnson, N.L. and Kotz, S. (1972). *Distributions in Statistics: Continuous Multivariate Distributions*. New York: Wiley.

Johnson, R.D., Rennie, R.D. and Wells, G.L. (1991). Outcome trees and baseball: A study of expertise and list-length effects. *Organizational Behavior and Human Decision Processes*, **50**, 324–342.

Jones, S.K., Jones, T.K. and Frisch, D. (1995) Biases of probability assessment: A comparison of frequency and single-case judgments. *Organizational Behavior and Human Decision Processes*, **61**, 109–122.

Juslin, P. (1993). An explanation of the hard-easy effect in studies of realism of confidence in one's general knowledge. *European Journal of Cognitive Psychology*, **5**, 55–71.

Juslin, P. (1994). The overconfidence phenomenon as a consequence of informal experimenter-guided selection of almanac items. *Organizational Behavior and Human Decision Processes*, **57**, 226–246.

Juslin, P., Olsson, H. and Björkman, M. (1997). Brunswikian and Thurstonian origins of bias in probability assessment: On the interpretation of stochastic components in judgment. *Journal of Behavioral Decision Making*, **10**, 189–209.

Juslin, P., Winman, A. and Olsson, H. (2000). Naive empiricism and dogmatism in confidence research: A critical examination of the hard-easy effect. *Psychological Review*, **107**: 384–396.

Juslin, P., Winman, A. and Olsson, H. (2003). Calibration, additivity, and source independence of probability judgments in general knowledge and sensory discrimination tasks. *Organizational Behavior and Human Decision Processes*, **92**, 34–51.

Kadane, J.B. (1994). An application of robust Bayesian analysis to a medical experiment. *Journal of Statistical Planning and Inference*, **40**, 221–232.

Kadane, J.B., Chan, N.H. and Wolfson, L.J. (1996). Priors for unit root models. *Journal of Econometrics*, **75**, 99–111.

Kadane, J.B., Dickey, J.M., Winkler, R.L., Smith, W.S. and Peters, S.C. (1980). Interactive elicitation of opinion for a normal linear model. *Journal of the American Statistical Association*, **75**, 845–854.

Kadane, J.B. and Schum, D.A. (1996). *A Probabilistic Analysis of the Sacco and Vanzetti Evidence*. New York: John Wiley and Sons.

Kadane, J.B. and Wolfson, L.J. (1998). Experiences in elicitation. *Statistician*, **47**, 1–20 (with discussion, 55–68).

Kahneman, D. (2003). A perspective on judgment and choice: Mapping bounded rationality. *American Psychologist*, **58**, 697–720.

Kahneman, D. and Lovallo, D. (1993). Timid choices and bold forecasts: A cognitive perspective on risk taking. *Management Science*, **39**, 17–31.

Kahneman, D., Slovic, P. and Tversky, A. (1982). *Judgment Under Uncertainty: Heuristics and Biases*, Cambridge: Cambridge University Press.

Kahneman, D. and Tversky, A. (1972). Subjective probability: A judgment of representativeness. *Cognitive Psychology*, **3**, 430–454.

Kahneman, D. and Tversky, A. (1973). On the psychology of prediction. *Psychological Review*, **80**, 237–251.

Kahneman, D. and Tversky, A. (1982a). The psychology of preferences. *Scientific American*, **246**, 136–142.

Kahneman, D. and Tversky, A. (1982b). Variants of uncertainty. *Cognition*, **11**, 143–157.

Kaplan, S. (1992). 'Expert information' vs 'expert opinions': Another approach to the problem of eliciting/combining/using expert knowledge in PRA. *Reliability Engineering and System Safety*, **35**, 61–72.

Kass, R.E. and Wassermann, L. (1996). The selection of prior distributions by formal rules. *Journal of the American Statistical Association*, **91**, 1343–1370.

Keren, G. (1985). On the calibration of experts and lay people. Unpublished manuscript cited in Keren (1987) Facing uncertainty in the game of bridge. *Organizational Behavior and Human Decision Processes*, **39**, 98–114.

Keren, G. (1987). Facing uncertainty in the game of bridge. *Organizational Behavior and Human Decision Processes*, **39**, 98–114.

Kerr, N.L., MacCoun, R.J. and Kramer, G.P. (1996). Bias in judgement: Comparing individuals and groups. *Psychological Review*, **103**, 687–719.

King, R. and Brooks, S.P. (2001). Prior induction in log-linear models for general contingency table analysis. *The Annals of Statistics*, **29**, 715–747.

Kleindorfer, P.R., Kunreuther, H.C., and Schoemaker, P.J.H. (1993). Decision Sciences: An Integrative Perspective. Cambridge: Cambridge University Press.

Kleinmuntz, B. (1990). Why we still use our heads instead of formulas: Toward an integrative approach. *Pscychological Bulletin*, **107**, 296–310.

Kleinmuntz, D.N., Fennema, M.G. and Peecher, M.E. (1996). Conditioned assessment of subjective probabilities: Identifying the benefits of decomposition. *Organizational Behavior and Human Decision Processes*, **66**, 1–15.

Koehler, J.J. (1996). The base rate fallacy reconsidered: Descriptive, normative and methodological challenges. *Behavioral and Brain Sciences*, **19**, 1–53.

Koehler, J.J. (2001a). When are people persuaded by DNA match statistics. *Law and Human Behavior*, **25**, 493–513.

Koehler, J.J. (2001b). The psychology of numbers in the courtroom: How to make DNA-match statistics seem impressive or insufficient. *Southern California Law Review*, **74**, 1275–1305.

Koehler, D.J. and Harvey, N. (2004). *Blackwell Handbook of Judgement and Decision Making*. Oxford, U.K.: Blackwell Publishers.

Koehler, D.J., White, C.M. and Grondin, R. (2003). An evidential support accumulation model of subjective probability. *Cognitive Psychology*, **46**, 152–197.

Koriat, A., Lichtenstein, S. and Fischoff, B. (1980). Reasons for confidence. *Journal of Experimental Psychology: Human Learning and Memory*, **6**, 107–118.

Kuipers, B., Moskowitz, A.J. and Kassirer, J.R. (1988). Critical decisions under uncertainty: Representation and structure. *Cognitive Science*, **12**, 177–210.

Kunda, Z. and Nisbett, R.E. (1986). The psychometrics of everyday life. *Cognitive Psychology*, **18**, 195–224.

Kurowicka, D. and Cooke, R. (2006). *Uncertainty Analysis with High Dimensional Dependence Modelling*. Chichester: Wiley.

Kurzenhauser, S. and Lucking, A. (2004). Statistical formats in Bayesian inference. In R.F. Pohl (Ed.), *Cognitive Illusions: A Handbook on Fallacies and Biases in Thinking, Judgement and Memory* (pp. 61–78). Hove, UK: Psychology Press.

Kynn, M. (2005) Eliciting Expert Knowledge for Bayesian Logistic Regression in Species Habitat Modelling. *PhD thesis*, Queensland University of Technology.

Lagnado, D.A. and Sloman, S.A. (2004). Inside and outside probability judgment. In D.J. Koehler and N. Harvey (Eds.), *Blackwell Handbook of Judgment and Decision Making* (pp. 157–176). Oxford: Blackwell.

Lahiri, K., Teigland, C. and Zaporowski, M. (1988). Interest rates and the subjective probability distribution of inflation forecasts. *Journal of Money, Credit, and Banking*, **20**, 233–248.

Langer, E.J. (1975). The illusion of control. *Journal of Personality and Social Psychology*, **32**, 311–328.

Larrick, R.P. (2004). Debiasing. In D.J. Koehler and N. Harvey (Eds.), *Blackwell Handbook of Judgement and Decision Making*. Oxford: Blackwell, pp. 296–310.

Lathrop, R.G. (1967). Perceived variability. *Journal of Experimental Psychology*, **73**, 498–502.

Lau, A.H.L., Lau, H.S. and Zhang, Y. (1996). A simple and logical alternative for making PERT time estimates. *IIE Transactions*, **28**, 183–192.

Lau, H.S., Lau, A.H.L. and Zhang, Y. (1997). An improved approach for estimating the mean and standard deviation of a subjective probability distribution. *Journal of Forecasting*, **16**, 83–95.

Lau, H.S. and Somarajan, C. (1995). A proposal on improved procedures for estimating task-time distributions in PERT. *European Journal of Operational Research*, **85**, 39–52.

Laud, P.W. and Ibrahim, J.G. (1995). Predictive model selection. *Journal of the Royal Statistical Society, Series B*, **57**, 247–262.

Laws, D.J. and O'Hagan, A. (2002). A hierarchical Bayes model for multilocation auditing. *The Statistician*, **51**, 431–450.

Lenox, M.J. and Haimes, Y.Y. (1996). The constrained extremal distribution selection method. *Risk Analysis*, **16**, 161–176.

León, C.J., Vázquez-Polo, F.J. and González, R.L. (2003). Elicitation of expert opinion in benefit transfer of environmental goods. *Environmental and Resource Economics*, **26**, 199–210.

Lichtenstein, S. and Fischhoff, B. (1980). Training for calibration. *Organizational Behavior and Human Performance*, **26**, 149–171.

Lichtenstein, S., Fischhoff, B. and Phillips, L.D. (1977). Calibration of probabilities: The state of the art. In H. Jungerman and G. de Zeeuw (Eds.), *Decision Making and Change in Human Affair* (pp. 275–324). Dordrecht: Reidel.

Lichtenstein, S., Fischhoff, B. and Phillips, L.D. (1982). Calibration of probabilities: The state of the art to 1980. In D. Kahneman, P. Slovic and A. Tversky (Eds.), *Judgment Under Uncertainty: Heuristics and Biases.*

Lichtenstein, S. and Newman, J.R. (1967). Empirical scaling of common verbal phrases associated with numerical probabilities. *Psychonomic Science*, **9**, 563–564.

Lindley, D.V. (1982). The improvement of probability judgements. *J R Statist Soc A*, **148**, 117–126.

Lindley, D.V. (1983). Reconciliation of probability distributions. *Operations Research*, **31**, 866–880.

Lindley, D.V. (1985). Reconciliation of discrete probability distributions (with discussion). In J.M. Bernardo, M.H. DeGroot, D.V. Lindley and A.F.M. Smith (Eds.), *Bayesian Statistics* (Vol. 2, pp. 375–390). Amsterdam: North-Holland.

Lindley, D.V. (1991). Subjective probability, decision analysis and their legal consequences. *Journal of the Royal Statistical Society. Series A (Statistics in Society)*, **154**, 83–92.

Lindley, D.V. and Singpurwalla, N.D. (1986). Reliability (and fault tree) analysis using expert opinions. *J Amer Statist Assoc*, **81**, 87–90.

Lindley, D.V., Tversky, A. and Brown, R.V. (1979). On the reconciliation of probability assessments. *Journal of the Royal Statistical Society, Series A*, **142**, 146–180.

Linstone, H.A. and Turoff, M. (Eds.) (1975). *The Delphi Method: Techniques and Applications*. Reading, MA: Addison-Wesley.

Littlepage, G., Robison, W. and Reddington, K. (1997). Effects of task experience and group experience on group performance, member ability, and recognition of expertise. *Organizational Behavior and Human Decision Processes*, **69**, 133–147.

Lloyd, A., Hayes, P., Bell, P.R.F. and Naylor, A.R. (2001). The role of risk and benefit perception in informed consent for surgery. *Medical Decision Making*, **21**, 141–149.

Lopes, L.L. (1991). The rhetoric of irrationality. *Theory and Psychology*, **1**, 65–82.

MacGregor, D., Lichtenstein, S. and Slovic, P. (1988). Structuring knowledge retrieval: An analysis of decomposed quantitative judgments. *Organizational Behavior and Human Decision Processes*, **42**, 303–323.

MacLeod, A.K. and Tarbuck, A.F. (1994). Explaining why negative events will happen to oneself: Parasuicides are pessimistic because they cannot see any reason not to be. *British Journal of Clinical Psychology*, **33**, 317–326.

Matheson, J.E. and Winkler, R.L. (1976). Scoring rules for continuous probability distributions. *Management Science*, **22**, 1087–1096.

May, R.S. (1986a). Interferences, subjective probability and frequency of correct answers: A cognitive approach to the overconfidence phenomenon. In B. Brehmer, H. Jungermann, P. Lourens and G. Sevo'n, (Eds.), *New Directions in Research on Decision Making*. Amsterdam: North Holland.

May, R.S. (1986b). Overconfidence as a result of incomplete and wrong knowledge. In R.W. Sholtz (Ed.), *Current Issues in West German Decision Research.* Frankfurt: Lang.

McClelland, A.G.R. and Bolger, F. (1994). The calibration of subjective probabilities: Theories and models 1980-94. In G. Wright and P. Ayton (Eds.), *Subjective Probability.* Chichester: John Wiley.

McClish, D.K. and Powell, S.H. (1989). How well can physicians estimate mortality in a medical intensive care unit? *Medical Decision Making*, **9**, 125–132.

McConway, K.J. (1981). Marginalization and linear opinion pools. *Journal of the American Statistical Association*, **76**, 410–414.

McDaniels, T.L. (1995). Using judgement in resource management: A multiple objective analysis of a fisheries management decision. *Operations Research*, **43**, 415–426.

McKendrick, I.J., Gettinby, G., Gu, Y., Reid, S.W.J. and Revie, C.W. (2000). Using a Bayesian belief network to aid differential diagnosis of tropical bovine diseases. *Preventive Veterinary Medicine*, **47**, 141–156.

McKenzie, C.R.M. (1998). Taking into account the strength of an alternative hypothesis. *Journal of Experimental Psychology: Learning, Memory and Cognition*, **24**, 771–792.

McKenzie, C.R.M. (1999). (Non)Complementary updating of belief in two hypotheses. *Memory and Cognition*, **27**, 152–165.

Meinhold, R.J. and Singpurwalla, N.D. (1987). A Kalman-filter smoothing approach for extrapolation in certain dose-response, damage assessment and accelerated-life-testing studies. *American Statistician*, **41**, 101–106.

Merkhofer, M.W. (1987). Quantifying judgmental uncertainty: methodology, experiences, and insights. *IEEE Transactions on Systems, Man and Cybernetics*, **SMC-17**, 741–752.

Miklas, M.P., Norwine, J.R., DeWispelare, A.R., Herren, L.T. and Clemen, R.T. (1995). Future climate at Yucca Mountain, Nevada proposed high-level radioactive waste repository. *Global Environmental Change –Human and Policy Dimension*, **5**, 221–234.

Monti, S. and Carenini, G. (2000). Dealing with the expert inconsistency in probability elicitation. *IEEE Transactions on Knowledge and Data Engineering*, **12**, 499–508.

Morgan, M.G. and Henrion, M. (1990). *Uncertainty: A Guide to Dealing with Uncertainty in Quantitative Risk and Policy Analysis.* Cambridge: Cambridge University Press.

Morgan, M.G. and Keith, D.W. (1995). Climate change – subjective judgements by climate experts. *Environmental Science and Technology*, **29**, 468A–476A.

Morgan, M.G., Pitelka, L.F. and Shevliakova, E. (2001). Elicitation of expert judgements of climate change impacts on forest ecosystems. *Climatic Change*, **49**, 279–307.

Morris, P.A. (1974). Decision analysis expert use. *Management Science*, **20**, 1233–1241.

Mosleh, A., Bier, V. and Aspostolakis, G. (1987). A critique of current practice for the use of expert opinions in probabilistic risk assessment. *Reliability Engineering and System Safety*, **20**, 63–85.

Murphy, A.H. (1970). The ranked probability score and the probability score: A comparison. *Monthly Weather Review*, **98**, 917–924.

Murphy, A.H. (1971). A note on the ranked probability score. *Journal of Applied Meteorology*, **10**, 155–156.

Murphy, A.H. (1973). A new vector partition of the probability score. *Journal of Applied Meteorology*, **12**, 596–600.

Murphy, A.H. and Brown, B.G. (1984). A comparative evaluation of objective and subjective weather forecasts in the United States. *Journal of Forecasting*, **3**, 369–393.

Murphy, A.H. and Daan, H. (1984). Impacts of feedback and experience on the quality of subjective probability forecasts: Comparison of results from the first and second years of the Zierikzee experiment. *Monthly Weather Review*, **112**, 413–423.

Murphy, A.H. and Winkler, R.L. (1970). Scoring rules in probability assessment and evaluation. *Acta Psychologica*, **34(2/3)**, 273–286.

Murphy, A.H. and Winkler, R.L. (1974). Credible interval temperature forecasting: Some experimental results. *Monthly Weather Review*, **102**, 784–794.

Murphy, A.H. and Winkler, R.L. (1977). Reliability of subjective probability forecasts of precipitation and temperature: Some preliminary results. *Applied Statistics*, **26**, 41–47.

Murphy, A.H. and Winkler, R.L. (1984). Probability forecasting in meteorology. *Journal of the American Statistical Association*, **79**, 489–500.

Must, A., Phillips, S.M., Stunkard, A.J. and Naumova, E.N. (2002). Expert opinion on body mass index percentiles for figure drawings at menarche. *International Journal of Obesity*, **26**, 876–879.

Nash, H. (1964). The judgement of linear proportions. *American Journal of Psychology*, **77**, 480–484.

Nau, R.F. (1985). Should scoring rules be 'effective'? *Management Science*, **31**, 527–535.

Nau, R.F. (1992). Joint coherence in games of incomplete information. *Management Science*, **38**, 374–387.

Nisbett, R.E., Zukier, H. and Lemley, R.E. (1981). The dilution effect: Nondiagnostic information weakens the implications of diagnostic information. *Cognitive Psychology*, **13**, 248–277.

Normand, S.L.T., Frank, R.G. and McGuire, T.G. (2002). Using elicitation techniques to estimate the value of ambulatory treatments for major depression. *Medical Decision Making*, **22**, 245–261.

Northcraft, G.B. and Neale, M.A. (1987). Experts, amateurs and real-estate: Anchoring and adjust perspective in property pricing decisions. *Organizational Behavior and Human Decision Processes*, **39**, 84–97.

Oakley, J.E. and O'Hagan, A. (2005). Uncertainty in Prior Elicitations: A Nonparametric Approach. *Research Report No. 521/02 (revised)*, Department of Probability and Statistics, University of Sheffield.

Oberkampf, W.L., Helton, J.C., Joslyn, C.A., Wojtkiewicz, S.F. and Ferson, S. (2004). Challenge problems: Uncertainty in system response given uncertain parameters. *Reliability Engineering and System Safety*, **85**, 11–19.

Ofir, C. (2000). Ease of recall vs recalled evidence in judgment: Experts vs laymen. *Organizational Behavior and Human Decision Processes*, **81**, 28–42.

O'Hagan, A. (1988). *Probability: Methods and Measurement.* London: Chapman and Hall.

O'Hagan, A. (1994). Robust modelling for asset management. *Journal of Statistical Planning and Inference*, **40**, 245–259.

O'Hagan, A. (1998). Eliciting expert beliefs in substantial practical applications. *The Statistician*, **47(1)**, 21–35.

O'Hagan, A. (2005). Elicitation. *Significance*, **2**, 84–86.

O'Hagan, A. and Forster, J. (2004). *Bayesian Inference* (Vol. 2B, 2nd ed.). *Kendall's Advanced Theory of Statistics.* Arnold.

O'Hagan, A., Glennie, E.B. and Beardsall, R.E. (1992). Subjective modelling and Bayes linear estimation in the UK water industry. *Journal of Applied Statistics*, **41**, 563–577.

O'Hagan, A. and Oakley, J.E. (2004). Probability is perfect, but we cannot elicit it perfectly. *Reliability Engineering and System Safety*, **85**, 239–248.

Olson, M.J. and Budescu, D.V. (1997). Patterns of preference for numerical and verbal probabilities. *Journal of Behavioral Decision Making*, **10**, 117–131.

Oman, S.D. (1985). Specifying a prior distribution in structured regression problems. *Journal of the American Statistical Association*, **80**, 190–195.

Önkal, D. and Muradoğlu, G. (1994). Evaluating probabilistic forecasts of stock prices in a developing stock market. *European Journal of Operational Research*, **74**, 350–358.

Önkal, D. and Muradoğlu, G. (1996). Effects of task format on probabilistic forecasting of stock prices. *International Journal of Forecasting*, **12**, 9–24.

Oremus, M., Collet, J-P., Corcos, J. and Shapiro, S.H. (2002). A survey of physician efficacy requirements to plan clinical trials. *Pharmacoepidemiology and Drug Safety*, **11**, 677–685.

Oskamp, S. (1965). Overconfidence in case study judgments. *Journal of Consulting Psychology*, **29**, 261–265.

Osherson, D., Shafir, E., Krantz, D.H. and Smith, E.E. (1997). Probability bootstrapping: Improving prediction by fitting extensional models to knowledgeable but incoherent beliefs. *Organizational Behavior and Human Decision Processes*, **69**, 1–8.

Paese, P.W. (1993). Uncertainty Assessment Accuracy and Resource Allocation Outcomes: An Empirical Test of a Presumed Relation. *The Journal of Psychology*, **127**, 443–450.

Paese, P.W. and Sniezek, J. (1991). Influences on the appropriateness of confidence in judgment: Practice, effort, information and decision-making. *Organizational Behavior and Human Decision Processes*, **48**, 100–130.

Parducci, A. (1963). The range-frequency compromise in judgment. *Psychological Monographs*, **77**(2, Whole No. 565).

Parent, E. and Bernier, J. (2003). Encoding prior experts judgments to improve risk analysis of extreme hydrological events via POT modeling. *Journal of Hydrology*, **283**, 1–18.

Parenté, F. J. and Anderson-Parenté, J.K. (1987). Delphi inquiry systems. In G. Wright and P. Ayton (Eds.), *Judgemental Forecasting* (pp. 129–156). Chichester: Wiley.

Patel, V.L. and Groen, G.J. (1991). The general and specific nature of medical expertise: A critical look. In K.A. Ericsson, and J. Smith (Eds.), *Toward a General Theory of Expertise.* Cambridge: Cambridge University Press.

Pattillo, C. (1998). Investment, uncertainty and irreversibility in Ghana. *International Monetary Fund Staff Papers*, **45**, 522–553.

Perrier, A., Bounameaux, H., Morabia, A., de Moerloose, P., Slosman, D., Didier, D., Unger, P.F. and Junod, A. (1996). Diagnosis of pulmonary embolism by a decision analysis-based strategy including clinical probability, D-dimer levels, and ultrasonography: A management study. *Archives of Internal Medicine*, **156**, 531–535.

Peterson, C.R. and Beach, L.R. (1967). Man as an intuitive statistician. *Psychological Bulletin*, **68**, 29–46.

Peterson, C.R. and Miller, A. (1964). Mode, median and mean as optimal strategies. *Journal of Experimental Psychology*, **68**, 363–367.

Peterson, C.R., Snapper, K.J. and Murphy, A.H. (1972). Credible interval temperature forecasts. *Bulletin of the American Meteorological Society*, **53**, 966–970.

Pfeifer, P.E. (1994). Are we overconfident in the belief that probability forecasters are overconfident? *Organizational Behavior and Human Decision Processes*, **58**, 203–213.

Pham-Gia, T., Turkkan, N. and Duong, Q.P. (1992). Using the mean deviation in the elicitation of the prior distribution. *Statistics and Probability Letters*, **13**, 373–381.

Phillips, L.D. (1999). Group elicitation of probability distributions: Are many heads better than one? In J. Shanteau, Barbara A. Mellers, and David A. Schum (Eds.), *Decision Science and Technology: Reflections on the Contributions of Ward Edwards* (pp. 313–330). Norwell, MA: Kluwer Academic Publishers.

Phillips, L.D. (2004). There's a high probability we'll end up uncertain. *The Independent*, 7th February, p.43

Phillips, L. and Edwards, W. (1966). Conservatism in a simple probability inference task. *Journal of Experimental Psychology*, **72**, 346–354.

Phillips, L.D. and Wisbey, S.J. (1993). The Elicitation of Judgemental Probability Distributions from Groups of Experts: A Description of the Methodology and Records of Seven Formal Elicitation Sessions Held in 1991 and 1992. *Report NSS/R282*, Nirex UK, Didcot.

Pill, J. (1971). The Delphi method: Substance, context, a critique, and an annotated bibliography. *Socio-Economic Planning Sciences*, **5**, 57–71.

Pitz, G.F. (1965). Response variables in the estimation of relative frequency. *Perceptual and Motor Skills*, **21**, 867–873.

Pitz, G.F. (1966). The sequential judgement of proportion. *Psychonomic Science*, **4**, 397–398.

Plous, S. (1993). *The psychology of Decision Making*. New York: McGraw Hill.

Poses, R.M. and Anthony, M. (1991). Availability, wishful thinking, and physicians' diagnostic judgments for patients with suspected bacteremia. *Medical Decision Making*, **11**, 159–168.

Poses, R.M., Bekes, C., Copare, F.J. and Scott, W.E. (1990). What difference do two days make? The inertia of physicians' sequential prognostic judgements for critically ill patients. *Medical Decision Making*, **10**, 6–14.

Poses, R.M., Cebul, R.D., Collins, M. and Fager, S.S. (1985). The accuracy of experienced physicians' probability estimates for patients with sore throats. *Journal of the American Medical Association*, **254**, 925–929.

Poses, R.M., Cebul, R.D. and Wigton, R.S. (1995). You can lead a horse to water – improving physicians knowledge of probabilities may not affect their decisions. *Medical Decision Making*, **15**, 65–75.

Poses, R.M., Cebul, R.D., Wigton, R.S., Centor, R.M., Collins, M. and Fleischli, G. (1992). Controlled trial using computerized feedback to improve physicians' diagnostic judgments. *Academic Medicine*, **67**, 345–347.

Poulton, E.C. (1989). *Bias in Quantifying Judgments*. Hillsdale, NJ: Erlbaum.

Price, P.C. (1998). Effects of a relative-frequency elicitation question on likelihood judgment accuracy: The case of external correspondence. *Organizational Behavior and Human Decision Processes*, **76**, 277–297.

Qian, S.S. and Reckhow, K.H. (1998). Modeling phosphorus trapping in wetlands using nonparametric Bayesian regression. *Water Resources Research*, **34**, 1745–1754.

Rakow, T., Vincent, C., Bull, K. and Harvey, N. (2005). Assessing the likelihood of an important clinical outcome: New insights from a comparison of clinical and actuarial judgment. *Medical Decision Making*, **25**, 262–282.

Ramachandran, G., Banerjee, S. and Vincent, J.H. (2003). Expert judgement and occupational hygiene: Application to aerosol Speciation in the nickel primary production industry. *Annals of Occupational Hygiene*, **47**, 461–475.

Rapoport, A. and Wallsten, T.S. (1972). Individual decision behavior. *Annual Review of Psychology*, **23**, 131–179.

Reagan-Cirincione, P. (1994). Improving the accuracy of group judgement: A process intervention combining group facilitation, social judgement analysis, and information technology. *Organizational Behavior and Human Decision Processes*, **58**, 246–270.

Reagan, R.T., Mosteller, F. and Youtz, C. (1989). Quantitative meanings of verbal probability expressions. *Journal of Applied Psychology*, **74**, 433–442.

Redelmeier, D., Koehler, D.J., Liberman, V. and Tversky, A. (1995). Probability judgment in medicine: Discounting unspecified alternatives. *Medical Decision Making*, **15**, 227–230.

Regazzini, E. and Sazonov, V.V. (1999). Approximation of laws of multinomial parameters by mixtures of Dirichlet distributions with applications to Bayesian inference. *Acta Applicandae Mathematicae*, **58**, 247–264.

Remus, W., O'Connor, M. and Griggs, K. (1996). Does feedback improve the accuracy of recurrent judgmental forecasts? *Organizational Behavior and Human Decision Processes*, **66**, 22–30.

Render, B. and Stair, R.M. (2000). *Quantitative Analysis for Management*. London: Prentice Hall.

Renooij, S. and Witteman, C. (1999). Talking probabilities: Communicating probabilistic information with words and numbers. *International Journal of Approximate Reasoning*, **22**, 169–194.

Risbey, J.S., Kandlikar, M. and Karoly, D.J. (2000). A protocol to articulate and quantify uncertainties in climate change detection and attribution. *Climate Research*, **16**, 61–78.

Rockette, H.E., Gur, D. and Metz, C.E. (1992). The use of continuous and discrete confidence judgements in receiver operating characteristics studies of diagnostic imaging techniques. *Investigative Radiology*, **27**, 169–172.

Ronis, D.L. and Yates, J.F. (1987). Components of probability judgment accuracy: Individual consistency and effects of subject matter and assessment method. *Organizational Behavior and Human Decision Processes*, **40**, 193–218.

Roosen, J. and Hennessy D.A. (2001). Capturing experts' uncertainty in welfare analysis: An application to organophosphate use Regulation in U.S. apple production. *American Journal of Agricultural Economics*, **83**, 166–182.

Root, H.E. (1962). Probability statements in weather forecasting. *Journal of Applied Meteorology*, **1**, 163–167.

Rothbart, M. (1970). Assessing the likelihood of threatening events. *Journal of Personality and Social Psychology*, **15**, 109–117.

Rottenstreich, Y. and Tversky, A. (1997). Unpacking, repacking, and anchoring: Advances in Support Theory. *Psychological Review*, **104**, 406–415.

Rowe, G. (1992). Perspectives on expertise in the aggregation of judgments. In G. Wright and F. Bolger (Eds.), *Expertise and Decision Support* (pp. 155–180). New York: Plenum Press.

Rowe, G. and Wright, G. (1999). The Delphi technique as a forecasting tool: Issues and analysis (with discussion). *International Journal of Forecasting*, **15**, 353–381.

Saaty, T.L. (1977). A scaling method for priorities in hierarchical structures. *Journal of Mathematical Psychology*, **15**, 234–281.

Saaty, T.L. (1980). *The Analytic Hierarchy Process.* New York: McGraw-Hill.

Saleh, K.J., Wood, K.C., Gafni, A., Gross, A.E. (1997). Immediate surgery versus waiting list policy in revision total hip arthroplasty – An economic evaluation. *The Journal of Arthroplasty*, **12**, 1–10.

Sanders, F. (1963). On subjective probability forecasting. *Journal of Applied Meteorology*, **2**, 191–201.

Savage, L.J. (1954). *The Foundations of Statistics* (2nd edition 1972; New York: Dover) New York: Wiley; and London: Chapman and Hall.

Savage, L.J. (1971). Elicitation of personal probabilities and expectations. *Journal of the American Statistical Association*, **66**, 783–801.

Schaefer, R.E. and Borcherding, K. (1973). The assessment of subjective probability distributions: A training experiment. *Acta Psychologics*, **37**, 117–129.

Schilling, S.G., and Hogge, J.H. (2001). Hierarchical linear models for the nomothetic aggregation of idiographic descriptions of judgment. In K.R. Hammond and T.R. Stewart (Eds.), The Essential Brunswik: Beginnings, Explications, Applications. (pp. 332–341). New York: Oxford University Press.

Schum, D.A., Goldstein, I.L., Howell, W.C. and Southard, J.F. (1967). Subjective probability revision under several cost-payoff arrangements. *Organizational Behavior and Human Performance*, **2**, 84–104.

Schwarz, N. (1999). Self-reports: How the questions shape the answers. *American Psychologist*, **54**, 93–105.

Schwarz, N. and Strack, F. (1991). Context effects in attitude surveys: Applying cognitive theory to social research. In W. Stroebe and M. Hewstone (Eds.), *European Review of Social Psychology*. (Vol. 2, pp. 31–50). Chichester: Wiley.

Sears, D.O. (1986). College sophomores in the laboratory: Influences of a narrow data base on psychology's view of human nature. *Journal of Personality and Social Psychology*, **51**, 513–530.

Seaver, D.A., von Winterfeldt, D. and Edwards, W. (1978). Eliciting subjective probability distributions on continuous variables. *Organizational Behavior and Human Performance*, **21**, 379–391.

Sedlmeier, P., and Betsch, T. (Eds.) (2002). *Etc. – Frequency processing and cognition*, Oxford: Oxford University Press.

Sexsmith, R.G. (1999). Probability-based safety analysis –value and drawbacks. *Structural Safety*, **21**, 303–310.

Shackle, G.L.S. (1969). *Decision, Order and Time in Human Affairs*. Cambridge: Cambridge University Press.

Shanteau, J. (1988). Psychological characteristics and strategies of expert decision makers In B. Rohrmaum, L. Beach, C. Vlek, and S. Watson (Eds.), *Advances in Decision Research.* Amsterdam: North Holland.

Shanteau, J. (1992a). Competence in experts: The role of task characteristics. *Organizational Behavior and Human Decision Processes*, **53**, 252–266.

Shanteau, J. (1992b). The Psychology of experts: An alternative view. In G. Wright and F. Bolger (Eds.), *Expertise and Decision Support.* New York: Plenum Press.

Shanteau, J. and Stewart, T.R. (1992). Why study expert decision making? Some historical perspectives and comments. *Organizational Behavior and Human Decision Processes*, **53**, 95–106.

Shephard, G.G. and Kirkwood, C.W. (1994). Managing the Judgmental Probability Elicitation Process: A Case Study of Analyst/Manager Interaction. *IEEE Transactions on Engineering Management*, **41**, 414–425.

Sherrick, B.J. (2002). The accuracy of producer's probability beliefs: Evidence and implications for insurance valuation. *Journal of Agricultural and Resource Economics*, **27**, 77–93.

Shuford, E.H. (1961). Percentage estimation of proportion as a function of element type, exposure type, and task. *Journal of Experimental Psychology*, **61**, 430–436.

Siegel-Jacobs, K. and Yates, J.F. (1996). Effects of procedural and outcome accountability on judgment quality. *Organizational Behavior and Human Decision Processes*, **65**, 1–17.

Simon, H.A. and Chase, W.G. (1973). Skill in chess. *American Scientist*, **61**, 394–403.

Simpson, W. and Voss, J.F. (1961). Psychophysical judgements of probabilistic stimulus sequences. *Journal of Experimental Psychology*, **62**, 416–422.

Slovic, P. (1972). From Shakespeare to Simon: speculations–and some evidence–about man's ability to process information. *Oregon Research Institute Research Monograph*, **12(2)**.

Slovic, P., Finucane, M., Peters, E. and MacGregor, D.G. (2002). The affect heuristic. In T. Gilovich, D. Griffin and D. Kahneman (Eds.), *Heuristics and biases: The psychology of intuitive judgment* (pp. 397–420). New York: Cambridge University Press.

Slovic, P., Fischhoff, B. and Lichtenstein, S. (1979). Rating the risks. *Environment*, **21**, 14–20.

Slovic, P. and Lichtenstein, S.C. (1968). The relative importance of probabilities and payoffs in risk taking. *Journal of Experimental Psychology Monograph Supplement*, **78(3)**, Part 2, 1–18.

Slovic, P. and Lichtenstein, S.C. (1971). Comparison of Bayesian and regression approaches to the study of information processing in judgement. *Organizational Behavior and Human Performance*, **6**, 649–744.

Slovic, P., Monahan, J. and MacGregor, D.G. (2000). Violence risk assessment and risk communication: The effects of using actual cases, providing instruction and employing probability versus frequency formats. *Law and Human Behavior*, **24**, 271–296.

Smedslund, J. (1963). The concept of correlation in adults. *Scandinavian Journal of Psychology*, **4**, 165–173.

Smith, J.Q. (1998). Discussion note to the papers on 'Elicitation' . *The Statistician*, **47**, 55–68.

Smith, J.Q. and Faria, A.E. (2000). Bayesian Poisson models for the graphical combination of dependent expert information. *Journal of the Royal Statistical Society, Series B*, **62**, 525–544.

Smith, J.F. and Kida, T. (1991). Heuristics and biases: Expertise and task realism in auditing. *Psychological Bulletin*, **109**, 472–489.

Smith, J. and Mandac, A.M. (1995). Subjective versus Objective Yield Distributions as Measures of Production Risk. *American Journal of Agricultural Economics*, **77**, 152–161.

Smith, G.C. and Wilkinson, D. (2002). Modelling disease spread in a novel host: Rabies in the European badger *Meles meles*. *Journal of Applied Ecology*, **39**, 865–874.

Sniezek, J.A. (1990). A comparison of group techniques with shared information. *Group and Organization Studies*, **15**, 5–19.

Sniezek, J.A. (1992). Groups under uncertainty: An examination of confidence in group decision making. *Organizational Behavior and Human Decision Processes*, **52**, 124–155.

Sniezek, J.A. and Henry, R.A. (1990). Revision, weighting and commitment in consensus group judgement. *Organizational Behavior and Human Decision Processes*, **45**, 66–84.

Soll, J.B. (1996). Determinants of overconfidence and miscalibration: The roles of random error and ecological structure. *Organizational Behavior and Human Decision Processes*, **65**, 117–137.

Soll, J.B. and Klayman, J. (2004). Overconfidence in interval estimates. *Journal of Experimental Psychology: Learning, Memory and Cognition*, **30**, 299–314.

Solomon, I. (1982). Probability Assessment by Individual Auditors and Audit Teams: An Empirical Investigation. *Journal of Accounting Research*, **20**, 689–710.

Sørensen, J.T., Østergaard, S., Houe, H. and Hindhede, J. (2002). Expert opinions of strategies for milk fever control. *Preventive Veterinary Medicine*, **55**, 69–78.

Sostman, H.D., Coleman, R.E., DeLong, D.M., Newman, G.E. and Paine, S. (1994). Evaluation of Revised Criteria for Ventilation-Perfusion Scintigraphy in Patients with Suspected Pulmonary Embolism. *Radiology*, **193**, 103–107.

Spencer, J. (1961). Estimating averages. *Ergonomics*, **4**, 317–328.

Spencer, J. (1963). A further study of estimating averages. *Ergonomics*, **6**, 255–265.

Spetzler, C.S. and Staël von Holstein, C.A.S. (1975). Probability encoding in decision analysis. *Management Science*, **22**, 340–358.

Spiegelhalter, D.J., Dawid, A.P., Lauritzen, S.L. and Cowell, R.G. (1993). Bayesian analysis in expert systems. *Statistical Science*, **8**, 219–247.

Spiegelhalter, D.J., Freedman, L.S. and Parmar, M.K.B. (1994). Bayesian approaches to randomized trials. *Journal of the Royal Statistical Society, Series A*, **157**, 357–416.

Spiegelhalter, D.J., Abrams, K.R. and Myles, J.P. (2004). *Bayesian Approaches to Clinical Trials and Health-Care Evaluation*. Wiley: Chichester.

Staël von Holstein, C.A.S. (1970a). *Assessment and Evaluation of Subjective Probability Distributions*. Stockholm: Economic Research Institute.

Staël von Holstein, C.A.S. (1970b). Measurement of subjective probability. *Acta Psychologica*, **34**, 146–159.

Staël von Holstein, C.A.S. (1971). An experiment in probabilistic weather forecasting. *Journal of Applied Meteorology*, **10**, 635–645.

Staël von Holstein, C.A.S. (1972). Probabilities forecasting: An experiment related to the stock market. *Organizational Behavior and Human Performance*, **8**, 139–158.

Staël von Holstein, C.A.S. (1977). The continuous ranked probability score in practice. In H. Jungermann and G. de Zeeuw (Eds.), *Decision Making and Change in Human Affairs* (pp. 263–273). Dordrecht, Holland: D. Reidel.

Stanovich, K.E. and West, R.F. (2000). Individual differences in reasoning: Implications for the rationality debate? *Behavioral and Brain Sciences*, **23**, 645–665.

Stärk, K.D.C., Wingstrand, A., Dahl, J., Møgelmose, V. and Lo Fo Wong, D.M.A. (2002). Differences and similarities among experts' opinions on *Salmonella entrica* dynamics in swine pre-harvest. *Preventive Veterinary Medicine*, **53**, 7–20.

Stasser, G. and Titus, W. (1985). Effects of information load and percentage of common information on the dissemination of unique information during group discussion. *Journal of Personality and Social Psychology*, **53**, 81–93.

Steffey, D. (1992). Hierarchical Bayesian Modelling with Elicited Prior Information. *Communications in Statistics –Theory and Methods*, **21**, 799–821.

Stephens, M.E., Goodwin, B.W. and Andres, T.H. (1993). Deriving parameter probability density functions. *Reliability Engineering and System Safety*, **42**, 271–291.

Stevens, S.S. (1957). On the Psychophysical law. *Psychological Review*, **64(1)**, 53–81.

Stevens, S.S. and Galanter, E.H. (1957). Ratio scales and category scales for a dozen perceptual continua. *Journal of Experimental Psychology*, **54**, 377–411.

Stevens, J.W. and O'Hagan, A. (2002) Incorporation of genuine prior information in cost-effectiveness analysis of clinical trial data. *International Journal of Technology Assessment in Health Care*, **18**, 782–790.

Stewart, T.R. and Lusk, C.M. (1994). Seven components of judgmental forecasting skill: Implications for research and the improvement of forecasts. *Journal of Forecasting*, **13**, 575–599.

Stewart, T.R., Moninger, W.R., Grassia, J., Brady, R.H., and Merrem, F.H. (1989). Analysis of expert judgment in a hail forecasting experiment. *Weather and Forecasting*, **4**, 24–34.

Steyerberg, E.W., Eijkemans, M.J.C., Harrell, F.E. and Habbema, J.D.F. (2001). Prognostic modeling with logistic regression analysis: In search of a sensible strategy with small data sets. *Medical Decision Making*, **21**, 45–56.

Stiber, N.A., Pantazidou, M. and Small, M.J. (1999). Expert system methodology for evaluating reductive dechlorination at TCE sites. *Environmental Science and Technology*, **33**, 3012–3020.

Stone, E.R. and Opel, R.B. (2000). Training to improve calibration: The effects of performance and environmental feedback. *Organizational Behavior and Human Decision Processes*, **83**, 282–309.

Subbotin, V. (1996). Outcome feedback effects on under- and overconfident judgments (general knowledge tasks). *Organizational Behavior and Human Decision Processes*, **66**, 268–276.

Sutherland, H.J., Lockwood, G.A., Tritchler, D.L., Sem, F., Brooks, L. and Till, J.E. (1991). Communicating probabilistic information to cancer patients: Is there 'noise' on the line? *Social Science and Medicine*, **32**, 725–731.

Tan, S-B., Chung, Y-F.A., Tai, B.C., Cheung, Y.B. and Machin, D. (2003). Elicitation of prior distributions for a phase III randomized controlled trial of adjuvant therapy with surgery for hepatocellular carcinoma. *Controlled Clinical Trials*, **24**, 110–121.

Tape, T.G., Heckerling, P.S., Ornato, J.P. and Wigton, R.S. (1991). Use of clinical judgement analysis to explain regional variations in physicians' accuracies in diagnosing pneumonia. *Medical Decision Making*, **11**, 189–197.

Taylor, S.E. (1981). The interface of cognitive and social psychology. In J. Harvey (Ed.), *Cognition, Social Behavior, and the Environment* (pp. 182–211). Hillsdale, NJ: Erlbaum.

Teigen, K.H. (2001). When equal chances = good chances: Verbal probabilities and the equiprobability effect. *Organizational Behavior and Human Decision Processes*, **85**, 77–108.

Teigen, K.H., Martinussen, M. and Lund, T. (1996a). Linda versus World Cup: Conjunctive probabilities in three real-life fictional and real-life predictions. *Journal of Behavioral Decision Making*, **9**, 77–93.

Teigen, K.H., Matinussen, M. and Lund, T. (1996b). Conjunction errors in the prediction of referendum outcomes: Effects of attitude and realism. *Acta Psychologica*, **93**, 91–105.

Thornbury, J.R., Fryback, D.G. and Edwards, W. (1975). Likelihood ratios as a measure of excretory urogram information. *Radiology*, **114**, 561–565.

Tierney, W.M., Fitzgerald, J., McHenry, R., Roth, B.J., Psaty, B., Stump, D.L. and Anderson, F.K. (1986). Physicians estimates of the probability of myocardial infarction in emergency room patients with chest pain. *Medical Decision Making*, **6**, 12–17.

Timmermans, D.R.M. (1999). What clinicians can offer: Assessing and communicating probabilities for individual patient decision making. *Hormone Research*, **51**(Suppl. 1), 58–66.

Timmermans, D., Kievit, J. and van Bockel, H. (1996). How do surgeons' probability estimates of operative mortality compare with a decision analytic model? *Acta Psychologica*, **93**, 107–120.

Todd, P.M. and Gigerenzer, G. (2000). Précis of 'Simple heuristics that make us smart'. *Behavioral and Brain Sciences*, **23**, 727–780.

Tollcott, M.A. and Marvin, F.F. (1988). *Reducing the Confirmation Bias in an Evolving Situation*. Technical Report No. 88–11. Reston, VA: Decision Science Consortium, Inc.,

Tollcott, M.A., Marvin, F.F. and Bresnick, T.A. (1989a). Situation assessment and hypothesis testing in an evolving situation. Technical Report No. 89–3. Reston, VA: Decision Science Consortium, Inc.,

Tollcott, M.A., Marvin, F.F. and Lehner, P.E. (1989b). Expert decision making in evolving situations. *IEEE Transactions on Systems, Man and Cybernetics*, **19**, 606–615.

Treadwell, J.R. and Nelson, T.O. (1996). Availability of information and the aggregation of confidence in prior decisions. *Organizational Behavior and Human Decision Processes*, **68**, 13–27.

Tversky, A. and Kahneman, D. (1971). Belief in the law of small numbers. *Psychological Bulletin*, **2**, 105–110.

Tversky, A. and Kahneman, D. (1973). Availability: A heuristic for judging frequency and probability. *Cognitive Psychology*, **4**, 207–232.

Tversky, A. and Kahneman, D. (1974). Judgment under uncertainty: Heuristics and biases. *Science*, **185**, 1124–1131.

Tversky, A. and Kahneman, D. (1982). Evidential impact of base rates. *In:* D. Kahneman, P. Slovic and A. Tversky (Eds.) *Judgment Under Uncertainty: Heuristics and Biases*. Cambridge: Cambridge University Press.

Tversky, A. and Kahneman, D. (1983). Extensional versus intuitive reasoning: The conjunction fallacy in probability judgment. *Psychological Review*, **90**, 293–315.

Tversky, A. and Koehler, D.J. (1994). Support theory: A nonextensional representation of subjective probability. *Psychological Review*, **101**, 547–567.

van Batenberg, P.C., O'Hagan, A. and Veenstra, R.H. (1994). Bayesian discovery sampling in financial auditing: A hierarchical prior model for substantive test sample sizes. *The Statistician*, **43**, 99–110.

van der Fels-Klerx, I.H.J., Goossens, L.H.J., Saatkamp, H.W. and Horst, S.H.S. (2002). Elicitation of quantitative data from a heterogeneous expert panel: Formal process and application in animal health. *Risk Analysis*, **22**, 67–81.

van der Fels-Klerx, H.J., Saatkamp, H.W., Verhoeff, J. and Dijkhuizen, A.A. (2002b). Effects of bovine respiratory disease on the productivity of dairy heifers quantified by experts. *Livestock Production Science*, **75**, 157–166.

van der Gaag, L.C., Renooij, S., Witteman, C.L.M., Aleman, B.M.P. and Taal, B.G. (2002). Probabilities for a probabilistic network: A case study in oesophageal cancer. *Artificial Intelligence in Medicine*, **25**, 123–148.

van der Pligt, J. (1988). Applied decision research and environmental policy. *Acta Psychologica*, **68**, 293–311.

van der Pligt, J., Eiser, J.R. and Spears, R. (1987). Comparative judgments and preferences: The influence the number of response alternatives. *British Journal of Social Psychology*, **26**, 269–280.

van Lenthe, J. (1993a). ELI: An interactive elicitation technique for subjective probability distributions. *Organizational Behavior and Human Decision Processes*, **55**, 379–413.

van Lenthe, J. (1993b). A blueprint of ELI, a new method for eliciting subjective probability distributions. *Behavior Research Methods, Instruments and Computers*, **25**, 425–433.

van Lenthe, J. (1994). Scoring-rule feedforward and the elicitation of subjective probability distributions. *Organizational Behavior and Human Decision Processes*, **59**, 188–209.

van Schie, E.C.M. and van der Pligt, J. (1994). Getting and anchor on availability in causal judgment. *Organizational Behavior and Human Decision Processes*, **57**, 140–154.

Varis, O. and Kuikka, S. (1997). BeNe-EIA: A Bayesian approach to expert judgement elicitation with case studies on climate change impacts on surface waters. *Climatic Change*, **37**, 539–563.

Vertinsky, P., Kanetkar, V., Vertinsky, I. and Wilson, G. (1986). Prediction of wins and losses in a series of field hockey games: A study of probability assessment quality and cognitive information-processing models of players. *Organizational Behavior and Human Decision Processes*, **38**, 392–404.

Vescio, M.D. and Thompson, R.L. (2001). Subjective tornado probability forecasts in severe weather watches. *Weather and Forecasting*, **16**, 192–195.

Villejoubert, G. and Mandel, D.R. (2002). The inverse fallacy: An account of deviations from Bayes's theorem and the additivity principle. *Memory and Cognition*, **30**, 171–178.

Vo, T.V., Heasler, P.G., Doctor, S.R., Simonen, F.A. and Gore, B.F. (1991). Estimates of rupture probabilities for nuclear power plant components: Expert judgement elicitation. *Nuclear Technology*, **96**, 259–271.

Walker, K.D., Catalano, P., Hammitt, J.K. and Evans, J.S. (2003). Use of expert judgment in exposure assessment –Part 2. Calibration of expert judgments about personal exposures to benzene. *Journal of exposure analysis and environmental epidemiology*, **13**, 1–16.

Walker, K.D., Evans, J.S. and MacIntosh, D. (2001). Use of expert judgment in exposure assessment–Part I. characterization of personal exposure to benzene.

Journal of Exposure Analysis and Environmental Epidemiology, **11**, 308–322.

Walley, P. (1991). *Statistical Reasoning with Imprecise Probabilities.* London: Chapman and Hall.

Walley, P. (1996). Measures of uncertainty in expert systems. *Artificial Intelligence*, **83**, 1–58.

Walls, L. and Quigley, J. (2001). Building prior distributions to support Bayesian reliability growth modelling using expert judgement. *Reliability Engineering and System Safety*, **74**, 117–128.

Wallsten, T.S. (1981). Physician and medical student bias in evaluating diagnostic information. *Medical Decision Making*, **1**, 145–164.

Wallsten, T.S. and Budescu, D.V. (1983). Encoding subjective probabilities: A psychological and psychometric review. *Management Science*, **29**, 151–173.

Wallsten, T.S., Budescu, D.V., Rapoport, A., Zwick, R. and Forsyth, B. (1986). Measuring the vague meanings of probability terms. *Journal of Experimental Psychology: General*, **115(4)**, 348–365.

Wallsten, T.S., Budescu, D.V., Zwick, R. and Kemp, S.M. (1993). Preferences and reasons for communicating probabilistic information in verbal or numerical terms. *Bulletin of the Psychonomic Society*, **31(2)**, 135–138.

Wallsten, T.S., Fillenbaum, S. and Cox, J.A. (1988). Base rate effects on the interpretations of probability and frequency expressions. *Journal of Memory and Language*, **25**, 571–587.

Ward, W.C. and Jenkins, H.M. (1965). The display of information and the judgement of contingency. *Canadian Journal of Psychology*, **19**, 231–241.

Weber, E.U., Shafir, S. and Blais, A.R. (2004). Predicting risk sensitivity in humans and lower animals: Risk as variance or coefficient of variation. *Psychological Review*, **111**, 430–445.

Weiler, H. (1965). The use of incomplete beta functions for prior distributions in binomial sampling. *Technometrics*, **7**, 335–347.

Weinstein, N.D. (1980). Unrealistic optimism about future life events. *Journal of Personality and Social Psychology*, **39**, 806–820.

Welch, R.D., Zalenski, R.J., Shamsa, F., Waselewsky, D.R., Kosnik, J.W. and Compton, S. (2000). Pretest probability assessment for selective rest sestamibi scans in stable chest pain patients. *American Journal of Emergency Medicine*, **18**, 789–792.

West, M. and Crosse, J. (1992). Modelling probabilistic agent opinion. *Journal of the Royal Statistical Society, Series B.*, **54**, 285–299.

Whitecotton, S.M., Sanders, D.E. and Norris, K.B. (1998). Improving predictive accuracy with a combination of human intuition and mechanical decision aids. *Organizational Behavior and Human Decision Processes*, **76**, 325–348.

Whitehead, J.C. (1992). Ex ante willingness to pay with supply and demand uncertainty: implications for valuing a sea turtle protection programme. *Applied Economics*, **24**, 981–988.

Wiggins, N. and Hoffman, P.J. (1968). Three models of clinical judgement. *Journal of Abnormal Psychology*, **73**, 70–77.

Wilkie, M.E. and Pollack, A.C. (1996). An application of probability judgment accuracy measures to currency forecasting. *International Journal of Forecasting*, **12**, 25–40.

Wilson, A.G. (1994). Cognitive Factors Affecting Subjective Probability Assessment. *ISDS Discussion Paper*, 94–02.

Windschitl, P.D. (2000). The binary additivity of subjective probability does not indicate the binary complementarity of perceived certainty. *Organizational Behavior and Human Decision Processes*, **81**, 195–225.

Windschitl, P.D. and Wells, G.L. (1996). Measuring psychological uncertainty: Verbal versus numeric methods. *Journal of Experimental Psychology: Applied*, **2**, 343–364.

Winkler, R.L. (1967). The assessment of prior distributions in Bayesian analysis. *Journal of American Statistical Association*, **62**, 776–880.

Winkler, R.L. (1981). Combining probability distributions from dependent information sources.*Management Science*, **27**, 479–488.

Winkler, R.L. and Murphy, A.H. (1968). Evaluation of subjective precipitation probability forecasts. In *Proceedings of the first National Conference on Statistical Meterology*, 148–157, American Meterological Society, Boston.

Winkler, R.L. and Poses, R.M. (1993). Evaluating and combining physicians' probabilities of survival in an intensive care unit. *Management Science*, **39**, 1526–1543.

Winkler, R.L., Smith, W. and Kulkarni, R. (1978). Adaptive forecasting models based on predictive distributions. *Management Science*, **24**, 977–986.

Winkler, R.L., Wallsten, T.S., Whitfield, R.G., Richmond, H.M., Hayes, S.R. and Rosenbaum, A.S. (1995). An assessment of the risk of chronic lung injury attributable to long-term ozone exposure. *Operations Research*, **43**, 19–28.

Winman, A., Hansson, P. and Juslin, P. (2004). Subjective probability intervals: How to reduce overconfidence by interval evaluation. *Journal of Experimental Psychology: Learning, Memory and Cognition*, **30**, 1167–1175.

Wittenbaum, G.M. and Stasser, G. (1996). Management of information in small groups. In J.L. Nye and A.M. Brower (Ed.), *What's Social about Social Cognition? Social Cognition Research in Small Groups*. Thousand Oaks, CA: Sage Publications.

Wolfson, L.J., Kadane, J.B. and Small, M.J. (1996). Bayesian environmental policy decisions: Two case studies. *Ecological Applications*, **6**, 1056–1066.

Woloshin, S., Schwartz, L.M., Byram, S., Fischhoff, B. and Welch, H.G. (2000). A new scale for assessing perceptions of chance: A validation study. *Medical Decision Making*, **20**, 298–307.

Wong, K.F.E. and Kwong, J.Y.Y. (2000). Is 7300 m equal to 7.3 km? Same semantics but different anchoring effects. *Organizational Behavior and Human Decision Processes*, **82**, 314–333.

Wood, L.E. and Ford, J.M. (1993). Structuring interviews with experts during knowledge elicitation. In K.M. Ford and J.M. Bradshaw (Eds.), *Knowledge Acquisition as Modeling* (Vol. 1, pp. 71–90). New York: Wiley.

Woods, C.M., Frost, R.O. and Steketee, G. (2002). Obsessive Compulsive (OC) Symptoms and Subjective Severity, Probability, and Coping Ability Estimations of Future Negative Events. *Clinical Psychology and Psychotherapy*, **9**, 104–111.

Wright, G. (1982). Changes in the realism and distribution of probability assessments as a function of question type. *Acta Psychologica*, **52(1 Suppl 2)**, 165–174.

Wright, G.W. (1984). *Behavioural Decision Theory*. New York: Penguin Books.

Wright, W.F. and Anderson, U. (1989). Effects of situation familiarity and financial incentives on use of anchoring and adjustment heuristic for probability assessment. *Organizational Behavior and Human Decision Processes*, **44**, 68–82.

Wright, G. and Ayton, P. (1986). Subjective confidence in forecasts: A response to Fischhoff and MacGregor. *Journal of Forecasting*, **5**, 117–123.

Wright, G. and Ayton, P. (1987). Task influences on judgemental probability forecasting. *Scandinavian Journal of Psychology*, **28**, 115–127.

Wright, G. and Ayton, P. (1992). Judgemental probability forecasting in the immediate and medium term. *Organizational Behavior and Human Decision Processes*, **51**, 344–363.

Wright, G., Rowe, G., Bolger, F. and Gammack, J. (1994). Coherence, calibration and expertise in judgmental probability forecasting. *Organizational Behavior and Human Decision Processes*, **57**, 1–25.

Wright, G., Saunders, C. and Ayton, P. (1988). Consistency, coherence and calibration in subjective probability forecasting. *Journal of Forecasting*, **7**, 185–199.

Wright, G. and Wisudha, A. (1982). Distribution of probability assessments for almanac and future event questions. *Scandinavian Journal of Psychology*, **23**, 219–224.

Wright, G. W. (1984). *Behavioural Decision Theory*, New York: Penguin Books.

Yamagishi, K. (1997a). When a 12.86% mortality is more dangerous than 24.14%: Implications for risk communication. *Applied Cognitive Psychology*, **11**, 495–506.

Yamagishi, K. (1997b). Upward versus downward anchoring in frequency judgments of social facts. *Japanese Psychological Research*, **39**, 124–129.

Yaniv, I. and Foster, D.P. (1995). Graininess of judgment under uncertainty: An accuracy informativeness trade-off. *Journal of Experimental Psychology: General*, **124**, 424–432.

Yaniv, I. and Foster, D.P. (1997). Precision and accuracy of judgmental estimation. *Journal of behavioral decision making*, **10**, 21–32.

Yates, J.F. (1990). *Judgment and Decision Making*. Englewood Cliffs, NJ: Prentice Hall.

Yates, J.F. (1994). Subjective probability analysis. In G. Wright and P. Ayton (Eds.), *Subjective Probability*, (pp. 382–410.) John Wiley: London.

Yates, J.F. and Curley, S.P. (1985). Conditional distribution analyses of probabilistic forecasts. *Journal of Forecasting*, **4**, 61–73.

Yates, J.F., McDaniel, L.S. and Brown, E.S. (1991). Probabilistic forecasts of stock prices and earnings: The hazards of nascent expertise. *Organizational Behavior and Human Decision Processes*, **49**, 60–79.

Zellner, A. (1972). On assessing informative prior distributions for regression coefficients. *Unpublished mimeo*.

Zukier, H. (1982). The dilution effect: The role of the correlation and the dispersion of predictor variables in the use of nondiagnostic information. *Journal of Personality and Social Psychology*, **43**, 1163–1174.

Author Index

Uncertain Judgements – Eliciting Experts' Probabilities A. O'Hagan, C. E. Buck, A. Daneshkhah, J. R. Eiser, P. H. Garthwaite, D. J. Jenkinson, J. E. Oakley and T. Rakow

Index

Uncertain Judgements – Eliciting Experts' Probabilities A. O'Hagan, C. E. Buck, A. Daneshkhah, J. R. Eiser, P. H. Garthwaite, D. J. Jenkinson, J. E. Oakley and T. Rakow

Statistics in Practice

Human and Biological Sciences

Berger – Selection Bias and Covariate Imbalances in Randomized Clinical Trials
Brown and Prescott – Applied Mixed Models in Medicine
Chevret (Ed) – Statistical Methods for Dose-Finding Experiments
Ellenberg, Fleming and DeMets – Data Monitoring Committees in Clinical Trials: A Practical Perspective
Lawson, Browne and Vidal Rodeiro – Disease Mapping with WinBUGS and MLwiN
Lui – Statistical Estimation of Epidemiological Risk
*Marubini and Valsecchi – Analysing Survival Data from Clinical Trials and Observation Studies
O'Hagan – Uncertain Judgements: Eliciting Experts' Probabilities
Parmigiani – Modeling in Medical Decision Making: A Bayesian Approach
Senn – Cross-over Trials in Clinical Research, Second Edition
Senn – Statistical Issues in Drug Development
Spiegelhalter, Abrams and Myles – Bayesian Approaches to Clinical Trials and Health-Care Evaluation
Whitehead – Design and Analysis of Sequential Clinical Trials, Revised Second Edition
Whitehead – Meta-Analysis of Controlled Clinical Trials
Willan and Briggs – Statistical Analysis of Cost-effectiveness Data

Earth and Environmental Sciences

Buck, Cavanagh and Litton – Bayesian Approach to Interpreting Archaeological Data
Glasbey and Horgan – Image Analysis in the Biological Sciences
Helsel – Nondetects and Data Analysis: Statistics for Censored Environmental Data
McBride – Using Statistical Methods for Water Quality Management
Webster and Oliver – Geostatistics for Environmental Scientists

Industry, Commerce and Finance

Aitken and Taroni – Statistics and the Evaluation of Evidence for Forensic Scientists, Second Edition

Balding – Weight-of-evidence for Forensic DNA Profiles
Lehtonen and Pahkinen – Practical Methods for Design and Analysis of Complex Surveys, Second Edition
Ohser and Mücklich – Statistical Analysis of Microstructures in Materials Science
Taroni, Aitken, Garbolino and Biedermann – Bayesian Networks and Probabilistic Inference in Forensic Science

*Now available in paperback